Industrial Assembly

Industrial Assembly

Shimon Y. Nof
School of Industrial Engineering
Purdue University
West Lafayette, IN, USA

Wilbert E. Wilhelm
Department of Industrial Engineering
Texas A&M University
College Station, TX, USA

and

Hans-Jürgen Warnecke
Fraunhofer Institute
Munich, Germany

CHAPMAN & HALL

London · Weinheim · New York · Tokyo · Melbourne · Madras

Published by Chapman & Hall, 2–6 Boundary Row, London SE1 8HN, UK

Chapman & Hall, 2–6 Boundary Row, London SE1 8HN, UK

Chapman & Hall GmbH, Pappelallee 3, 69469 Weinheim, Germany

Chapman & Hall USA, 115 Fifth Avenue, New York, NY 10003, USA

Chapman & Hall Japan, ITP-Japan, Kyowa Building, 3F, 2-2-1 Hirakawacho, Chiyoda-ku, Tokyo 102, Japan

Chapman & Hall Australia, 102 Dodds Street, South Melbourne, Victoria 3205, Australia

Chapman & Hall India, R. Seshadri, 32 Second Main Road, CIT East, Madras 600 035, India

First edition 1997

© 1997 Shimon Y. Nof, Wilbert E. Wilhelm and Hans-Jürgen Warnecke

Typeset in 10 on 12 pt Times by Best-set Typesetter Ltd., Hong Kong
Printed in Great Britain by St Edmundsbury Press Ltd, Bury St Edmunds, Suffolk

ISBN   0 412 55770 3

A catalogue record for this book is available from the British Library

Library of Congress Catalog Card Number: 96-85302

∞ Printed on permanent acid-free text paper, manufactured in accordance with ANSI/NISO Z39.48-1992 and ANSI/NISO Z39.48-1984 (Permanence of Paper).

To our families

Contents

Preface

Industrial assembly is a rapidly changing field with significant importance in production. The purpose of this book is threefold:

1. to provide a comprehensive coverage of technological, engineering and management aspects of this field;
2. to present multi-disciplinary approaches to rationalization, design, planning and control of assembly operations and systems;
3. to explain quantitative models, information technologies and engineering techniques that have been practised effectively in industrial assembly, as well as theoretical foundations and emerging trends that shape its future.

The three of us have been attracted to the area of assembly for many years (a total of about a century between us), because of its unique role in industry and its challenging complexity, involving both human and technological concerns. Our objective in this book has been to compile and organize the vast amount of relevant information and knowledge that has been developed and published in this field, and the worldwide experience of many researchers and workers in projects related to assembly. This book is timely: we believe that assembly has matured and at the same time still holds promise for significant innovations and productivity improvements in service and manufacturing industries.

The book is written for senior undergraduate, graduate and continuing education students of engineering and operations management; for managers, engineers and consultants working in industry on acquisition, development, design and implementation of assembly facilities; and for researchers in this rich and interesting field.

The organization of the book follows the general lifecycle of assembly, from history of assembly in Chapter 1 to its expected future in Chapter 10; from fundamentals to design and then to operations. The ten chapters can thus be viewed in three logical parts, as follows:

Assembly History and Fundamentals	Design of Products and Systems	Planning, Operations and Control
1. Introduction and Fundamental Concepts of Assembly	3. Design for Assembly	7. Sequencing and Scheduling of Assembly Operations
2. Assembly Tasks and Technology	4. Design of Assembly Systems (Technology)	8. Time-Managed Material Flow Control
	5. Assembly System Design and Planning	9. Quality and Inspection in Assembly
	6. Performance Evaluation of Stochastic Assembly Systems	10. Emerging Trends in Assembly

Finally, we wish to acknowledge the following people for their help and contributions in preparing this book: Mark Hammond, our publisher, Richard Owen and David Hemsley, our production editors, and Sara Hulse, our proof-reader. Dr Michael Boasson, Jasmin Nof, Moriah Nof, Nava Nof, Dr Venkat Rajan, Michael Seidman and Dr James Witzerman (S.Y.N.); Dennis Allen, Michele Bork and Shay Sanders (W.E.W.); Dr Manfred Schweizer and Claus Scholpp (H.J.W.).

1

Introduction and fundamental concepts of assembly

1.1 Introduction

The purpose of Chapter 1 is to provide the reader with an overview of the scope of assembly. The chapter begins by describing the fundamental concepts and characteristics of assembly, followed by a review of its historical development (section 1.2) and economic significance (section 1.3). The introduction continues with engineering and management techniques and approaches for rationalizing and improving assembly operations and systems in industry (sections 1.4, 1.5). Most of these rationalization strategies are explained in detail later in the book. The first part of Chapter 1 is concluded (in section 1.6) with a general taxonomy of assembly operations and systems.

The next three sections complete the chapter by introducing and illustrating at some depth the nature of the rest of the book. In section 1.7, the key design issues are presented. Fundamental models to design and analyse assembly are described in section 1.8, and in section 1.9 assembly operational planning issues are introduced and illustrated.

1.2 Nature and role of assembly in industry

Assembly is part of the **production system**. Industrially-produced final products consist mainly of several individual parts and subassemblies that have mostly been manufactured at different times, possibly in separate locations. Assembly tasks thus result from the requirement to build together certain individual parts, subassemblies and substances such as lubricants and adhesives into final assemblies of higher complexity in a given quantity or within a given time period. Assembly represents a cross-section of the problems within the whole of production engineering, with different assembly activities and processes being performed in various branches of industry.

Assembly (plural 'assemblies', abbreviation 'ass'y') is defined as:

> The aggregation of all processes by which various parts and subassemblies are built together to form a complete, geometrically designed assembly or product (such as a machine or an electronic circuit) either by an individual, batch or a continuous process.

Some examples of common assemblies are:

- an electric motor, which is an assembly of the stationary parts of the stator and the subassembly that turns, called the rotor;
- a power train, which is an assembly of gears and associated parts by which power is transmitted from an engine to a driving axle;
- a fuselage, the central body of an aircraft, to which the wings and tail assembly are assembled and which accommodates the crew, passengers and cargo;
- furniture, which is an assembly of several wood, fabric, metal, etc. subassemblies;
- a computer, which is an assembly of numerous electronic circuits.

Two examples of assembled products are shown in Fig. 1.1. The paper trimmer is assembled from metal and wood components, and the electronic board is assembled from a printed circuit board mounted with various electronic components and modules.

Three common terms associated with industrial assembly are **modular components**, **assembler** and **assembly line**.

- **Modular component.** This is a standardized, often interchangeable component of a system or product that is designed for easy assembly or variable use, e.g. the handle module of the paper trimmer, or a cabinet consisting of two end modules and three center modules. In electronic manufacturing, a module is a self-contained assembly of electronic components and circuitry, such as a controller circuit card, that is installed as a unit.
- **Assembler.** This is a worker who assembles by building together components of a product being manufactured. (This usage should be distinguished from the computer science context, in which an assembler means a program operating on high-level, symbolic input data to produce the equivalent, executable computer program.)
- **Assembly line.** This is an arrangement of workers, machines and equipment in which the product being assembled passes consecutively from one specialized operation to the next until completed. It is also called a production line. During the twentieth century, the term has become a metaphor for any process in which finished products are turned out in a mechanically efficient, though impersonal, manner. Well known are the humorous movie scenes of Charlie Chaplin turning into an automaton from the repeated assembling of mechanical parts, and Lucy having to swallow chocolates while trying to keep up with assembling chocolate

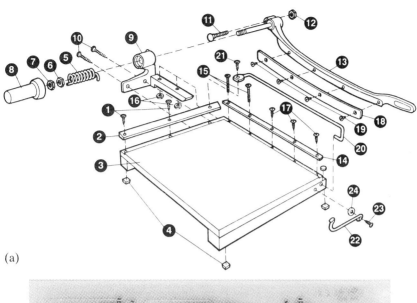

(a)

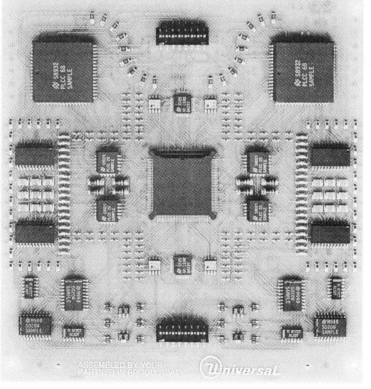

(b)

Fig. 1.1 A sampling of assembled products: (a) paper trimmer assembly (numbers indicate types of individual component parts and fasteners) (courtesy of Quartet Manufacturing Co.); (b) electronic board assembly (courtesy of Universal Co.).

candy boxes on a fast-moving conveyor. The latter part of the twentieth century has seen the development of flexible assembly, assembly work-stations and assembly cells.

Industrial assembly is distinguished from non-industrial assembly (e.g. for a hobby) by its goals of efficiency, productivity and cost-effectiveness. This book focuses on repetitive assembly. Construction site assembly, e.g. prefabricated house building, shipbuilding and facility construction, will not be treated in depth. This book describes and explains the general subject of industrial assembly with an integrated approach to:

- selection and design of the appropriate assembly method;
- design and planning of products for assembly;
- assembly techniques; and
- assembly system planning and operation.

Before discussing the details of these topics, it may be of interest to explore the historical developments related to assembly.

1.2.1   A brief history of assembly

The history of assembly can be divided roughly into three main periods (Fig. 1.2):

- pre-industrial;
- continuous assembly; and
- flexible assembly.

Before the Industrial Revolution, manual assembly was performed as part of the crafts-oriented production of jewelry, household utensils, clothing items, weapons, religious artifacts, etc.

The two essential features influencing industrial assembly evolution were

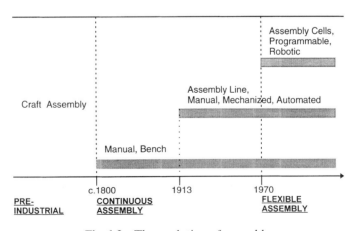

Fig. 1.2   The evolution of assembly.

M'CORMICK'S PATENT VIRGINIA REAPER.

Though the farmer may put his Reaper together by the above cut, I give the following:

Assembly Directions

I. Lap the in hound and cross piece which has two holes for the axle, for high or low stubble. Lower hole placed on the axle for higher stubble.

II. Bolt the angular board (S, in the cut) marked thus -----, to its place with the 8 inch bolt in the back end, which rests close to the platform.

III. Put the axle in place.

IV. Place the wheel frame on the other end of the axle lapping the finger beam above or below - suiting the higher or lower cut at Q.

V. Put on the main brace D, marked thus ==, one side of driver's seat on same bolt.

VI. Separator W, (marked O) to its corresponding mark, there is a little wooden pin to be removed to give place to a bolt.

VII. dividing iron marked thus x. If the reel be required very low, this iron is taken off, for small wheat.

VIII. Side board (with brace nailed to it) marked U, to its place, removing a small block to give place to it.

IX. Two small posts erected and cloth bar on top marked T.

X. Reel bearer V.

XI. Reel shaft Z to its place, arms put in with blocks (on each end) forward as the reel turns, and after braces are in and tightened by a cross pin in the one with long tenon (there is one brace for each end with long tenon) the boards are then nailed on the blocks. See d in cut. The boards have to be strained into a shape that will fit, as the arms are not parallel. Each of these boards having a block on the end (and the end without the block is next the wheels) put on, so as to pass $3/4$ of an inch from the angular board.

XII. If fingers do not fit right, they can be knocked out with a punch, and with the wedges trained right with very little trouble.

XIII. The small ground wheel has a third and lower sett by bolting the block on the upper instead of the lower side, and the side next the horses may be raised or lowered some two inches by changing the position of the tongue which may be done if required by boring another hole in the tongue.

XIV. The square washer may be placed on either end of the axle to gear deeper or shallower and other washers or leather can be added.

XV. Put in the driver (9 or 12 for lower or higher stubble) and bolt on the guard (small cast-iron piece) to keep it in place, see that the guard does not tighten on the driver. The block in the driver to be tightened on the crank, when it wears.

Bolts having beviled heads belong to corresponding places, heads of tongue bolts to the left. The 8-inch bolt (in the box) goes through the back end of angle board (SS in the cut), cross piece and in bound C. The $5^3/4$ bolt through cross piece and in bound near the same place. The two 7-inch bolts through in bound and cross piece near the axle open side of small ground what outward.

Operation instruction follow.

C. H. M'CORMICK, PATENTEE.   **CHICAGO,   1851**

Fig. 1.3 Cyrus M'Cormick's instructions for reaper assembly; also included are an illustration ('cut') and operation instructions. (Adapted from Hounshell, 1984.) Reapers were invented in 1831.

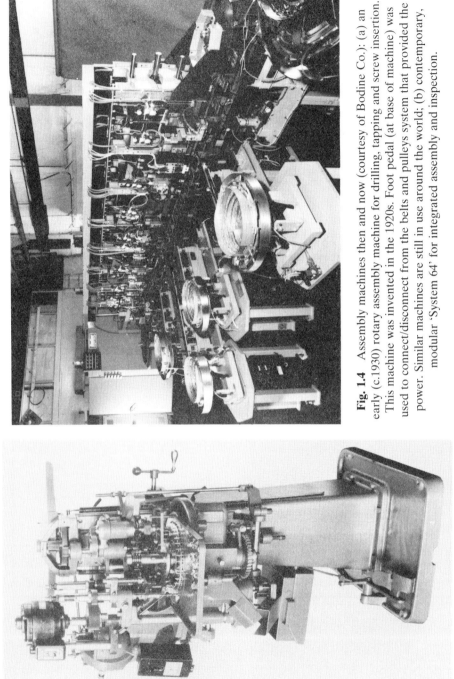

(b)

Fig. 1.4 Assembly machines then and now (courtesy of Bodine Co.): (a) an early (c.1930) rotary assembly machine for drilling, tapping and screw insertion. This machine was invented in the 1920s. Foot pedal (at base of machine) was used to connect/disconnect from the belts and pulleys system that provided the power. Similar machines are still in use around the world; (b) contemporary, modular 'System 64' for integrated assembly and inspection.

(a)

parts interchangeability and **conveyors**. The use of interchangeable parts, which became available at the turn of the eighteenth century in Europe and the US, enabled the manual assembly of various products such as rifles, clocks, sewing machines, bicycles and agricultural machinery (Fig. 1.3). However, assembly was performed at stationary benches. The system of conveyor belts and chutes introduced in 1784 by Oliver Evans in his Delaware flour mill was a concept adopted gradually by many industries during the nineteenth century to provide continuous parts and materials mobility.

In the early twentieth century, the Ford Motor Co. demonstrated the effectiveness of the **continuous assembly line** for mass assembly by combining the conveyor and the use of interchangeable parts. In parallel, **automatic assembly machines**, e.g. rotary and multiple-spindle assembly machines, became common in industry and provided more precise and faster assembly cycles (Fig. 1.4). Thus the era of purely manual assembly changed to the era of mixed manual and automatic assembly. The assembly line approach was also adopted by industries other than the automotive for high-demand products that require relatively precise assembly, such as radios, washing machines and refrigerators.

Market pressures for flexibility in design and production during the 1970s introduced the current era of **flexible assembly**. The two essential features here have been the availability of **computers** for assembly design, planning and control, and **robots** for reprogrammable assembly. Increasingly, industry is shifting to flexible assembly in the production of cars, machine tools, computers, airplanes, consumer electronics, household appliances and so on.

Development of **assembly theory** is in progress. Various operations research, artificial intelligence techniques and information technologies are currently being harnessed to provide new, powerful methods for rationalizing the complete assembly system. New materials and assembly technologies are also being developed. These efforts are in response to emerging global markets, increasing demands for higher quality products at competitive price, decreasing product lifecycles, and growing social awareness of environmental concerns and the need for recycling. The following timeline is a brief summary of interesting events in the evolution of industrial assembly.

Significant dates in the evolution of industrial assembly

c. 2500 BC   Noah's Ark is the first recorded assembly plan (Genesis 6:8–22) including the ark's material, design and dimensions.

16th century   In Venice, a reserve military fleet of 100 galleys was stored in prefabricated sections, ready for quick assembly during an emergency.

1799   Following the Industrial Revolution in Europe, the two principles of division of labor and of interchangeable components are applied by the US Armory in Springfield. The new assembly methods and work organization enable production to increase from 80 to 442 muskets per month.

Interchangeable parts are manufactured on machines with consistent and better quality compared with manual production. Accuracy is achieved by dimensioning from a single datum and checking accuracy with gauges at each manufacturing stage. Secondary and tertiary gauges are used to check the bench gauges.

1801   Eli Whitney, a pioneer of mass production, demonstrates in his factory at New Haven, Connecticut, assembly of musket locks with parts he picks at random from several heaps.

1824   The interchangeability requirement for all US government weapons is proven when 100 rifles from different armories are brought together, disassembled and reassembled at random, successfully.

1853   Complete assembly and adjustment of a single big bicycle wheel requires 60 minutes; assembly, despite use of interchangeable parts, is the biggest bottleneck.

One operator can assemble 75 clock movements per day – which is considered slow for the annual assembly of over 3000 clocks per week. British observers of US technology are impressed by the ease of musket assembly but fail to comment on clock assembly.

1858   Eight thousand sewing machines are assembled in the US by Wheeler & Wilson Co., compared with only 800 in 1853. By applying jigs, fixtures and gauges, the Brown & Sharp Co. eliminates intensive part-fitting by Wheeler & Wilson's highly skilled assemblers. In 1859 Wheeler & Wilson increases production to over 21 000 sewing machines, and by 1876 to about 109 000.

1875   Disassembly lines in Cincinnati and Chicago meat slaughterhouses become famous for using overhead conveyors and gravity flow-lines.

1885   Can production and assembly line for canned food, using special purpose machinery and conveyors, are introduced by Norton Co.

1901   Significant improvement is achieved in assembly accuracy: Carl Johanson develops in Sweden precision gauge blocks for use as quick 'go–no go' inspection decisions by unskilled assemblers.

1913   The first assembly line is developed by Henry Ford and Charles Sorenson at the Ford flywheel magneto assembly department. Flywheels rest on sliding surfaces mounted on a pipe frame, instead of workers standing at individual workbenches. Each assembly requires 16 permanent magnets, supports and clamps, 16 bolts and other components. Work is broken down into repeated elements. The breakthrough: from 29 assemblers producing 35–40 magnetos per day (per bench), i.e. one every 20 minutes, now they assemble 1188 on the line, i.e. one every 13 minutes and 10 seconds.

New problems appear: backaches – corrected by raising the line's working surface 6–8 inches, which results in a further cycle-time reduction to 7 minutes; pace imbalance between fast and slow assemblers – solved by a paced, continuous-driven chain. Within one year the number of assemblers is down from 29 to 14, with a yield of 1335 units, which means that each magneto now costs only 5 operator-minutes compared with the original 20 minutes.

Next to employ assembly lines are transmission, engines, axles, dashboard and upholstery departments.

1916	Time-and-motion studies and behavioral studies are performed by Frank and Lillian Gilbreth to improve manual assembly methods.
1920	Ford assembles half of the cars in the world. Fifteen million Model T cars are assembled before the line is discontinued in 1927.
	Early assembly machines are invented in the US in response to laws restricting child labor.
1941	Henry Kaiser establishes seven California shipyards using pre-fabrication and assembly line methods to set new speed records in shipbuilding.
1957	Pre-cut, pre-drilled Bakelite and glass–epoxy resin boards become available for electronic assembly of printed wiring boards (PWB) and inspire the design of integrated circuits (IC) and printed circuit boards (PCB).
1958	Automatic insertion machines apply fixed insertion heads over movable tables for electronic assembly of axial lead components. Panasonic and TDK in Japan later invent assembly machines to accommodate radial leads.
1975	Olivetti Sigma robot is used in Italy for assembly operations with two hands.
1976	Remote Center Compliance (RCC) device and Part Mating Theory are developed at C.S. Draper Labs near Boston for adaptive component insertion in assembly.
1978	PUMA (Programmable Universal Machine for Assembly) is introduced by Unimation for assembly, following research at General Motors.
1979	SCARA robot (Selective Compliance Arm for Robotic Assembly) is developed at Yamanashi University in Japan. Robots with advanced control and better accuracy become practical for board handling and electronic assembly.

	First Handbook of Product Design for Assembly is published at the University of Massachusetts.
1980	Of all the mechanical products assembled in the US, about 6% are assembled by automatic machines, mostly in high-volume factories that produce 20% of all manufactured products.
1983	APAS (Adaptable-Programmable Assembly System) pilot project by Westinghouse and the US National Science Foundation develops a flexible robotic assembly cell.
1990	Assembly robots are routinely used in industry with repeatability within 0.005 cm at speeds of 5 m/sec; visual sensing is used for component assembly from conveyors and bowl feeders; automated guided vehicles are used for flexible assembly.
1993	It is estimated that 30% of all semiconductor assembly is by surface mount technology (SMT) as opposed to the previously common technology of through-the-hole component insertion.
1995	Over 20% of all industrial robots in use are for assembly applications, compared with only 4% in 1975.

1.3   The economic significance of assembly

Assembly of manufactured goods accounts for over 50% of total production time according to various surveys (e.g. Nevins and Whitney, 1978) and according to a recent survey (Martin-Vega *et al.*, 1995) for 20% of the total unit production cost (Fig. 1.5). Typically, about one-third of a manufacturing company's labor is involved in assembly tasks. In the automotive industry, 50% of the direct labor costs are in the area of assembly, and in precision instruments it is between 20 and 70% (Lotter, 1986; Warnecke *et al.*, 1992). These statistics indicate the relative importance of assembly in terms of time and cost of assembled products. They also point to the potential savings that can be generated by efforts to understand and improve assembly technology and systems.

It is interesting to examine next the relative significance of industries involved with assembly. Highlights of the economic significance of assembly are shown in Fig. 1.6. Figure 1.6(a), (b) and (c) display the annual production, annual value-added and annual employment in 10 economically developed market economy countries. The annual statistics are based on the period 1988–92 (OECD, 1994). In our analysis based on these statistics, industry sectors are divided into two groups: (1) 'No–Lo assembly' – sectors with no assembly at all, such as civil works or production of minerals and chemicals, and sectors with relatively low assembly contents, such as food processing, structural metal, glass and paper production. This group includes OECD classification codes 31–37; (2) 'Med–Hi assembly' – sectors with medium or high assembly contents, such as mechanical, electronic and instrument production. This group, which can be called 'assembly industries', includes OECD classification codes 38–39.

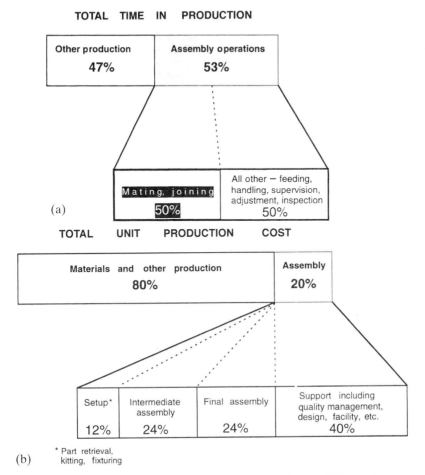

Fig. 1.5   Typical average breakdown of (a) production time and (b) production cost of industrial products.

The following observations can be made from the statistics:

1.   Countries with a relatively larger production volume tend also to have a relatively higher percentage of assembly. For instance, the three leading economies, the US, Japan and Germany, all have above 40% of their production in industries with medium-to-high percentage of assembly.
2.   In most cases, the percentage of assembly in total value-added is higher than its percentage of total production, indicating a higher relative value-added by sectors with assembled products.
3.   The percentage of employment in assembly is consistently similar to the percentage of value-added by assembly industries.

A consideration of the growth (or decline) of value-added and employment over one decade in the sectors with assembly compared to those with none or

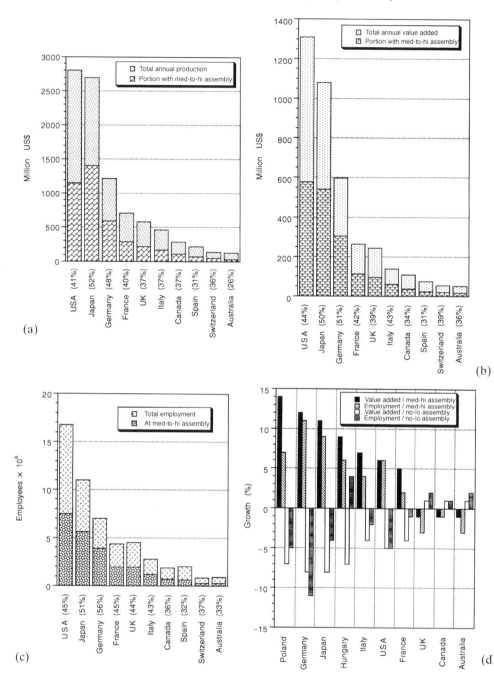

Fig. 1.6 Assembly statistics from selected countries: (a) annual production; (b) annual value-added; (c) annual employment; (d) percent growth in manufacturing industries 1975–85. (Prepared from data in UNIDO, 1988; OECD, 1994.)

Table 1.1 Share (%) of assembly value-added worldwide in 1985 vs. 1965

% assembly in industry	Developing countries		Centrally planned economies		Developed market economies	
	1965	1985	1965	1985	1965	1985
None-Low	14	19	16	24	70	57
Med.-High	6	8	12	19	82	73

Source: UNIDO (1988).

little as shown in Fig. 1.6(d), shows that significant growth is indicated in assembly industries. Poland and Hungary are included in the comparison as economies that had central planning during the time period considered, in which there is also a significant increase in assembly industries. Table 1.1 summarizes, in terms of value-added, where assembly industries are distributed around the world.

1.4 Assembly engineering and rationalization

In industrial production systems the assembly function is related to other functions, such as material procurement and marketing, via the flow of materials and information (Fig. 1.7(a)). The assembly system consists, as does the total production system, of different subsystems, such as assembly workstations and machines, basic construction units and interlinking equipment, which are also joined by a flow of materials and information. Figure 1.7(b) depicts, in an analogy to the other main types of manufacturing operations, the subfunctions of assembly and its related handling, joining and inspection activities.

The early history of assembly process development is closely related to the history of the development of mass production methods. Thus, the pioneers of mass production are also considered the pioneers of modern assembly. Their ideas and concepts significantly improved the manual and automated assembly methods employed in large-volume production. In the past decade efforts have been directed at reducing assembly costs and improving quality by the application of flexible automation and modern techniques.

Rationalization of assembly implies the efforts and investments to improve assembled products' quality and reduce their cost. Rationalization can be accomplished by a variety of engineering and management methods, including development of new materials, time-and-motion studies, methods analysis and improvement, new manufacturing and joining techniques, product development and design, mechanization and automation. Other approaches include design, planning and control models and decision-support methods, and or-

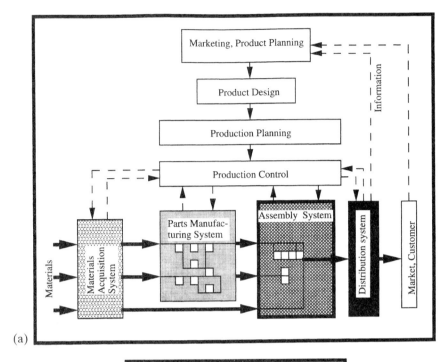

(a)

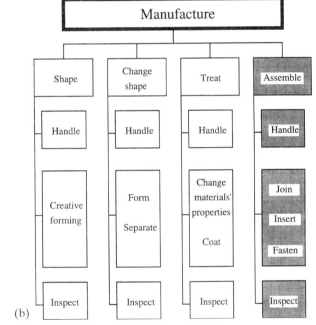

(b)

Fig. 1.7 The role of assembly: (a) assembly as a subsystem of production; (b) assembly activities as part of manufacturing.

ganizational and management systems. These techniques and methods are discussed in further detail in the rest of the book.

In the course of industrial production development, mechanization and automation have reached a high level of sophistication in parts manufacturing, permitting the efficient production of individual parts with a relatively low proportion of labor costs. In the field of assembly, by contrast, automation remained limited to high-volume production. In medium and short-run production, rationalization measures were mainly taken in the area of work structuring and workstation design. Automation measures were slowly undertaken here because of the following reasons.

1. Assembly is highly product specific, with highly variable tasks of handling, joining, adjusting, testing. Previously found solutions can, therefore, be applied only with great difficulty to other products or companies, in contrast to parts manufacturing.
2. Assembly, as the final production stage, must cope with continuously shifting market requirements in regard to timing, batch sizes and product design or style.
3. Assembly automation has, until recently, been difficult to justify economically, or infeasible for lack of programmable equipment.

As a result, part manufacturing improvements may be hindered by the increasing proportion of assembly costs. These costs can amount to between 20 and 70% (of total production cost) according to the product and level of production (Fig. 1.8).

Studies in the automotive industry indicate the relative lag of automation in assembly compared to other production areas (Fig. 1.9). In the areas of part manufacturing, spot welding and press shop, the level of automation is between 60 and 90%, whereas in assembly, it is less than 15% (apart from a few pilot projects). About 15% of all fasteners are assembled automatically. The least degree of automation in assembly is found in final assembly – not more than 5%. A more recent study of electrical and mechanical assembly in the US (Martin-Vega *et al.*, 1995) found that 19% of assembly operations are fully automated.

As a direct result of the low degree of automation, 50% of all direct labor costs in the automotive industry today are incurred in the area of assembly. Next to increasing material costs, the growing proportion of assembly labor costs in the total production cost of a product is a major motivation for assembly rationalization.

Human factors are also decisive in assembly rationalization. Worker performance in terms of speed, stamina and accuracy is limited. Extensive use of the division of labor principle has had negative repercussions, particularly in repetitive and large-volume assembly. Manual assembly work has been increasingly improved by better design of the workplace and substitution of muscle power by other energy sources. But the large fluctuations in assembly

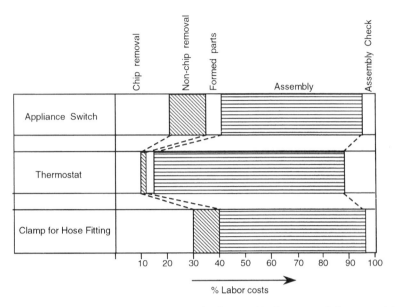

Fig. 1.8 Proportion of direct labor costs in the production cost of three precision products (Lotter, 1986).

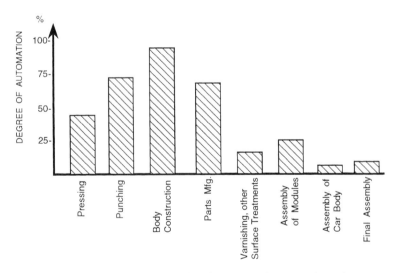

Fig. 1.9 Degree of automation in automotive manufacturing (Warnecke *et al.*, 1992).

employment, particularly in seasonal production, the proliferating training costs and other human-related interruptions in the assembly process all contribute to the interest in assembly rationalization that will result in cost reduction and product quality improvement.

1.4.1   Present state of industrial assembly products

A large variety of products are assembled in industry. There are also many different systems, requirements and conditions for assembly. To discuss the present state and identify emerging trends for rationalization, the general area of assembly must be analysed by its main distinguishing characteristics, namely **product complexity** (number of parts per product), **product turnover** and **industry type** (Table 1.2). In all cases, however, it is clear that the highest motivation for rationalization would be with assemblies that have a higher turnover and larger annual production volume.

When annual volumes of products to be assembled are studied, differences are found between different industries. Consider the frequency distribution of assembly of the fastest moving products in the machine tool industry (Fig. 1.10). It is evident that the largest proportion of products in groups 2 (31–500 parts) and 3 (over 500 parts) calls for relatively less rationalization effort if a quantity of more than 50 000 units/year is needed for justification. The quantity of 50 000 units/year is taken as a rough estimate for justifying rationalization investments, e.g. in flexible automation. Similarly, there is less motivation for rationalization efforts for products of group 1 (less than 31 parts) where higher volumes can be expected. Of the high turnover assemblies surveyed, those parts constitute only 8%.

In the electronic industry, there is a relatively high potential for rationalization (Fig. 1.11). Here, even in group 3 products, 30% of the products with the greatest turnover are produced in volumes of 50 000 units/year. Simple products that offer the best justification for automation represent a share of 20%. About 90% of them are presently manufactured in annual volumes of over 50 000 units.

Besides the annual volume to be assembled, the cycle time represents an important measure in evaluating the potential for rationalization. The main justification for automated assembly applications is within the operation cycle range of 5 s to 3 min. More flexible systems with assembly robots are justified when cycle times are greater than 30 s. In the machine tool industry 84% of all workstations have a cycle time of more than 3 min, hence there is less motivation to automate. On the other hand, in the automotive and electronic industries, cycle times indicate a high potential for the application of automated flexible assembly systems. In industrially developed countries, efforts made by companies have brought cycle times for 42% of all assembly tasks to fall below 1.5 min and for 26% to fall even below 30 s. Above-average high proportions of short-cycle tasks are to be found particularly in the electronic industry and precision instruments industry.

1.4.2   Trends and strategies in the rationalization of assembly

Distribution of assembly-related activities in industrial companies is shown in Fig. 1.12, based on a representative survey. About 50% of the total assembly

Table 1.2   Characteristics of product assembly

	Group 1	Group 2	Group 3
Product → Industry ↓	Simple products <31 parts	31–500 parts →increasing no. of parts (complexity)	Complex products >500 parts
Machine tool	Example: bearings	Transmissions	Agricultural machinery
Automotive	Automotive parts	Engines	Vehicles
Electrical equipment	PCBs, lamps, electric motors, electrical switching and control elements	Small domestic appliances	Radio, TV and audio equipment
Precision instruments	Mechanical measuring and control instruments	Photographic equipment	Clocks, projectors
Office equipment	Pens, pencils	Calculators	Office machines, computers

Source: Warnecke *et al.* (1992).

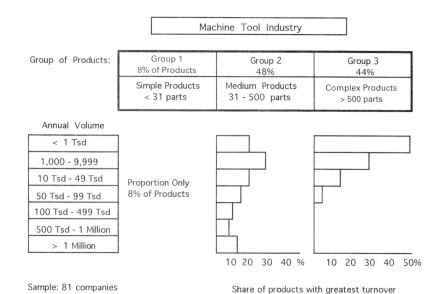

Fig. 1.10   Frequency distribution of annual production (tsd = thousands) in the machine tool industry for products with greatest turnover (Warnecke *et al.*, 1992).

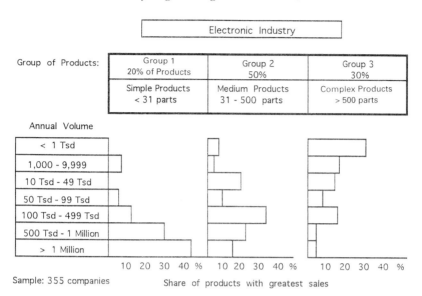

Fig. 1.11 Frequency distribution of annual production (tsd = thousands) in the electronic industry for products with greatest turnover (Warnecke *et al.*, 1992).

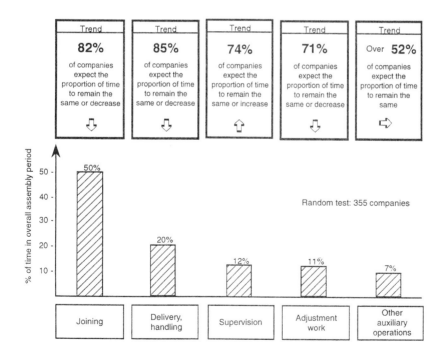

Fig. 1.12 Distribution of assembly operations and emerging trends (Warnecke *et al.*, 1992).

time is spent on actual joining operations. The rest of the assembly time is mainly in part-feeding, inspection and adjusting activities. This breakdown reveals the significance of peripheral mechanisms, for example sensors, testers and feeding devices, for future rationalized assembly. On the other hand, often the non-joining peripheral tasks such as visual checking and inspection cannot yet be automated economically.

When an assembly process is being rationalized, there are organizational and personnel issues to consider (discussed in the next section), as well as engineering considerations. The latter may relate both to the product design for assembly and to higher mechanization/automation of production equipment. A brief outline of several effective rationalization strategies is given here.

Plant investment will continue to increase in coming years for rationalization of assembly operations. On average 25% of all industrial investments in the 1980s were in the area of assembly (electronic industry, 31%; machine tool, 17%). Presently, approximately 30% of total industrial investment is in assembly operations (Warnecke *et al.*, 1992).

Expenditures for automation are increasing in disproportion to total investment in assembly. Investment in assembly automation started seriously in the late 1970s. A decade later it reached about half of the total investment in assembly. Savings expectations vary for the various strategies from sector to sector in industry according to survey results (Fig. 1.13).

Interestingly, product design is considered by all sectors in the survey as having the greatest potential for savings in assembly time (between 11.5% and 21.5% savings). For other rationalization strategies, there are differences between the sectors. The electronic and precision instruments industries both consider the next largest potential for obtaining benefits to be from mechanization and automation (19.5% time savings). In contrast, the machine tool industry anticipates the next greatest potential for benefits from optimizing employee work organization and labor structure (16% time savings). Section 1.5 addresses labor-related rationalization.

It can be said, in summary, that rationalization of assembly cannot depend on a single strategy or on technology alone. Instead, the various available rationalization strategies must be examined and evaluated by a cost–benefit analysis for the factors specific to a given industrial sector.

1.5 Management issues and assembly organizations

1.5.1 Assembly employment

Several general observations can be made about assemblers based on surveys in the German industry (Warnecke *et al.*, 1992) and experience in other industrial countries:

- The skill structure of assemblers is highly dependent on the industry sector. For instance, skilled workers are the major segment, on average 50% or more, in the machine tool and precision instrument industry

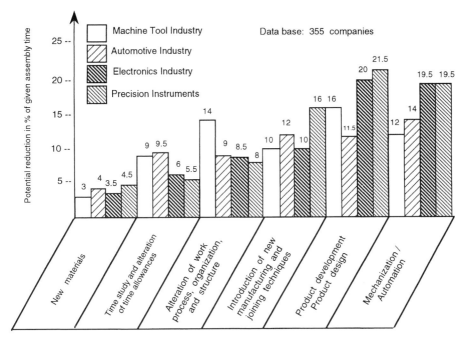

Fig. 1.13 Potential assembly time reduction by various rationalization strategies (Warnecke *et al.*, 1992).

which is typically greater than the proportions of unskilled and of semi-skilled labor. In the automotive and electronic industries, the proportion of unskilled/semi-skilled labor amounts to 80 to 85% while the proportion of skilled workers is only 16 and 10%, respectively.

- Assembly personnel are generally younger than the average employee; the 'under 35' assembler age group is twice as large as the average.
- Typically, there are twice as many female workers in assembly work as in other types of work.
- Group piecework is often found as the basis for payment in the field of assembly. An exception is the machine tool industry, where payment is most typically based on time worked, probably because of the relatively loose structure of assembly workstations and the relatively high proportion of skilled labor.

1.5.2 Organizational rationalization

Considerable success in rationalization, particularly by organizational measures, can be achieved even in lot-size-of-one and small-lot assembly. Experts in the field of assembly know well that interruptions, faults and defects that may not appear in prior production stages will suddenly surface when the parts or subassemblies converge in the assembly area. For instance, single components of acceptable quality as individual parts may be found unsuitable for assembly; components acceptable for assembly may arrive in insufficient quantity, or they may arrive late and cause delays in assembly. Any assembly rationalization effort must therefore first come to terms with the problem of trouble-free materials availability. Only when all the interrelationships, or interactions, of all production and supply areas have been adequately addressed can the assembly process proceed optimally.

Careful planning of the assembly operation by the production control function can guarantee appropriate quantities and timely availability of necessary component parts with the required quality. Costly interruptions in assembly due to part shortage can thus be avoided.

Computer-supported performance problem-solving methods that are commonly and effectively applied in high-volume manufacturing and mass production for shopfloor control, can also be applied as an effective logistic, organizational measure for assembly system rationalization. Apart from the steady-state improvement in efficiency and workload schedule for the operators, a reduction of setup and changeover time can be achieved when new product designs, new methods and new systems are introduced.

1.5.3   Labor-related rationalization

It has been widely recognized that the general motivation of a human operator will determine his or her ability to achieve the volume and quality goals of given assembly work. Investigations have shown that, next to wages, the improvement of the working environment can significantly raise workers' motivation. Creating a work environment that suits human needs is mandatory. Every effort should be made to prevent any physical and psychological harm to the operator. Thus, for example, actions to lower the noise level or to improve accident prevention through safety programs at the assembly workplace are given high priority.

Advantages and disadvantages of paced assembly line work for human assemblers have been discussed in all industrialized countries for many years. The rigidity and lack of freedom for self-expression by the operator in this working environment is generally considered to be adverse and counter-productive in the long run. Alternative improvements to the assembly work environment by various forms of collaboration have been tried and are still being tested and refined. They include, for example:

- work structuring;
- job enlargement;
- job rotation;
- job enrichment;
- quality circles;
- autonomous cell groups;
- teamwork;
- employee empowerment.

All of these approaches (Kochan, Cutcher-Gershenfeld and MacDuffie, 1992), some of which are devised within the just-in-time methodology, seek to avoid workplace monotony and reduce psychological stress. For example, small autonomous groups may be formed on the assembly line; within the group the operators can themselves determine their work method and have the right to be included in discussions about the equipment in their workstations, thereby bearing greater collective responsibility. Such organizational restructuring also aims to promote the assemblers' awareness of responsibility for their product.

Assembly work structuring in cells frequently leads to the creation of autonomous, flexible working groups oriented to the product, or to single workstations which are usually separated from automated sections by buffers. Such work structures may, in general, be safer and more flexible with regard to demand fluctuations; they also permit a more human-centered layout than those which conventionally interlink assembly lines.

In surveys, about 25% of all assembly workstations have been affected by certain types of work structuring since 1975. In about two-thirds of all surveyed companies, work structuring measures have been carried out. The predominant reason is the reduction of non-productive times. Other prominent reasons are flexibility increases with regard to variations in demand and a better work environment for employees.

1.5.4   Automation of assembly

A variety of mechanisms, components and facilities are presently available for assembly automation (Warnecke and Walther, 1982; Lotter, 1989; Boothroyd, 1992). They differ in degrees of automation (semi-automatic, fully automatic), in their complexity (automatic station, assembly cell, assembly line or system) and degree of flexibility (single-purpose automatic mechanism, flexible assembly system, assembly system with industrial robots).

Flexible assembly automation is considered an important rationalization strategy for the future, because:

1.   intensified competition and shorter product life demand more flexibility in completion of orders with shorter process times, quicker resetting and equipment reusability;
2.   elements for flexible layout of automated assembly systems are available as a result of technological advances in control engineering, material handling engineering and sensor technology, enabling cost-effective assembly automation even without mass production.

1.6   General taxonomy of assembly operations and systems

Assembly tasks include two basic categories: **parts mating** and **parts joining**. In parts mating two (or more) parts are brought into contact or alignment with each other. Parts joining means that after parts are mated, fastening is applied to hold them together. Mating tasks include: (1) peg in hole, (2) hole on peg, (3) multiple peg in hole and (4) stacking. Joining or fastening tasks involve: (1) fastening screws, (2) retainers, (3) press fits, (4) snap fits, (5) welding and related metal-based joining methods, (6) adhesives, (7) crimpings and (8) riveting.

In a recent survey of 24 product lines in the US (Martin-Vega *et al.*, 1995) the following assembly operation frequencies were observed:

Electronic assembly		**Mechanical assembly**	
Wiring	27%	Fastening by screw or bolt	38%
Surface mount technology (SMT)	15%	Riveting	26%
Soldering	12%	Pressing	6%
Through-hole insertion	7%	Miscellaneous operations	30%
Cabling	5%		
Cleaning	4%		
Miscellaneous operations	30%		

The three leading mechanical assembly operations shown are traditional and have been well established for years. Interestingly, in electronic assembly a relatively new technique, SMT, is listed as second, and is already used twice as frequently as through-hole insertion which it is replacing. This observation indicates the highly dynamic nature of assembly in general and electronic assembly in particular. Further details of these assembly tasks and operations are discussed in Chapter 2.

The total **assembly cost** of a product is a function of **product design** and the **assembly system** used for production. The lowest assembly cost can be achieved by designing the product so that it can be economically assembled by the most appropriate system. Assembly systems can be classified according to **facility organization** or **level of automation**. In terms of organization, these are: (1) assembly workstation, (2) assembly line, (3) assembly cell.

An **assembly workstation** is any area in which manual and/or non-manual work is performed, e.g. a bench, press, assembly machine, etc. Several workstations in sequence comprise an **assembly line** (section 1.1) where the assembly work progresses from one workstation to the next. In an **assembly cell** operations and processes are grouped to include all the assembly work needed for a complete subassembly or product.

Another way to classify assembly is by the type of assembler (Fig. 1.14): (1) manual assembly, (2) special-purpose assembly, (3) flexible, adaptable or pro- grammable assembly.

In **manual assembly**, control of motion, decision-making capability and flexibility, assuming well-trained operators, are superior to current machines. At times, it is economical to provide operators with mechanical assistance (fixtures, gauges, computer displays, etc.) to reduce assembly time and errors.

Special-purpose assembly machines are often one of a kind, customized for a specific product. They consist of a **transfer device** with single-purpose

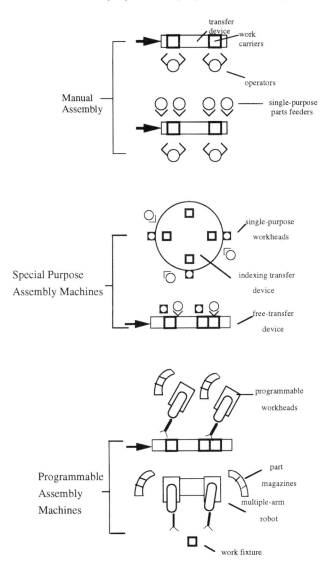

Fig. 1.14   Assembly systems by type of assemblers.

workheads and **feeders** at successive workstations. A transfer device can operate on **synchronous indexing** (as in Fig. 1.4) or on asynchronous indexing, also known as **free transfer**.

Flexible   assembly   is   achieved   by   **adaptable**   and/or   **programmable workheads**, or by **assembly robots**. It allows multiple operations to be performed at each workstation and provides flexible transfer of parts and tools between machines and cells. There is also flexibility in production volume and adaptability to product design changes and to different product styles. Advanced sensors and artificial intelligence techniques provide more accuracy.

There are three categories of **automation level and type**: (1) automation of plant level, (2) automation of shop level, and (3) automation of workstation/ line/cell level. Selection of automation level is discussed in Chapter 4.

Through **design for assembly (DFA)** techniques, one can determine whether a product design is (or can be made) compatible with automated assembly. The type of automation varies with the type of work transfer that is used. The four main transfer system types are: continuous, synchronous, asynchronous and stationary.

With the **continuous method of transfer**, parts are indexed continuously at constant speed. As a result, the workheads must also move during processing to keep up with the moving parts. Examples of continuous transfer are in bottling operations, packaging, manual assembly where the human operator can move with the moving line, and relatively simple automatic assembly tasks. The advantage is that the assembled product is kept moving at a steady speed, therefore there are no acceleration/deceleration problems.

In **intermittent transfer** the workstations' position is fixed. Parts are moved with an intermittent or discontinuous motion between the stations and then registered at the proper locations for processing. All parts are moved at the same time, hence the term **synchronous transfer** system which is also used to describe this method of workpiece transport. Examples of intermittent transfer applications can be found in machining operations, pressworking operations and mechanized assembly.

The **asynchronous transfer** system, often used as a 'power-and-free system', allows each work part to advance to the next station when processing at the current station has been completed. Each part moves independently of other parts, increasing flexibility. The flexibility of the system can be advantageous in certain situations. Another advantage is that in-process storage of workpieces can be incorporated into the asynchronous systems with relative ease. Power-and-free systems can also adjust, by allowing queues to form between stations, for line balancing problems where there are significant differences in process times between the stations.

Asynchronous lines are often preferred where there are manually operated stations, and cycle-time variations can lead to problems on either the continuous or synchronous systems. A disadvantage of the power-and-free systems is that the cycle rates are generally slower than for the other types.

In the **stationary** base part system, the base part (to which other components are added) is placed in a fixed location where it remains during the entire assembly process. This approach would be preferred when the assembled product is bulky or otherwise difficult to handle, for example airplanes, truck containers or recreational vehicles.

Another way to classify assembly systems is by physical configuration. Possible configurations of assembly machines include:

Table 1.3 Alternative work transfer systems for different assembly configurations

Assembly system configuration	Work transfer system			
	Base part	Continuous	Synchronous	Asynchronous
Dial-type	No	Yes	Yes	No
In-line	No	Yes	Yes	Yes
Carousel	No	Yes	Yes	Yes
Single-station	Yes	No	No	No

- dial-type;
- in-line;
- carousel;
- single-station.

In **dial-type** assembly base parts are loaded into fixtures or nests that are attached to a circular dial. Components are added and/or fastened at various workstations located around the periphery of the dial. Dial-type assembly machines are often designed to use a continuous motion rather than an intermittent motion.

The **in-line** assembly machine consists of a series of automatic workstations located along an in-line transfer system. It is the automated version of the manual assembly line. Continuous, synchronous or asynchronous transfer systems can be used with the in-line configuration.

The **carousel** assembly system is a hybrid between the circular work flow associated with a dial assembly machine and the straight work flow of an in-line system. The carousel configuration, with asynchronous transfer of work, is often used in partially automated assembly systems.

In a **single-station** assembly machine, operations are performed at a single, stationary location. Typically, the operation involves the placement of a base part at the workstation. The various components are added or fastened to the base part. In this case the assembled components must be delivered to the single station by one or more feeding mechanisms, while one or several workheads perform the various assembly and fastening operations. Table 1.3 lists the possible combinations of work transfer systems and assembly system configurations.

1.7 Assembly information representation

Assembly information representation is important: it is an essential first step to planning systems and sequences, topics that are discussed in detail later in the book; it serves for documentation and replanning; it involves interpretation of design data to identify relationships between assembled components.

Representation methods are categorized as manual or computerized. Two typical manual methods, widely used in industry, are the **assembly tree** and **assembly process chart** (Fig. 1.15(a) and (b)). Both convert drawings to logical steps and provide a graphical summary. Computerized methods have increased in popularity because they are useful for automated analysis and programming (Delchambre, 1992). Various CAD methods are available, e.g. wire frame or sweep representations, boundary representation (often called B-Rep), parametrized shapes and constructive solid geometry.

Three ways of representing computerized assembly information exist in addition to product CAD models (Rajan and Nof, 1996): (1) language-based; (2) data structures; (3) graph-based.

Language-based schemes have been developed for reasoning about assemblies. The first language, called AUTOPASS, was developed by Lieberman and Wesley (1977). It uses a world model (i.e. the world in which a system has to operate) that captures information on components and the spatial relationships between them. AUTOPASS is used to describe the way an assembly is built. It incorporates state change, tool and fastener statements. The second language, RAPT, was developed by Popplestone, Ambler and Bellos (1978). It adapts the numerical control language APT to robotic assembly. RAPT adds object geometry specification by additional commands such as FACE and BODY.

Representing assemblies with **data structures** (Lee and Gossard, 1985), relationships between components and subassemblies are captured by a 'virtual link' structure. The link specifies the nature of mating between components. Individual component models can be in the form of CAD representations. The benefit of this approach is that the spatial locations of components can be derived. However, in a similar way to language-based structures, the user must predetermine the assembly sequence. It is difficult to reason about valid sequences using such representation.

The more recent and popular representations for assemblies are **graph-based**. These schemes are more suitable for the generation of valid assembly sequences. Homem de Mello and Sanderson (1991) suggest an AND/OR graph representation with which a robot can pick a course of action based on any current state. The representation is similar to a state-transition graph (a graph in which states are shown as nodes, and links between the nodes represent possible transitions) with each node representing a stable subassembly. The representation can be used to generate both the optimal assembly and disassembly sequences. De Floriani and Nagy (1989) and Khosla and Mattikali (1989) describe the generation of an AND/OR graph given a solid model or a 3-D solid model of an assembly. Correctness and completeness of five different, yet interrelated, assembly representations are discussed by Homem de Mello and Sanderson (1991): directed graph, AND/OR graph, establishment conditions, precedence assembly relations and precedence

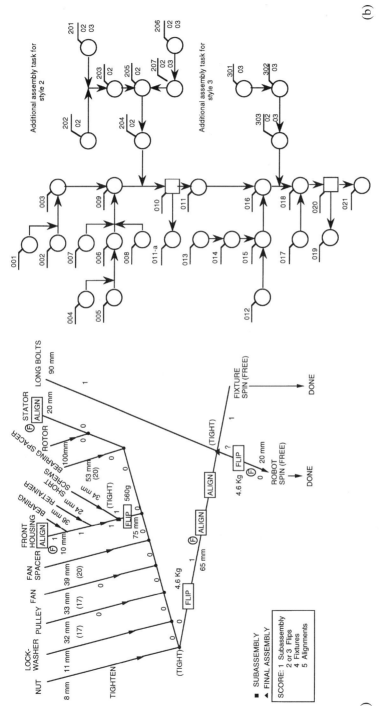

Fig. 1.15 Examples of assembly information representation: (a) assembly tree (Whitney, 1985); (b) assembly process chart (Csakvary, 1985).

component-mating relations. A computerized assembly representation is important for computer-aided design and planning methods, which are discussed in Chapter 3.

1.8   Fundamental models and measures of assembly

The significant economic impact of assembly on product cost discussed earlier in this chapter and the need for rationalization require the use of analytic models. A variety of models for assembly system design, optimization, planning and control are discussed in the following chapters. As an introduction, several fundamental models and measures are presented in this chapter. However, it is first interesting to observe the total number of parts that are included in an assembled product. Naturally, products that are assembled from more parts will have more costly and more complex assembly operations.

1.8.1   How many parts are assembled in a product?

What is the total number of component parts that are included in an assembled product? This question may be easier to answer for simple products than for complex ones. Consider the storage stacker shown in Fig. 1.16. It has 22 = 8 + 8 + 2 + 2 + 2 parts, assuming the two dowel pins are not needed. For a chess game, the total number of parts is also relatively easy to count: 2 kings + 2 queens + 4 rooks + 4 bishops + 4 knights + 16 pawns + 1 board + 1 instruction sheet + 1 box + 1 cover = 36. Table 1.4 illustrates several assemblies.

1.8.2   Assembly models and effectiveness measures

One approach to assembly analysis is by the five fundamental assembly models (Nof, 1995): chain, base, arithmetic, geometric and combinatorial assembly

Table 1.4   Total number of parts in various assemblies

Assembled product	Total number of component parts
Two-shelf storage stacker	22
Chess game	36
Average printed circuit board	~250
Missile	~25 000
Communication satellite	~200 000
Passenger car	~350 000
Passenger airplane (Boeing 777*)	~1 500 000
Passenger airplane (Boeing 747)	~4 500 000

* Major reductions in component numbers were achieved by extensive design for assembly efforts, and two jet engines are used instead of four.

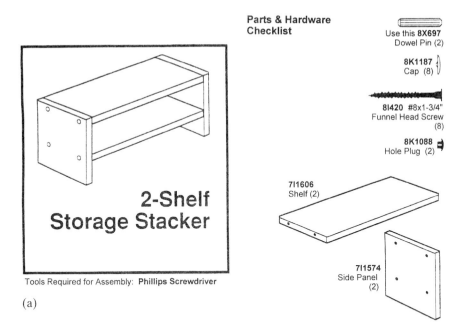

Parts & Hardware Checklist

Use this **8X697** Dowel Pin (2)

8K1187 Cap (8)

8I420 #8x1-3/4" Funnel Head Screw (8)

8K1088 Hole Plug (2)

7I1606 Shelf (2)

7I1574 Side Panel (2)

2-Shelf Storage Stacker

Tools Required for Assembly: **Phillips Screwdriver**

(a)

How to Assemble Your Stacker

• Fasten side panels and shelves together (*finished side of shelves facing up*), using the pre-drilled holes in the side panels and the starter holes of the shelves, as shown. Tighten screws flush with the surface of the side panels. Then, cover screwheads with caps; tap them into heads of screws with back of screwdriver handle.

• If you intend to stack units one on top of the other, insert a dowel pin into the top edge of each side panel, as shown. Use the screwdriver handle to tap the pins in. Carefully place another unit over the stacking pins. If the pins are not used, plug holes with small hole plugs.

Use **Dowel Pin** for stacking units. Otherwise, cover hole with small hole plug.

Cover Screwheads with Caps.

Use this **8I420** #8x1-3/4" Funnel Head Screw

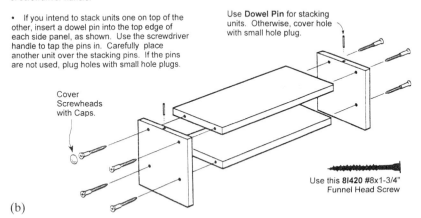

(b)

Fig. 1.16   Two-shelf storage stacker assembly (courtesy of Hirsh Co.): (a) product and parts; (b) manual assembly instructions.

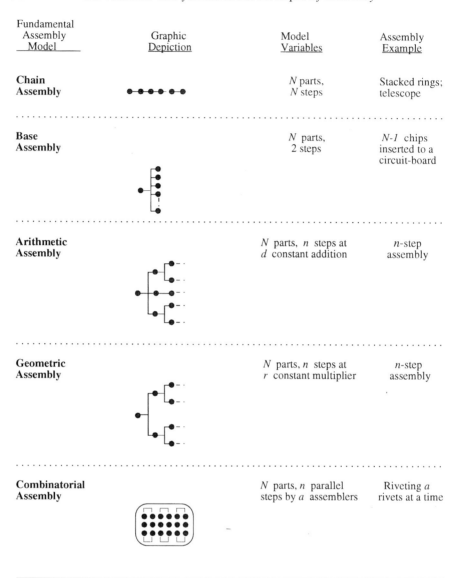

Fundamental Assembly Model	Graphic Depiction	Model Variables	Assembly Example
Chain Assembly		N parts, N steps	Stacked rings; telescope
Base Assembly		N parts, 2 steps	N-1 chips inserted to a circuit-board
Arithmetic Assembly		N parts, n steps at d constant addition	n-step assembly
Geometric Assembly		N parts, n steps at r constant multiplier	n-step assembly
Combinatorial Assembly		N parts, n parallel steps by a assemblers	Riveting a rivets at a time

Fig. 1.17   The five fundamental models of assembly (Nof, 1995).

(Fig. 1.17). These models are theoretical; nevertheless, they can be useful to the study of basic assembly characteristics. Practical assembly will usually follow a combination of these basic models. For all five models assume:

N = total number of components to be assembled;
n = number of assembly steps or operations.

Chain assembly In this model, component parts are added or stacked one at a time, or one in each operation, each to all the previously assembled items. Obvious examples are products that are chain-like, telescoping or modules added one on top of another. However, the chain model is not necessarily for the chain appearance but for the logic of adding one component at a time. Fixing a cup of tea may be an example, as may assembling a mechanical pencil by hand or sequentially inserting components on an assembly line. Assuming that the first component is fixtured as the first step, then for the chain model:

$$n = N \tag{1.1}$$

Base assembly Here the assembly involves one base part to which all the other components are added at the same step or operation. For instance, if all the electronic components are inserted at once from a magazine to a base printed circuit board, or inserted at the same SMT machine, the assembly follows the base model. Other examples are placement of parts on a kit tray (the tray is the base), or riveting an airplane window frame in a single workstation. The base model represents the shortest possible assembly operation, and in it always:

$$n = 2 \tag{1.2}$$

Arithmetic assembly An assembly is considered arithmetic when components are added to it at an arithmetic progression. In this case:

$$N = (n/2)(2a_1 + (n-1)d) \tag{1.3}$$

where a_1 = the number of components placed in the first assembly step and d = the number of parts (assumed to be equal) added in each subsequent step. For instance, for $a_1 = 1$ and $d = 2$ the assembly of nine components will require three steps ($1 + 3 + 5$) but so would the assembly of five, six, seven or eight components (e.g. $7 = 1 + 3 + 3$). In general, for:

$$a_1 = 1 \text{ and } d = 1 \quad n = -1/2 + 1/2\sqrt{(1+8N)}$$
$$a_1 = 1 \text{ and } d = 2 \quad n = \sqrt{N}$$
$$a_1 = 1 \text{ and } d = 3 \quad n = 1/6 + 1/6\sqrt{(1+24N)} \tag{1.4}$$

and so on. Observe that for $a_1 = 1$ and $d = 0$ the arithmetic model is equivalent to the chain model, and for $a_1 = 1$ and $d = N - 1$ it is equivalent to the base model. Thus an arithmetic assembly of $N = 10$ components will take ten steps for $d = 0$; four steps for $d = 1$ or 2; three steps for $d = 3$ to 7, and two steps for $d = 9$.

How many components can be assembled arithmetically in four steps? The answer for small products ($N = 10$) is that four steps can be applied with $d = 1$ or 2; for medium-size products ($N = 30$) d will have to be between 5 and 8. To

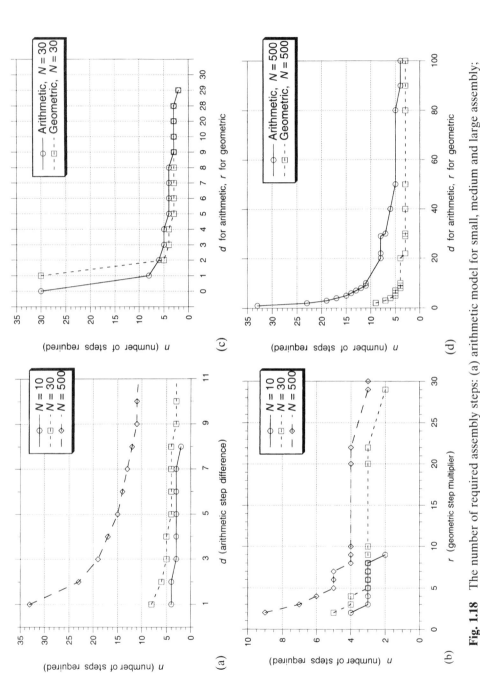

Fig. 1.18 The number of required assembly steps: (a) arithmetic model for small, medium and large assembly; (b) geometric model for small, medium and large assembly; (c) arithmetic vs. geometric, medium assembly; (d) arithmetic vs. geometric, large assembly.

assemble large products (say, $N = 500$) in four steps, d must be around 100 or 150 (Fig. 1.18).

Geometric assembly   An assembly is considered geometric if components are added to it at a geometric progression. In general,

$$N = (1 - r^n)/(1 - r) \tag{1.5}$$

where the number of parts added in each step is equal to the number of parts added in the previous step, multiplied by a constant r ($r \geq 1$); r is called the constant multiplier. For instance, with $r = 3$, starting with one part, the geometric assembly will add 3, 9, 27, 81, etc., parts to the assembly in each step, respectively. In general, starting with a single component, for:

$$r = 1 \qquad n = N$$
$$r = 2 \qquad n = \log_2(N + 1)$$
$$r = 3 \qquad n = \log_3(2N + 1)$$
$$r = 4 \qquad n = \log_4(3N + 1)$$

and so on. Thus a geometric assembly of $N = 10$ components will require four steps with $r = 2$, and three steps with $r = 3$ or 4. How many components can be assembled geometrically in four steps? The answer for various size products is:

For
$$N = 10 \qquad r = 1$$
$$N = 30 \qquad r = 2$$
$$N = 500 \qquad r = 8$$

Combinatorial assembly   This assembly model is somewhat different from the four previous models in that it considers the question of how many assemblers are needed for a given assembly. The model assumes:

- a = the number of assemblers (manual, automatic heads and/or robots) that execute the assembly of N components by N assembly activities, for example inserting N components in a circuit board, or N rivets in an airplane wing panel;
- each assembler executes one assembly activity (on one component) at a time, so if all a assemblers work in parallel, a out of the total N are completed.

In other words, the combinatorial assembly can be defined as the execution of N assembly activities in n steps, 'combinations', where each step includes x out of the N components in parallel, without repeating, where $x = a$ in all the steps except the last step, in which $0 < x \leq a$. The adjective 'combinatorial' implies that the assembly involves x out of N combinations without repetition, and $x \leq a \leq N$. For instance (Fig. 1.19), three assembly heads can insert 15 pegs in

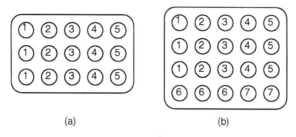

Fig. 1.19 Insertion of pegs in a board – combinatorial model (step numbers indicated): (a) five steps by three assembly heads; (b) seven steps by three assembly heads.

five steps; to insert 20 pegs they would require seven steps. The chain model, in comparison, assumes $a = 1$, and therefore, 15 pegs require 15 steps. The base model always implies two steps, which means that the first step is executed by one assembler to secure the base board, and the second step by 15 assemblers or assembly heads working in parallel. In arithmetic assembly and geometric assembly, the number of assemblers (or the ability of assemblers to handle a variable number of components in each assembly step) is variable.

Effectiveness measures   Several measures are used with the models throughout the book to assess the performance and cost-effectiveness of alternative assembly designs and decisions.

1.   **Assembly throughput** – the rate of good-quality assemblies produced by an assembly system. Often this rate is determined by the bottleneck resource of the system. Equivalent terms are system capacity, production rate or yield.
2.   **Assembly capacity** – the maximum throughput.
3.   **Assembly lead-time** – total time required to produce one assembly including all waiting time. Equivalent terms are throughput time, flow time or unit time.
4.   **In-process inventory** – total amount of material or products in the system, either waiting or being processed. Equivalent terms are WIP and work-in-process.
       During a steady-state or over a relatively long period,

$$\left(\text{in-process inventory}\right) \approx \left(\text{throughput}\right)$$
$$\times \left(\text{mean assembly lead-time}\right). \tag{1.6}$$

The goal in effective assembly is to have a shorter lead-time, higher throughput and lower in-process inventory.
5.   **Availability** – the percentage of time the system (or cell or workstation) is available and not down because of failures, part delays or poor plan-

ning. This measure indicates the readiness of the facility for useful work. In modern production it should not be confused with idle time. For instance, when a cell is idle for a period of time and it is not assigned any work due to poor production planning, it is indeed considered idle. On the other hand, if a cell is kept idle intentionally for excess (reserve) capacity, the objective is flexibility or reliability, and the cell is considered available. Availability is usually the complement of downtime.

6.   **Flexibility** – the difference (over time) between the system's capacity and the demand, as a ratio of the capacity:

$$\text{Flexibility} = \left(\max. \left(\text{capacity} - \text{demand};\ 0\right)\right)\big/\text{capacity} \qquad (1.7)$$

There are many possible measures of flexibility. The simple one defined here focuses on the excess capacity which would provide the flexibility to accommodate increases in demand, either in assembly product and volume variety, or due to time spent on assembly design changes and experimentation with new products. The flexibility measure in expression 1.7 is zero if the capacity cannot fulfill the total demand over time. A high flexibility measure may mean that there is too much excess capacity.

A more detailed measure of flexibility in assembly, the versatility index (Makino, 1990), relates the frequency of model changes that can be accommodated by a given assembly process over time, at a desired efficiency, to the production volume. For instance, a system with a versatility index of 100 at 90% efficiency would allow model changes only once every 20 hours or more. A versatility index of 1000 indicates that the system can accommodate model changes every 10 minutes.

7.   **Quality** – Several measures are associated with the quality of assembly:

(a)   **scrap rate** or **defect rate**, which is the complement of the yield;
(b)   total **rework time** per assembly;
(c)   total **inspection and test time** per assembly;
(d)   **responsiveness** to customers, also called service level, measured as the percentage of time customer orders are filled correctly and on time.

8.   **Cost per assembly** – the total cost of each acceptable assembly produced by the system. This cost is generally the sum of material cost, operations cost, cost of quality such as rework cost and cost of system downtime.

Other measures of effectiveness will be defined later. Several measures can be calculated indirectly with the five fundamental models. For instance, the number of steps per assembly, n, indicates throughput and lead-time. A capacity measure is indicated by the **parallelism** of each model, since parallel assembly tasks require less lead-time.

In combinatorial assembly, the number of assemblers, a, determines the degree of parallelism. In arithmetic assembly, higher values of d imply higher degrees of parallelism, and geometric assembly with $r = d$ has an even higher relative degree of parallelism, as shown in Fig. 1.18. Clearly, the base assembly model represents ideal parallelism, with the least possible number of steps ($n = 2$). Naturally, a cost evaluation of alternative models is necessary to analyse the influence of cost per step for each model.

1.9   Assembly planning issues

The five fundamental models in section 1.8 illustrate analysis of assembly processes and systems. Another family of models described in the book deals with operational planning and control. This type of model can be introduced by the general **Assembly Plan Problem** (following Drezner and Nof, 1984). In this section, the Assembly Plan Problem and its models are defined and several practical cases are illustrated.

1.9.1   Characteristics of assembly operations

Assembly operations depend on three main characteristics: product variety and mix, variety and quantity of components and sequence of tasks. Depending on product variety and mix, assembly systems operate in one of three modes: (1) single product, with design changes that require only equipment-program changes; (2) changeable product, where minor resources such as fixtures and feeders are changed for each model switch; or (3) multiple products, where the facility must possess all required assembly capabilities at all times.

Concerning components' variety and quantity, certain products are assembled each with one component of each type, for instance a toolkit with three different tools. Other products each have a number of identical components, e.g. a board with four identical legs. Instead of picking individual components, a magazine containing several components can be applied. Inexpensive, small and frequently used components are kept in bulk, although not all of them are used for every product.

The third characteristic is task sequencing. Sometimes, the sequence of tasks (or motions) is dictated by the nature of components. For instance, components that accept other components must be placed first. If a sequence is completely defined, planning is reduced to optimizing the resource layout. Otherwise, the task sequence is also subject to optimization.

The following resources are typical (Fig. 1.20): one or more assemblers (operators or robot arms); input/output equipment, e.g. conveyors; component storage, e.g. feeders, bins, pallets; peripheral machines, e.g. presses; assembly accessories, e.g. powered screwdriver or special-purpose removable grippers.

1.9.2   Assembly Plan Problem definition

It is assumed that a product design provides specification of (1) the components and quantities to be assembled into each product, and (2) the physical

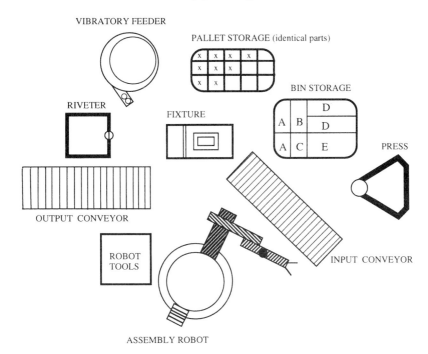

VIBRATORY FEEDER

PALLET STORAGE (identical parts)

BIN STORAGE

RIVETER

FIXTURE

A B | D
A C | E

PRESS

OUTPUT CONVEYOR

ROBOT TOOLS

INPUT CONVEYOR

ASSEMBLY ROBOT

Fig. 1.20 Example of a general assembly layout (Nof and Drezner, 1986).

placement and location of each component on the assembled product. For example, a printed circuit board design determines the specific electronic components and their exact placement on the board. With this input, the Assembly Plan Problem is defined as follows:

1. Find the optimal location of each resource, e.g. location of component feeders, and arrangement of components in bins.
2. Find the optimal sequence of assembly tasks if it is not pre-specified.

The Assembly Plan Problem and its solution are explained in detail by Nof and Drezner (1986) and are summarized here.

The Simple Assembly Plan problem (SAP) can be formulated when the number of bin-cells, b, is equal to the number of components in a product. (Note: bin-cells are smaller divisions within a bin.) The total movement time of the assembler arm, which is moving loaded with a component from a bin-cell and empty on the way back, is to be minimized. This problem can be solved by well-known algorithms for the Traveling Salesman Problem (TSP) (Crowder and Padberg, 1980). The solution involves definition of $2b$ 'cities' as b bin-cells and b locations on the assembled product. Assuming the assembler arm handles only one component (or magazine) at a time, the arm must alternate between bin-cells and insertion locations. Therefore, there is no connection between any two bin-cells or any two insertion locations. This guarantees that the arm moves from a bin-cell to an assembly location, and when it moves back empty, it must move to another bin-cell. The distances (or travel times) between the cities can depend on the part type because we know which part is located at each bin-cell. Also, distances (or travel times) for a loaded arm may be different from those for an empty arm.

Small problems can be solved optimally by complete enumeration. For large problems, one heuristic approach is to break the tour between the 2*b* cities into two separate problems: (1) tour of all loaded arm moves, termed the **Bin Assignment Problem** (BAP); (2) tour of all empty arm moves, termed the **Pick-Insert Sequencing** (PIS) problem. Each of these tours can be formulated as a linear assignment problem with *n* cities. For the loaded arm tour the assignment obtained is of parts to bin-cells. Given this assignment, the empty arm tour can be formulated as a TSP among *b* cities.

To illustrate, suppose six component types in six bin-cells are to be assembled to a base which is moving on a conveyor. Given the physical placement of components and the matrix of transfer times between each bin-cell and placement location and back, the SAP can be defined and solved. Solving by the TSP with 12 cities (2×6) yields the optimal layout, sequence and total assembly time. Solving for comparison by the BAP and PIS was found by Nof and Drezner (1986) to yield the same solution as the optimal in half the cases. The total cost of the heuristic solution never exceeded the optimal cost by more than 1%, and was only 0.2% higher on the average.

1.9.3   *Electromechanical assembly planning example*

An electromechanical product, assembled by a robot, comprises 10 components: two large frame-ends providing structural support and eight small parts. The 10 components are conveyed on 10-cell pallets, one pallet per assembly, to a fixed position. From this position they are picked up by the robot. Finished assemblies and empty pallets leave on the conveyor.

The SAP model can be applied as follows. Repeat components that are to be assembled in *K* different places are treated as *K* different components; movements which are not allowed (e.g. from small cells to assembly positions of large parts) are assigned very large distance values; the different speed of the loaded vs. unloaded arm is handled by multiplying the time matrix by a **load factor** relative to each carried part.

A TSP solution for a given problem (Nof and Drezner, 1986), using a weighted time matrix for 20 'cities', yields the optimal solution with a total arm movement of 787 units (= 471 units when loaded, plus 316 when empty). Solution by the heuristic procedure of BAP–PIS results in a close total of 788 units. (First the BAP procedure finds the best total movement of the loaded arm alone, 470, then the total distance of the empty arm is found by the PIS model as 318.) The results serve as a guideline in implementing the assembly cell.

Four practical extensions of particular assembly situations are considered and illustrated by Nof and Drezner (1986): single conveyor feeding of identical, repeated conponents; the number of bin-cells is greater than the number of component types stored; different products, with the objective to minimize the expected total transfer time; different products, with the objective to minimize the maximum transfer time. An extension of the Assembly Plan Problem to multiple, cooperating robots (assemblers) is developed by Nof and Drezner (1993). They propose a method to determine an optimal plan, which includes assignment of parts to each robot (assembler).

The Assembly Plan Problem has been described in this chapter as an introduction to a variety of useful planning models covered later in the book.

1.10 Summary

Almost any product of interest around us involves assembly. Industrial assembly has a significant economic role: about one-third of the labor force in manufacturing companies is employed in assembly, and over one-third to one-half of the annual value-added of industrialized countries is created by production activities with a medium-to-high assembly content.

Industrial assembly has evolved as an integral part of production from a purely manual to a partly mechanized activity, from being performed at stationary workbenches to assembly lines, and then to flexible assembly with a combination of manual assemblers, automated machines and robots. The fundamental concepts of these developments and approaches to rationalize the design of assembly operations and systems have been introduced and illustrated in this chapter.

Design of products for assembly has been the foremost contributor to time and cost savings in assembly. Other methods, including automation, improved work organization and other engineering and management techniques, have also been recognized as useful for effective and efficient assembly. Five fundamental models of assembly analysis and some general models for assembly planning have been described as introductory models. A variety of other models will be included in the following chapters.

After covering the assembly basics, it is time to delve into the assembly operation itself: how is it actually done? Chapter 2 is devoted to assembly tasks and technology.

1.11 Review questions

1. How can you explain the relatively high percentage of assembly workers in the more developed, industrial countries?
2. A tuna sandwich is an assembled product. Describe how it would have been assembled during each of the assembly evolution stages.
3. Try to develop an assembly tree and assembly process chart for the reaper assembly described in Fig. 1.3.

4. What changes in industrial assembly can be expected for countries that during the 1990s have moved from a centrally planned economy to a market economy?

5. For each of the following industries, suggest which assembly rationalization approaches would be most effective:
   (a) aerospace;
   (b) pharmaceutical;
   (c) fast food;
   (d) furniture;
   (e) plumbing fixtures.

6. For each of the possible system combinations in Table 1.3, identify the advantages and disadvantages for assembly of the storage stacker (Fig. 1.16).

7. Find the unique characteristics and specifications of the PUMA robot and of a SCARA robot, and explain what makes certain robots ideally suited for assembly.

8. For each one of the five fundamental models described in section 1.8 find examples of products that can be assembled according to the model.

9. Count the number of parts in the paper trimmer assembly (Fig. 1.1(a)), and then:
   (a) Try to organize them according to the five fundamental assembly models.
   (b) Which model would be the shortest in terms of steps? Which would be the longest?
   (c) Which model would be the most logical for this product?

10. Explain under what conditions it would make sense to apply combinatorial assembly for:
    (a) circuit boards;
    (b) television sets;
    (c) telephones;
    (d) chairs;
    (e) chocolate boxes.

11. Compare the size of products that can be assembled in five steps according to each of the five fundamental models.

12. Develop an assembly tree for:
    (a) a pen;
    (b) a pair of scissors;
    (c) a table.

13. Develop an assembly process chart for the products in question 12.

14. List the precedence relations for component mating in questions 12 and 13.

15. Use the Assembly Plan Problem models to design a manual assembly cell for a game box (such as chess) assembly.

(a) Describe the product components, assembly resources and layout plan.
(b) Describe the operational characteristics.
(c) Define the input information for each of the three possible production modes.
(d) Define design and planning problems that you can solve.
(e) Using hypothetical sets of values for the problem data, develop specific solutions.

References

Boothroyd, G. (1992) *Assembly Automation and Product Design*, Marcel Dekker, New York.

Crowder, H. and Padberg, M.W. (1980) 'Solving large scale symmetric traveling salesman problems to optimality', *Management Science*, Vol. 26, pp. 495–509.

Csakvary, T. (1985) 'Planning robot applications in assembly', *Handbook of Industrial Robotics* (ed. S.Y. Nof), Wiley, New York, Chapter 63, pp. 1054–83.

De Floriani, L. and Nagy, G. (1989) 'A graph model for face-to-face assembly', *Proceedings of the IEEE International Conference on Robotics and Automation*, Scottsdale, AZ, May 14–19, pp. 75–8.

Delchambre, A. (1992) *Computer-Aided Assembly Planning*, Chapman & Hall, London.

Drezner, Z. and Nof, S.Y. (1984) 'On optimizing bin picking and insertion plans for assembly robots', *IIE Transactions*, Vol. 16, No. 3, pp. 262–70.

Homem de Mello, L. and Sanderson, A.C. (1991) 'Representations of mechanical assembly sequences', *IEEE Transactions on Robotics and Automation*, Vol. 7, No. 2, pp. 211–27.

Hounshell, D.A. (1984) *From The American System to Mass Production 1800–1932, The Development of Manufacturing Technology in the U.S.*, The John Hopkins University Press.

Khosla, P.K. and Mattikali, R. (1989) 'Determining the assembly sequence from a 3-D model', *Journal of Mechanical Working Technology*, Vol. 20, pp. 153–62.

Kochan, T., Cutcher-Gershenfeld, J. and MacDuffie, J.P. (1992) 'Employee participation, work redesign, and new technology', *Handbook of Industrial Engineering*, 2nd edn (ed. G. Salvendy), Wiley, New York.

Lee, K. and Gossard, D. (1985) 'A hierarchical data structure for representing assemblies: Parts 1 and 2', *Computer-Aided Design*, Vol. 17, No. 1, pp. 15–24.

Lieberman, L.I. and Wesley, M.A. (1977) 'AUTOPASS: an automatic programming system for computer controlled mechanical assembly', *IBM J. Res. Develop.*, pp. 321–33.

Lotter, B. (1986) 'Automated assembly in the electrical industry', *Automated Assembly*, 2nd edn, Society of Manufacturing Engineers, pp. 49–56.

Lotter, B. (1989) *Manufacturing Assembly Handbook*, Butterworth, London.

Makino, H. (1990) 'Versatility Index – an indicator for assembly system selection', *CIRP Annals*, Vol. 39, No. 1, pp. 15–18.

Martin-Vega, L.A., Brown, H.K., Shaw, W.H. and Sanders, T.J. (1995) 'Industrial perspective on research needs and opportunities in manufacturing assembly', *Journal of Manufacturing Systems*, Vol. 14, No. 1, pp. 45–58.

Nevins, J.L. and Whitney, D.E. (1978) 'Computer-controlled assembly', *Scientific American*, Vol. 238, No. 2, pp. 62–74.

Nof, S.Y. (1995) 'Five fundamental models of assembly – definitions and calculations', Research Memo 95-26, Purdue University, West Lafayette, IN.

Nof, S.Y. and Drezner, Z. (1986) 'Part flow in the robotic assembly plan problem', *Robotics and Material Flow* (ed. S.Y. Nof), Elsevier Science, New York, pp. 197–205.

Nof, S.Y. and Drezner, Z. (1993) 'The multiple-robot assembly plan problem', *Journal of Intelligent and Robotic Systems*, Vol. 5, pp. 57–71.

OECD (1994) *Industrial Structure Statistics (1992)*, OECD, France.

Popplestone, R.J., Ambler, A.P. and Bellos, I.M. (1978) 'RAPT: a language for describing assemblies', *The Industrial Robot*, Vol. 5, No. 3, pp. 131–7.

Rajan, V.N. and Nof, S.Y. (1996) 'Minimal precedence constraints for integrated assembly and execution planning', *IEEE Transactions on Robotics and Automation*, Special Issue on Assembly and Task Planning, Vol. 12, No. 2, pp. 175–86.

UNIDO (1988) *Handbook of Industrial Statistics*, UN Industrial Development Organization, Vienna.

Warnecke, H.J. and Walther, J. (1982) 'Automatic assembly – state of the art', *Proceedings of the 3rd International Conference of Assembly Automation*, Boeblingen, pp. 1–14.

Warnecke, H.J., Schweizer, M., Tamaki, K. and Nof, S.Y. (1992) 'Assembly', *Handbook of Industrial Engineering*, 2nd edn (ed. G. Salvendy), Wiley, New York, Chapter 19.

Whitney, D.E. (1985) 'Part mating in assembly', *Handbook of Industrial Robotics* (ed. S.Y. Nof), Wiley, New York, pp. 1084–116.

2

Assembly tasks and technology

2.1 Introduction

Toddlers all over the world are often given toys to teach them, quite early in life, basic assembly skills. Some toys have pegs of different shapes to be inserted in matching openings in a box. Other toys have rings of different diameter, the purpose being to stack and insert them on a post according to size. Such toys help develop the child's skills and coordination. Naturally, they involve manual dexterity.

Industrial assembly and disassembly started with manual operations. Over the years, tasks have evolved from manual to mechanized and robotic ones. In the process, their characteristics have also changed. In addition, components are increasingly made of new materials such as composites, special alloys, ceramics, etc. Designers have to consider such developments. For instance, while screws and bolts have traditionally been the fasteners of choice, a designer may prefer adhesives or clinching for joining certain components or when planning to apply automation.

Because of the increasing influence of automation on assembly tasks, section 2.2 analyses the interrelations between flexibility goals and automation, the limits of automation, and its impact on companies and their workers. The objective is to determine the desirable degree of automation, which in turn determines task selection.

Sections 2.3 and 2.4 describe the requirements and characteristics of assembly and disassembly tasks, respectively. The technical details of the most common tasks for insertion and joining are covered in section 2.5. This section also discusses the challenges encountered in special assembly. For instance, different considerations are necessary for non-rigid components, e.g. those made of cloth, rubber or leather, and for small components, such as parts for precision instruments.

2.2   Levels of automation and flexibility in assembly

2.2.1   Flexibility

In many cases today the issue is not about automating a single, constant assembly process, but the creation of automated devices that are capable of performing different or, in some instances, similar tasks of assembly. So these devices have to be flexible, and this demand for flexibility is driven mainly by the market. With growing market saturation greater product variety is offered/demanded, coinciding with an increase in the critical behaviour of the customer, so that overall there has been a change from a seller's market to a customer's market. In order to be able to measure any increase in the flexibility of the assembly industry, relevant starting-points have to be established. We will therefore begin by examining the term flexibility more closely.

In order to perform different assembly tasks different states of an assembly system are required. If we consider first a single unit system, the ability to fulfill different assembly tasks depends on:

- how many different states can be assumed automatically;
- how long it takes to execute a change of state (the setup process).

The number of different possible situations and the type of changes required make it sensible to identify two kinds of flexibility, and to make an overall judgment on flexibility, both are connected. Both kinds concern the possible situations of the system, but only the different work situations of a manufacturing system during an undisturbed process run are of interest. Flexibility is interesting from two aspects of the manufacturing system: during the stable operation (process), and during transition and breakdown situations.

The number of different work situations a manufacturing system can be related to is called the **application flexibility**. This is the probability that an arbitrary manufacturing task inside a definite class of such tasks can be carried out. A comparison between the application flexibility of different assembly systems is only of use if it is applied to the same range of manufactured parts.

Adaptation flexibility, on the other hand, is a measure of the duration of time and the cost incurred for an assembly system at the transition from one definite work state to another. In quantitative terms, adaptation flexibility can only be indicated relatively, i.e. by comparing one assembly system with another and only for one defined change of state. The change of state is marked by one definite (work) possibility before and one after the change.

According to these observations, as a measure of the adaptation flexibility of an assembly system, the relation between what it requires to set up, taken as setup time T_R or setup cost K_R, and the corresponding dimensions T_{R_o} and K_{R_o} of a system considered optimal, has to be established.

A rough estimate of the two flexibility components for some cutting machine tools is shown in Fig. 2.1.

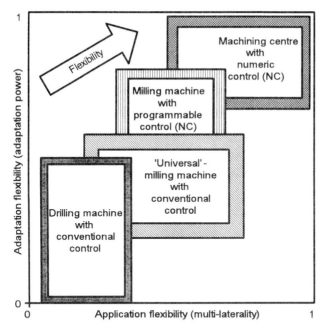

Fig. 2.1 Application flexibility and adaptation flexibility.

2.2.2 Degree of automation

Automation very rarely means that a plant works automatically in all its functions. According to DIN 19233:

> An area of automation is an automated area of a plant. A plant can have several areas of automation.

An automatic central lubricating oil supply for a steam turbine is an example of automation of a complete system; the automatic speed regulation control is another.

Figure 2.2 shows the progression in the automation of swivel-base production. The increase of automated partial functions on peripheral areas of the manufacturing process is clearly visible.

Again according to DIN 19233:

> The degree of automation describes the share of automated functions of the overall function of an installation. It can only be indicated for a system with known limits. Numerically, it is commonly indicated as the quotient of the number of automated functions to the total number of functions of a system, so that the numerical value is between 0 and 1.
>
> Partial automation of a system is where not all systems are automated, i.e. the degree of automation is less than 1.

Semi-automation as a general indication instead of partial automation should be avoided, because one could demand a degree of automation of

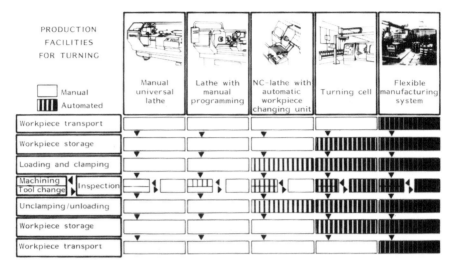

PRODUCTION FACILITIES FOR TURNING	Manual universal lathe	Lathe with manual programming	NC-lathe with automatic workpiece changing unit	Turning cell	Flexible manufacturing system
Workpiece transport					
Workpiece storage					
Loading and clamping					
Machining / Tool change / Inspection					
Unclamping/unloading					
Workpiece storage					
Workpiece transport					

☐ Manual ▥ Automated

Fig. 2.2   Gradual automation of swivel-base production.

exactly 0.5. But in operating practice a semi-automat is a machine that operates automatically according to a program, but the supply function for blanks and prefabricated units as well as the initiation of the manufacturing process is executed manually.

Despite the definition according to DIN 19233 there is no uniform method to determine the degree of automation. Therefore the data for different processes is not comparable. Specification of a degree of automation is of little use without an exact description of the process and the method of determination. As soon as a degree of automation is mentioned the system concerned has to be defined, e.g.:

- operation cycle;
- place of work;
- machine;
- group of machines;
- workshop;
- plant;
- branch;
- economy.

In determining the degree of automation of a manufacturing system it must be specified whether the following functions are to be considered or not:

- setup;
- handling;
- control;
- maintenance and repair;
- job planning;

- construction;
- administration.

For example, if our consideration of a copying lathe is limited only to the turning process, i.e. omitting (part) handling, dimensional inspection and failure recovery, then the process being considered has a degree of automation of 1. But if all the necessary partial functions are included, such as failure indication and recovery, then the degree of automation is far less than 1.

There are a number of methods to determine quantitatively the degree of automation. These can be divided into two groups:

- **Determination through comparison with a graded scale.** The graded scale contains all the functions of a defined assembly process with regard to a defined system and the corresponding degrees of automation. For the process concerned the degree of automation is found through comparison with the graded scale. This procedure is approximately the same as using Mohs' hardness scale or the Beaufort wind speed scale. Figure 2.3 shows an example of the definition of mechanization grades according to Bright (1958).
- **Determination of a quotient of two or more reference units weighted according to the circumstances.** The reference units, for example, may be:
  - amount of production;
  - manpower;
  - time units such as time of work, allowed time, etc.;

From the worker		Through control-mechanism, testing determined work sequences		Through variable influences in the environment				Origin of the check	
Variable		Fixed in the machine		React to signals	Reacts to the execution			Type of the machine reaction	
					Selects from determined processes	Changes actions itself inside influences			
Manual		Mechanical (not done by hand)						Energy source	
1 2 3	4	5	6	7	8	9 10 11	12	13 14	15 16 17	Step-No.
Manual	Machine tools, manual controlled			Machine system with remote control		Measures characteristics of the execution · Registers execution	Changes speed, position change and direction according to the measured signal	Identifies and selects operations	Corrects execution after the processing · Foresees the necessary working tasks, takes care of the execution	Steps of the mechanization

Fig. 2.3 Scale for comparing grades of automation.

- number of functions of motion;
- program steps;
- number of decisions made during the process (DIN 19233).

In DIN 19233 the degree of automation is determined as the quotient of the sum of estimated decisions made automatically to the decisions made in total – including those made with human assistance.

The calculation of the assessment should always indicate the weighting that has been used. For example, the total function of a system has 10 decisions, six made by technical devices and four by humans. But because these decisions are not equal, the degree of automation is not $6/(4 + 6) = 6/10 = 0.6$. The decisions can be weighted by the number of necessary program steps:

Decision no.	Assessment (program steps)
1	14
2	10
3	12
4	8
5	15
6	11
Sum automated:	70
7	31
8	40
9	72
10	55
Sum human:	198

$$\text{Degree of automation} = \frac{\text{Sum of decisions made by machine}}{\text{Total sum of all decisions}} = \frac{70}{70 + 198} = 0.27$$

In each case when quantitatively determining the degree of automation the following requirements have to be observed:

- the ability to reproduce the procedure;
- objectivity of the base of comparison (i.e. farthest possible independence from subjective estimates, operating influences, etc.);
- numerical values should be between 0 and 1 to simplify the determination.

2.2.3   Targets and restrictions for automation

The operational rationalization aims for automation in manufacturing are:

- improvement in time and cost structures in manufacturing in the sense of more economic production;

- improvement in productivity;
- improvement in working conditions for humans.

From these general aims, specific targets can be derived that will have different priorities in different companies. While one company may emphasize decreasing process times, another may stress quality and know-how. What and how much to automate is therefore influenced by these targets and their priorities.

Figure 2.4 shows the influence of the most important targets for rationalization on automation. The savings of main and idle machine time as well as of costs is mainly achieved by automation of the assembly process, i.e. of the main and additional functions. Reductions in setup and throughput time are achieved by automating the flow of parts, tools and information, as well as of accessory agents and waste. Considerable reductions in staff and therefore in employment costs can only be achieved with total automation of all operations.

Quality issues such as repetitive accuracy, surface finish and tolerances are mainly improved by the use of numerical control. Improvements in working conditions can be achieved in places where humans have to work close to the manufacturing process and the work is monotonous or overexerting.

The type of automation, how much is automated and when it is automated as well as the specific targets for automation are determined mainly by the assembly requirements of each company and the range of parts involved. The assembly requirements are determined by the assembly tasks or the assembly program.

Inhibitors of automation   Automation is not possible if the following requirements are not fulfilled:

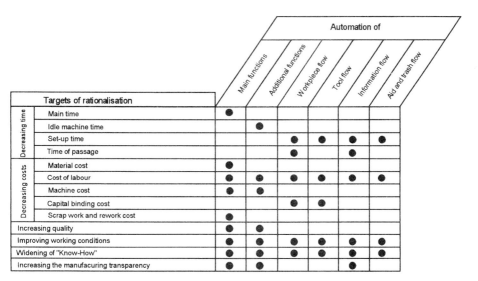

Fig. 2.4   Targets of automation in assembly (Westkämper, 1977).

- The process has to be capable of being defined by an algorithm. (An algorithm is an unambiguously defined sequence of operations that is either fixed from the start or depends on the results of preceding operations.)
- The process must be capable of being measured by technical devices (sensors). The lack of suitable sensors is the main problem for the automation of deburring and assembly procedures, but relevant devices are presently being tested.
- The operations that are to be automated must be repeatable. Automation is most successful at present for similar recurring actions (mass production). For small-lot or single-piece production automation is economically sensible only if there are recurring, similar complicated tasks and if flexible devices may be used. Repeatable operations can be achieved by:

  - standardization of the products;
  - simplification;
  - normalization;
  - standardization of assembly processes and sequences;
  - the high investment involved means extensive use of the automated devices is required, but this in turn requires that high volumes of product sales are achieved;
  - the investment for automation has to be available;
  - the assembly areas preceding and succeeding the automated area must be coordinated with the least possible expense. In many cases a technically and economically feasible automation fails because of the high costs in these areas;
  - the procedures considered for automation must be programmable, e.g. numerical control of machine tools or program control of industrial robots.

2.2.4   *The influence of automation*

Influence on humans   From the 1950s until the second half of the 1970s there was a mainly defensive attitude towards the latest technical achievements and automation. Many examples have shown that with automation jobs may well be 'rationalized away', but at the same time automation is a key to increasing productivity and therefore preserve competitiveness – which in turn preserves jobs. The concern that only large firms have the necessary capital to afford automation was proved to be only partly true:

- Automation methods and devices have been developed that are also suitable for smaller companies. A typical example is the advance of microelectronics in smaller firms.
- Small, flexible companies have developed into prosperous subcontractors for large companies.

- The expansion of economic trading to new geographical areas has counteracted the threat of monopolization.

The following developments in manufacturing companies with increasing automation can be observed:

- workers assume more responsibility for the company's operations;
- automation contributes to the design of human-oriented processes and human-centred systems.

Automation can influence the humanization of work in the following ways:

- reduction in the risks and safety hazards of the workplace;
- reduction in environmental hazards;
- reduction in physical stress;
- reduction in mental stress.

Influence on companies   As a result of increasing competition in national and international markets, many companies are being forced to look for new ways to manufacture high quality products more economically and to adapt their product line quickly to changes in customer demands. In production technology there is a need to increase both productivity and flexibility to stay competitive in the future.

Such structuring has not been possible before. Developments over the past decades have been marked mainly by an increase in the production capacity of plants. For example, in cutting, the cutting speed was increased by applying new tool-materials. Today the limits of such improvements have generally been reached. Now new problems are emerging and another range of conflicts must be resolved (Fig. 2.5).

The classic conflict in fulfilling the demands of the market is the difference between the lowest possible level of capitalization with regard to the means of production and distribution and the highest possible readiness to deliver, with products adaptable to the market. The capital needed is determined by the productivity of the production means and the use of their capacity. Maximizing the use of capacity and at the same time minimizing throughput time (the scheduling dilemma) is only possible with an increase in flexibility.

Productivity and flexibility can today be increased at the same time, and this is a paradox in itself. A highly productive installation like a production line or a special-purpose machine is, on the one hand, unbeatable regarding productivity but, on the other hand, the more rigid equipment is the least adaptable. With new technologies, the solution to this dilemma has emerged.

Thus the goal is not only an increase in the capacity of individual machines, but a better use of the capital invested. Present developments in production technology are characterized by this target. The capacity offered by a production plant (Fig. 2.6), that to date has been increased vertically by increasing the output per time unit, should now be increased through more usage over the

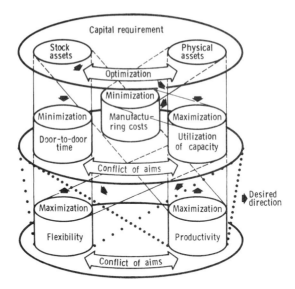

Fig. 2.5 Optimization of capital demand within the target system production.

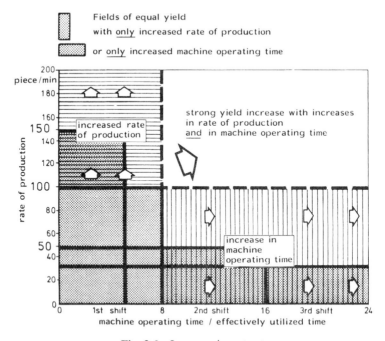

Fig. 2.6 Increase in output.

total time. The machine should not just work fractions of eight hours per day, but much longer than that.

The increase in capital productivity following an increase in machine time over the first, second and maybe even third shift is no longer dependent on an extremely intensified production rate per time unit. Indeed, less can be more, because the probability of failure in the system is decreasing hence the availability of machines is increasing.

In some factories highly automated assembly systems are run during unsupervised night operation at 50 to 65% of the capacity of day operation, to prevent breaking of tools.

'Capital productivity' may be defined as the ratio Turnover/Investment capital. However to increase capital productivity requires not just maximum usage of equipment but also the optimization of storage capacity. Companies are forced to optimize their stock holding. Experience shows that in custom production this target cannot be achieved by decreasing material costs (starting prices can only be decreased 5 to 10% on average), but rather by the possibilities of material flow scheduling to decrease costs and stock (by about 20%). A reduction in stock may also be achieved by the use of a suitable information system. To achieve such stock reduction, a company-specific information system can be set up to include not only the well known advantages of information processing, e.g. mass data processing and up-to-date and orderly representation of information, but also, for example, functions such as prognosis calculation, approximate plan simulation, customer order management (stocks of finished goods), consolidated order management, and calculation of order amounts and terms.

With these measures stock reductions can be achieved that increase capital productivity noticeably, without the need for high investment. Tightly linked to this idea is the consequent planning and regulation of transport, conveyance and stockpiling with regard to the use of the production factors of work and capital flow. Following this realization many companies have formed separate or outsourced logistics operations.

To achieve an increase in productivity, products and production processes must also be constantly optimized with regard to the use of material. For years an essential task for engineers has been to reduce waste materials and to design products so that materials are used optimally according to their characteristics. On the other hand, the characteristics of the materials themselves are also being continuously optimized.

The consumption of materials is also influenced, for example, by reducing waste. The transition from cutting and abrasive manufacturing processes to forming processes generally leads to a reduction in the consumption of materials. But this also requires high investment in machines, devices and manufacturing know-how that is only justified for a high-volume output. Another example has been material re-use in coach manufacturing of an automobile company. They have succeeded to reduce by about 11% the amount of materials required, by the re-use of waste for smaller plate parts. Put another way, a year's worth of materials for the whole coach manufacturing operation is saved every decade.

After the phases of rationalization in the 1960s and 1970s in the area of production, it is now the turn of information, its possession and its efficient use which have come to the forefront of companies' attention. Information has become a new production factor, the obtaining and processing of which requires a lot of energy, time and money. Better information requires more efficient use of new technologies such as word processing, communication networks and personal computers. Controlling first the information itself and then the utilization of information leads to better control of the future. Only the correct handling of information and the flow of information as a production factor on a par with the classic factors of labor, capital, production resources, energy and materials will lead to success. The effects on personnel are considerable, as has been illustrated by office automation.

Another concern arises in the division of working tasks. In a traditional company structure there is vertical organization of individual work areas and most of the time a technical division between design/development and manufacturing/assembly. Because design is responsible for many of the costs, close cooperation between a design department and a manufacturing department is necessary from the outset of construction to develop a product with reasonable costs and which is geared to market demands. The possibilities of influencing costs after design, manufacturing and assembly are finished are very limited. If it is recognized at this point that design has to be changed because of manufacturing reasons, high changeover costs will have to be incurred e.g. for tools, etc.

In general, the risks of a narrow view of automation may be prevented by a company-wide acceptance and education. The level of automation and flexibility determines the selection of tasks and tooling. In the next section, attention turns to specific assembly tasks.

2.3 Assembly tasks

2.3.1   Electronic assembly

In the electronic industry 50 to 75% of the total production costs of a product are in assembly (Lotter, 1986). Examples of assembly required in the electronic industry include computer components, controllers for various machines and instruments, audio and video electronic components, and instrumentation for vehicles, airplanes and so on. One of the most common component assemblies in this industry is the printed circuit board (PCB).

Effective, competitive electronics assembly is a capital-intensive business. For instance, a complete surface-mounted printed circuit board assembly line can cost upward of $300000 (Schwartz, 1988). A major process of electronic production is the assembly of the printed circuit board. The two main techniques are through-hole insertion of electronic components on the board and surface-mounted technology in which the components are glued to the surface of the PCB.

According to one study, almost 50% of all PCBs already have a surface-mounted part, compared with only 2% in 1984. Yet another study estimates that 70% of all PCBs are still through-hole only. (See also the survey results discussed in section 1.6.) Although somewhat contradictory, these data portray both the growth of surface-mounted technology (SMT) and the staying power of through-hole insertion (Bralower, 1989).

Manual assembly of through-hole PCBs was initially the more common technique. An improvement was introduced over the years which guided the assembly operator to the next insertion location by an automatic light beam. While this ensured the proper insertion location, it did not prevent errors in selecting the right component.

The next step was the use of universal NC (numerical control) insertion machines. Initially, special tapes holding the right components in the required sequence were assembled and spooled. Next, an NC inserter would automatically feed the tape and insert the next component in the sequence to the X–Y board position as specified in the NC program (Fig. 2.7).

Later, fully automatic and highly efficient NC insertion machines were

Parts no.	Quantity	Component name	Shape	Packing	Performing	Lead cut/clinch	Insertion shape	Diameter, mm Lead	Diameter, mm Hole
1	24	Relay		100 pcs/tray	None	Yes		0.65x0.3	1.2–1.35
2	8	Thyristor (transistor)		45 pcs/tray	Yes	"		0.45x0.3	1.2–1.35
3	8	"	"	"	"	"	"	0.45x0.3	1.2–1.35
4	16	Thermistor		Tape / 50 pcs/tray	None	"		0.6	0.95–1.1
5	8	Hybrid IC A	SIP		"	"		0.8	0.95–1.1
6A	8	Hybrid IC B	SIP	"	"	"	"	0.8	0.95–1.1
6B	(8)	Hybrid IC C	SIP	"	"	"	"	0.8	0.95–1.1
7	8	Fuse		10 pcs/tray	"	(Yes)		0.9 (diagonal)	1.2–1.35
8	16	Resistor		Tape	Yes	Yes		0.88	1.2–1.35

Fig. 2.7 Printed circuit board: inserted parts sample (Truman, 1988).

developed where a robotic arm retrieves the next component from a magazine and immediately inserts it into the board, all under the control of the NC program. The latest PCB technology in practice involves automatic SMT machines.

Minimizing the manual handling of PCBs can yield reductions of 70% or more in physical and electrostatic discharge damage. Proper positioning of assembly machines allows manual unloading of one PCB and loading of the next in one motion. Automatic loading elevators allow the transfer of a magazine instead of totes with boards. Automated conveyor systems eliminate all manual handling between machines.

Solder mask (also called solder resist) allows greater freedom of layout by preventing bridging, while simultaneously reducing solder consumption by confining solder only to the intended areas. The soldering process should be planned carefully: solder joint inspection is costly and inaccurate, while touchup usually conceals defects and almost always reduces joint reliability. Consistent, high-quality wave soldering requires solderable components, proper circuit board design and fabrication, the right flux and solder, and the correct preheat.

Reworking soldered SMT packages costs an order of magnitude more than reworking equivalent through-hole components. Therefore, a fully debugged and properly managed SMT assembly operation must have much lower defect rates at each process step than those typical of through-hole assembly. Automated optical inspection (AOI) systems have been developed and play an increasing role in ensuring high-quality results. Their use makes SPC (statistical process control) methods a viable means of correcting process problems. The electronics industry is using AOI in each process step for manufacturing SMT boards; for pattern error correction on screen printers; pattern error correction, part-to-part matching and lead inspection on placement machines; and solder joint inspection.

A number of analytical and simulation models have been developed for the analysis and design of electronic assembly systems (Badalamenti and Bao, 1986; Nof and Drezner, 1986). Some of the assembly system design and control issues that were analysed are as follows:

1.   How should the components be organized and loaded into a carousel, bin or magazine?
2.   What is the optimal bin-picking logic of the assembly, workhead or robot arm?
3.   What is the optimal configuration of the assembly system?
4.   What are the performance characteristics of the assembly during transition periods, when switching from one production order or mix to another?
5.   What is the influence of test and rework strategies?
6.   Priority release of kits.
7.   Lot sequencing and dispatching at workstations.

8. Generalized flow line routing.
9. Parallel machines at each assembly workstation.

Additional discussion of these issues can be found in other chapters.

2.3.2 *Automotive assembly*

One of the most extensive assembly automation projects to date is the auto-mated assembly of the VW Golf factory, where 25% of the final assembly work is carried out automatically, not only by robots but also by a number of special assembly machines designed for a specific purpose. The tasks com-pleted by robots include the automated insertion, tensioning and tightening of the V-belt (Fig. 2.8), the insertion of the spare wheel, and insertion and bolting of the battery. Volkswagen is the first company in the automotive industry to achieve such a high degree of automation at the final assembly stage. Other companies worldwide have between 4 and 6% automation in this area.

It is naturally difficult to draw conclusions from the automation as practiced in a major concern like VW with regard to possible applications for small businesses. The decisive points in the automation at Volkswagen were the possible influences on design and the actual design modifications to the prod-uct that resulted in automation-optimized design of the entire vehicle. Compli-cated tool motions were simplified; for example, all bolting operations on the

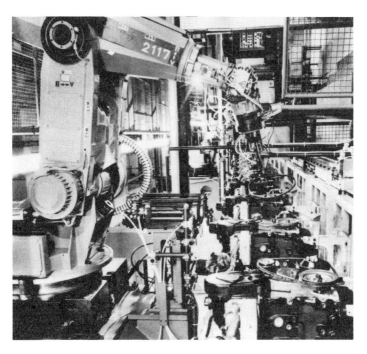

Fig. 2.8 Automated loading, stressing and tightening of a V-belt (courtesy VW).

floor of the car are now vertical, i.e. with linear joining motions. The brake cables were also modified so that they could be automatically connected. In addition, the entire vehicle concept was modified so that the fitting of the front is the last assembly task, with the front end staying open until all parts in the engine compartment have been mounted. This last operation is carried out by a linear joining motion. By then, the entire front section has been pre-assembled, including adjustment of the headlamps.

But it is not only VW who have made intensive efforts in this area; in all other automobile factories there are many examples of flexible assembly automation. KUKA use linear and rotation axis robots with a maximum load of 8 to 240 kg for assembly tasks, for the assembly of flywheels in car engines, e.g. the 60 kg robot (Fig. 2.9). This assembly cell installation is able to screw to crankshafts three different types of flywheel with the same punch image. The position of the crankshaft is sensed by a telecamera system.

Also in the assembly of car cylinder-heads the robot has to grip and screw with two separate tools, which it couples alternately by using an automatic tool-changing mechanism. This mechanism increases the flexibility and the utilization of the robot. First the gripper is used to place a cylinder-head on the engine block. Next the robot replaces the gripper with a twin screwdriver to screw down the cylinder-head screws in pairs. The screwdriver's torque is controlled and documented in order to ensure the quality. In other applications several grippers are applied, spot-welding guns and clinch pliers are exchanged and nozzle-heads for different kinds of adhesives and out-foaming materials are made available.

Two other examples show that a complex periphery is necessary in addition to a robot. For the automatic assembly of wheels to cars a solution allowing an assembly process on a moving car was developed. Front and rear tires are

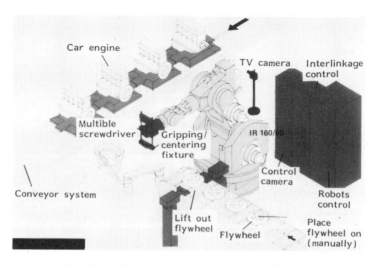

Fig. 2.9 Flywheel assembly (courtesy KUKA).

Fig. 2.10 Tire assembly (courtesy KUKA).

attached in pairs to the car, while the work-carrier does not have to adjust its unsteady movements (Fig. 2.10). The following modular units are therefore required:

- robots with heavy maximum load and synchronization with the work-carrier;
- television sensor system for synchronization of the car and robot movement as well as for the acquisition of the X- and Y-coordinates of the hub and the position of rotation of the nave image;
- robot gripper with integrated multiple screwdriver;
- a system of distance measurement for the acquisition of speed and position of the car;
- a transfer point including a wheel aligning fixture and a screw feeding device.

In modern vehicle concepts the automatic assembly of glass panes to passenger cars is accomplished by pasting. Robots position the panes spread with adhesive above the window cut, insert them and press them down (Fig. 2.11). The exact positioning is controlled by optical sensors that sense characteristic profiles which are made visible by laser beams. The sensor control sends correction signals to the robot control if the position of the panes deviates from the required position. This signal causes a correction motion of the robot.

The pre-cleaning of the glass panes and the application of the adhesive is also performed by robots. The panes are transported automatically. For an

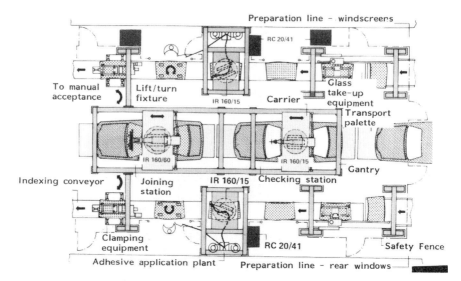

Fig. 2.11   Installation for the automatic assembly of car window panes, including
preparation (courtesy KUKA).

optimum time of exhausting the cleaning agent and the adhesive buffer sections are established. The success of an installation depends on the unique way
of integrating the system with robots, flow of materials, communication links
and manual jobs to decisively influence the assembly productivity.

2.4   Disassembly tasks

The aim of modern environmental policies and technology is to move from
waste disposal and burning to extensive recycling. An important factor are the
emerging stringent requirements of lawmakers with regard to waste disposal.
An explosive increase in the number of items to be recycled can be expected
in the 1990s, e.g. electrical equipment and components or old vehicles.

Today waste management is marked by a conspicuous 'imbalance between
a highly developed, well ordered supply system and a disposal management
system which is still under-developed and characterized by muddle and randomness'. Further development and transfer of the most modern methods
used in production engineering to the problems set by recycling are required.

We can distinguish between two recycling treatment processes (VDI 2243 E,
1990) – reprocessing and reuse. With product reuse the main thrust in production processes is in the recovery of workpieces for further use of a similar
nature. This method requires as the first step extensive non-destructive disassembly. Even during reprocessing, i.e. the industrial process for recovering
material for reuse, disassembly to separate material according to kind is the
first step in the recycling process.

Both for economic reasons and in order to improve the working environment in what are usually heavily contaminated working areas, extensive automation of the disassembly process is desirable. Therefore, for equipment flexibility to meet disassembly needs, the main thrust of development is the application of industrial robots.

2.4.1   Recycling and disassembly

Product recycling   According to VDI 2243 E, recycling means 'the reuse or reprocessing of products by establishing resource conservation cycles' (1984). Following the product lifecycle one can differentiate between the following types of cycles:

- recycling with reuse of products;
- production waste recycling; and
- recycling of old materials.

Furthermore, within the types of recycling, one can distinguish between recycling in the form of reuse or continued use of components while maintaining the recycled item in its original form, or continued utilization of material with the loss of its original form.

Before reuse, the products or material submitted for recycling undergo a treatment process which in all cases includes disassembly as the first step. With recycling for reuse, complete non-destructive disassembly, as comprehensive as possible, is the first stage of the five-stage preparation leading to restoration of the product properties (Steinhilper, 1988). Even in the preparation of product waste and recycling of old material, disassembly is increasingly being carried out first to separate the materials before they are crushed.

Disassembly   As the reverse of assembly, disassembly is defined as the totality of all processes which break down the structure of geometrically defined bodies (VDI 2860, 1990). Disassembly is always a sequence of functions. In addition to the main function of separation (Fig. 2.12) it includes, as with assembly, the auxiliary functions of manipulation and inspection.

2.4.2   Classification of disassembly systems

As with assembly the disassembly station is defined as the basic component in the disassembly system. Disassembly stations are differentiated according to the degree of automation into manual disassembly stations, automatic disassembly stations and flexible automated disassembly stations. Spur and Stöfele (1986) define manual disassembly as follows:

> A manual disassembly system is a work system, i.e. a work position with an operator at a specific location with specific tools and machinery for separating specific workpieces, or a work system group with several operators, but only a single end function applies.

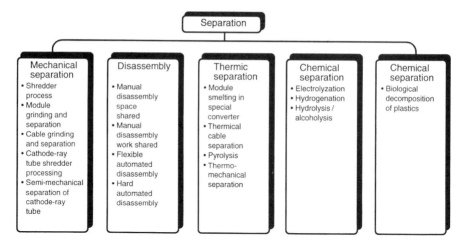

Fig. 2.12   Overview of separation methods.

In the following, 'manual disassembly station' should be understood to mean a module, complete in itself, of the disassembly system, i.e. a workstation. Because manual labour is used the manual disassembly station is characterized as having maximum flexibility.

An automatic disassembly station is an automated unit controlled by a fixed pre-set program with at least two options. The program comprises the complete instructions for performing a task. These plants are designed for special functions. A generally high degree of productivity is gained at the expense of relatively low flexibility with regard to product variety, usually with a minimum production variation.

A disassembly plant is considered a flexible automated disassembly station (FADS) if it is freely programmable and can react to any possible variations or product change. The basic element of a FADS is a disassembly robot, the so-called manipulation system, with a disassembly tool (generally a gripper or actuator) flanged to the disassembly robot. The combination of a freely programmable robot control system, disassembly tool and disassembly robot, similar to the definition of a flexible assembly station, comprises a flexible automated disassembly station.

Manual disassembly   Common solutions for the disassembly of products in the manufacture of replacement parts, for example in the telecommunications field or car engine reconditioning, are only sporadically applied in efficient production lines. Because of the relatively low number of items at the present time, these solutions are purely manual in form, with a limited degree of mechanization. They are based on the use of conventional assembly tools such as screwdrivers, hammers, etc., and occasionally of mechanical appliances such as compressed air screwdrivers. Also in the preparation of a product, e.g. in the case of electronic data processing equipment or the most recent disassembly tests (for example as carried out by BMW, VW), disassembly has so far been carried out through manual or partially mechanized work-stations.

Automatic disassembly   Solutions for the automation of disassembly, in the form of single or multi-spindle automatic unscrewing machines, exist at the present time only where large quantities of items with a minimum variety are concerned, e.g. in container recycling (Huber, 1991). Figure 2.13 shows as a further example a dedicated automated system for screw-top removal, a station for unscrewing bottle tops for the refilling of returnable bottles.

In addition, there are trends towards using robots for the disassembly of large volume products, for example in the overhaul of combustion engines for agricultural machines or diesel motors for locomotives (Zimmerman and Hartung, 1986; Kobrow *et al.*, 1990). Here three-axis manipulators are used in order to mechanize unscrewing operations during disassembly.

An automated installation for the dismantling of television and computer display units has been implemented by VICOR Recycling GmbH in Berlin. Figure 2.14 shows the rotary transfer table with two automatic and two manual stations for thermal/mechanical separation of picture tubes.

Examples of the industrial use of automated disassembly installation are characterized by the following:

- single-purpose stations are the most common in industrial facilities;
- the use of industrial robots is in general restricted to laboratory operations;

Fig. 2.13   Automatic machine for unscrewing bottle tops (courtesy Alcoa).

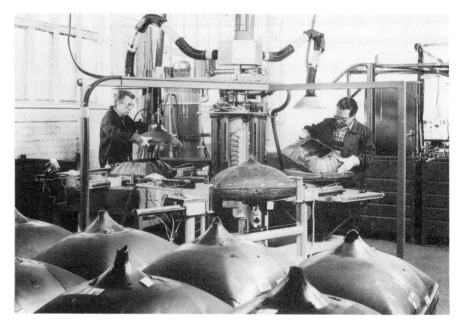

Fig. 2.14 Rotary transfer table for thermal/mechanical separation of picture tubes (courtesy VICOR).

- there is a high degree of automation only in the case of non-flexible facilities with a low product variety and with large quantities;
- there is a low degree of automation with low output in complex disassembly processes.

Flexible disassembly Automation of disassembly for most technological products requires a relatively higher degree of flexibility than assembly automation for the following reasons. The timespan that most products are in use is usually longer than the introduction cycles of new products, typically between 10 and 15 years. Hence, a number of generations of products have to be considered in the disassembly plant. Moreover, the shape of the products to be disassembled varies broadly as a result of ageing, corrosion, damage, etc. Therefore, an automated disassembly operation has to cope with a high need for adaptation both towards regular and irregular changes in the process.

In parallel with assembly automation the concept of a flexible disassembly cell is illustrated in Fig. 2.15. The flexible disassembly cell consists of the following main components:

- one or more industrial robots for the handling of parts and disassembly tools, typically with six degrees of freedom, capable of direct numerical control (DNC) for communication;
- the cell computer for control of the disassembly process and for commu-

nication both with a computer-integrated environment and the operator of the cell;

- flexible disassembly tools such as automated programmable screwdrivers, grippers, drilling tools, cutting tools (e.g. laser-based), stored in a tool magazine and ready to be engaged to the end-effector of the robot by a standard interface;
- sensors such as tactile sensors, vision tools or laser scanners for process control and to enable the robot-cell to adapt to different states of the product such as corrosion, damage, etc.;
- conveyors for the transport of boxes and palettes with disassembled parts out of the cell area for cleaning, sorting, repair and reassembly or granulation.

After disassembly, components are ready for the next cycle of assembly. The next section returns to the assembly tasks of insertion and joining.

2.5 Insertion and joining tasks

Establishment of connections is one of the most common assembly tasks. The design of this task is thus one of the central problems which has to be confronted in assembly rationalization.

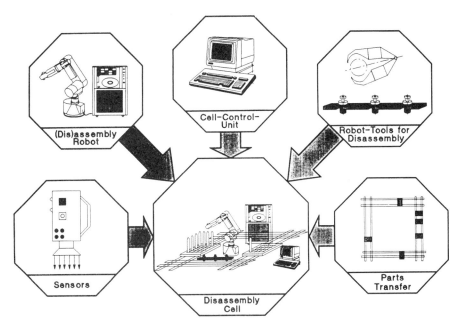

Fig. 2.15 Components of a flexible disassembly cell.

2.5.1   Screwing

The screw connection is one of the most frequently used connection types in the precision instrument industry and also in the automobile and machine building industry (Fig. 2.16) (Abele, 1984; Gießner, 1975; Schweizer, 1978; Warnecke and Walther, 1984). In the future, screwing will preserve its predominant position over other jointing processes (Weule, 1983a).

However, in comparison to welding, for example, the degree of automation in screwing is quite low. Today the work processes which monitor supply of groove screws or nuts in assembly lines are mostly manual. Only the tightening to a specific torque is performed by automated screw devices. Development work has concentrated mainly on this last operation in the past. There are numerous investigations into the problem of tightening screws and nuts and predicting the necessary prestressing force according to the workpiece and the environmental conditions to ensure safety. The correlation between prestressing and frictional force, friction torque, and starting and setting torque can be specified. The influence of factors such as surface texture and lubrication can also be considered and the rules for selection of a suitable starting process can be specified (Junker, 1974a, 1974b; Erhardt, 1981, 1982; Strelow, 1981; Pfaff and Thomala, 1982; Weule, 1983b; Muck, 1985).

The result of these efforts on the market today are the electronic or electropneumatic screw-devices with spindle-integrated sensors for the registration of the angle of rotation and torque. Also available are control units equipped with micro-processors and control-display for user-guidance and statistical process control. However, there are still problems in the automation of the screwing process itself which must be solved, beginning with the supply of

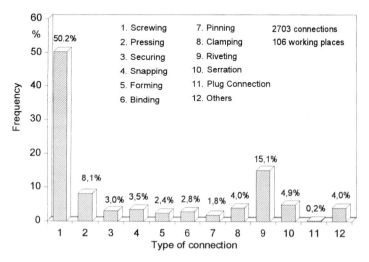

Fig. 2.16   Frequency of types of connection (DIN 8593) in the automobile and machine-tool industry.

workpieces, continuing with the distribution of frictional information, and ending with the setting conditions of sealing in multi-layer plate packages.

The pressure for 100% safety with 'life critical' screw connections has led to a multi-stage approach of manual assembly in the automobile industry: tightening is performed by conventional screwdrivers, where several screws are tightened (for actual tightening) by a knuckle wrench. This process is repeated by another worker and finally there is a sample testing by quality supervisors. This multi-stage process is an example of the rationalization potential.

What are the reasons that modern automation technology has not been applied? Investigations have suggested the following.

- The resetting of conventional piece supply systems is only partially feasible and associated with an extensive effort, i.e. the full flexibility of a programmable handling tool cannot be justified.
- The use of jointed links to compensate for positioning errors is not process-compatible, because traditional compliance devices (remote centre compliance – RCC) compensate for lateral and angular errors. But these devices because of their inherent forgiveness in turning, are unable to provide reliable tightening.
- Automated changing units for screw tools or screw spindles are only available with large-scale special designs; the changing of a tool generally requires a relatively large increase in installation complexity and in cycle time, so that economic application is not possible.
- Reliable systems for conveyor synchronization are not available, so automation in conveyor-based installations is inappropriate.

The connecting process of 'screwing' was studied in detail at the Fraunhofer Institute for Manufacturing Engineering and Automation (IPA) with a finite-element simulation. The application of flexible handling appliances as screw assembly systems was tested on typical industrial tasks (Fig. 2.17). The problems of recovery from positioning errors by controlled correction of the screwing tool and adapting to diameter variability for handling different screw sizes without changing tools have been solved through the development of special hardware modules. By combining these modules with a five-axis industrial robot, different screwing tasks can be handled (see also chapter 9).

The system offers the possibility, when assembling metric screws, of increasing the joining speed, of handling screws with different heads, of self-tapping in wood and/or plastic and of mounting hexagon head cap screws of different diameters without changing tools.

2.5.2 Riveting

Without doubt riveting is one of the 'classic' connection techniques and, together with the screw, is the most frequently applied. In aerospace engineering, riveting is employed most frequently – millions of rivets are used, for example, in the manufacture of large passenger aircraft. But permanent rivet-

Components:

- screw
- thread plate
- hole plate
- hexagonal nut

Fig. 2.17 Screw assembly system with industrial robot.

ing is used in other fields too. Recently, connection techniques using rivets have gained a marked growth. Influencing factors in this growth are the continuing development of different types of rivets, e.g. pop and blind rivets, carbon fiber-reinforced plastic rivets, etc., and the associated processing methods and tools. Another factor is the increasing replacement of steel and iron by aluminum, plastics or connecting materials which because of their different properties require alternative connecting techniques. For example, in the field of car bodies, stress-bearing aluminum parts are riveted, and in aerospace manufacturing fiber-reinforced components are connected using combined rivet–adhesive bonds and titanium bosses.

Generally, in riveting the connecting element with its pre-formed head is inserted through holes in the components to be connected. Finally the closing head is plastically formed from the projecting stem material. The rivet stem length I_s and the rivet stem excess length z (Fig. 2.18) are basically determined by the requirements for the rivet joint, the type of rivet used, the shape of the closed head to be formed and the relevant processing method. Among the important types of rivets are:

- full and hollow rivets;
- blind rivets; and
- pop rivet systems.

In addition to these three common types of rivet, alternative types have recently been developed, such as, for example, self-piercing or punched rivets, or rivets made of shape-memory alloys, as well as plastic rivets which are formed using ultrasonics or heat.

Until now, greater automation in riveting has been prevented due to the

lack of automatic means of production, poor accessibility at the rivet location
and poor reproducibility of the riveting process. Economic impediments exist
in small production runs and batch sizes, stemming from the excessive costs of
flexible automated plants, the great variation in rivets, and the high retooling
costs. A review of this situation is provided in Fig. 2.19.

The IPA has been developing a blind riveter for industrial robots based on

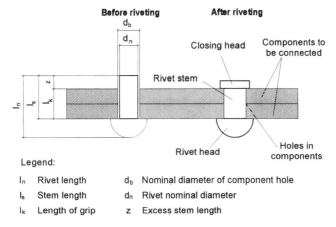

Before riveting   **After riveting**

Legend:

l_n	Rivet length	d_b	Nominal diameter of component hole
l_s	Stem length	d_n	Rivet nominal diameter
l_k	Length of grip	z	Excess stem length

Fig. 2.18   Terms used in connection techniques using rivets.

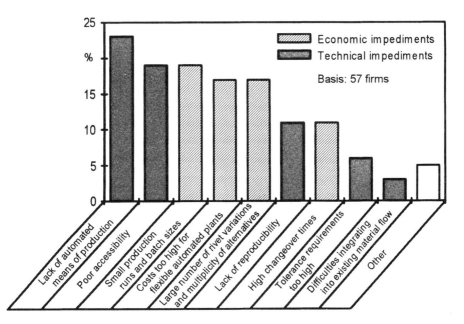

Fig. 2.19   Impediments to automation in riveting: overview.

a standard tool which, apart from its size and weight, is also notable for its enhanced flexibility. With the aid of a changeover device, for example, different rivet diameters and types of blind rivet can be handled.

The IPA has also studied the development of press riveting. A press riveting tool for SCARA robots has been developed for hollow or solid rivets of 1 mm to 6 mm in diameter with heads. This tool also allows the joining path and riveting speed and force to be programmed. An integrated sensor system monitors the pressing process and adjusts the process parameters if necessary.

2.5.3   Clinching

To meet the requirements of anti-corrosion protection and lightweight construction in many parts of the sheet metal industry, a steady rise in the use of surface-treated thin gauge sheet metal and aluminum sheet has been noted in recent years. However, plates of this type can be connected by conventional connecting methods only at great cost. Consequently, designers are at pains to discover new methods of joining which suitably supplement the connecting methods already known and also support the advance of automation. One element in this new technology is the technique of clinching as a joining method combined with forming. The joint element is formed directly out of the material of the parts to be joined, without the action of heat and without great power consumption.

The clinching method, introduced for the first time in 1981 as an economic and technically interesting joining technique, has taken a permanent position among existing connection methods. Moreover, since 1985 it has been included in the German standards under the generic term 'interleaved joints', in DIN 8593.

For structural reasons, with clinched joints (Fig. 2.20) the permanent joining element is manufactured in a single jointing stroke. The tool set required to make the clinched joint consists of a stamp and a die, the latter consisting of a fixed anvil, and laterally spring-loaded plates. During the first phase of the clinching operation the stamp is placed above the die so that it spans the plate joint sections lying across the plates. In the subsequent stamping motion, the stamp presses out the plate to form a web-shaped indentation. In a further phase of the same stamping motion, this web impacts on the anvil, where it is formed by cold forging between the pressure surface of the stamp and the crowned surface of the anvil so that its width increases. The widening of the web takes place against the resistance of the sprung plates and leads to the bracing of the joined plate material.

The clinching tool (Fig. 2.21) itself has relatively small dimensions, so that the clinching technique can also be used with small flange widths and under difficult conditions. A unique feature of this technique is that the tool-set used requires only small, easily operated presses as the operating units, which can be adapted to the particular application and the clinch-set used.

A robot-compatible, programmable tool has been developed for flexible

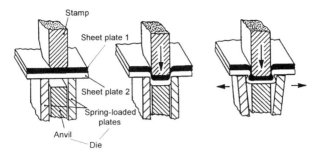

Fig. 2.20 Principle of the clinched joint (according to Liebig, 1985).

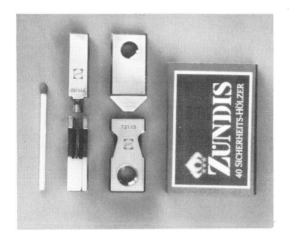

Fig. 2.21 Standard tool set for clinch joints consisting of stamp and cutting-bracing die: comparison of sizes (according to Dobrikow, 1992).

automated clinching, and is notable for its compact outer dimensions and low handling weight. With this robust but sensitive robot tool joining forces of up to 50 kN can be generated. Different sets of dies for a specific connecting task in question can also be used by means of the automated changeover device. The sensor system integrated in the tool enables customized process control. Rapid working strokes and success of the join at each point can be checked and documented (Fig. 2.22).

2.5.4 Assembly of non-rigid parts

The automated assembly of flexible parts (e.g. hoses, O-rings) has not been investigated to any great extent to date, probably because of the difficulties involved in such automation. The fact that 120 million hoses are assembled in the motor industry every year and a further 45 million in the household appliances sector indicates the significance of automation in this area. Problems which need to be solved here include: (a) the absorption of joining forces

Fig. 2.22 Clinching tool for industrial robot.

and moments by bodies with very little inherent rigidity (flexible parts); (b) the relatively high workpiece tolerance; and (c) the deformation of parts as a result of joining or gripping forces.

Hose assembly Both the theoretical and experimental aspects of hose assembly have been investigated by the IPA. Tests have been carried out with different hose materials and assembly conditions to verify analytical joining process calculations as well as three-dimensional finite-element simulation of dynamic joining processes. The design, development and testing of various system components for hose assembly, including hose grippers, systems for on-the-fly gripper or gripper jaw changeover aids and a clip assembly tool, have also been investigated.

Sensor solutions for hose assembly are of considerable significance. Experience has been gained with diverse systems such as laser scanners for free hose ends or the integration of a joining force monitoring system in the assembly tool. Various strategies with video processes (for monitoring and control) have also been investigated for hose assembly. Two pilot systems have been designed and built for the complete assembly of cooling water hoses, one of the main applications in this area. Of interest in this prototype is the successful evaluation of sensor signals for location of the spatially undefined hose position after the first joining operation, and the monitoring of the final join.

A cooler hose with three ends has been completely assembled in an automated assembly cell (Fig. 2.23) comprising:

- programmable industrial robot, gripper and gripper control;
- gripper jaw changeover system (on-the-fly change in one to three seconds, depending on the robot speed);
- laser scanner, sensor data processing.

The large tolerances occurring with elastomeric parts are compensated by the use of joining strategies which do not require sensors. In this example, the laser scanner basically has two functions:

- recognition of approximately pre-positioned hose ends – after the first hose-end has been assembled, the position and orientation of the other two ends can only be roughly determined; with this distance-measuring system the position and orientation of the hose-ends can be established;
- control of joining result.

The following main issues were studied:

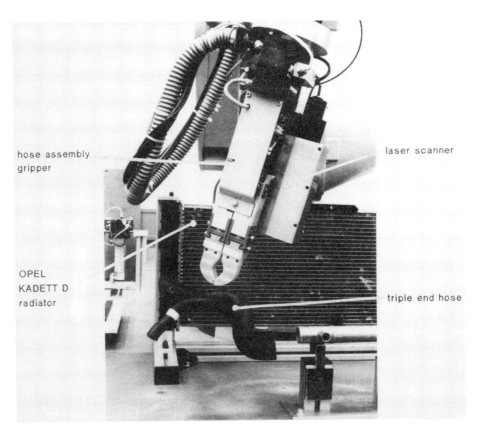

Fig. 2.23 Assembly of cooling water hoses.

- assembly-optimized design of connecting and hose elements;
- design and planning aids, e.g. use of a finite-element method for design of system parts;
- optimal gripper jaw design;
- determination of optimal assembly parameters within a computer-assisted knowledge database;
- concept for standard hose assembly operation;
- construction of two pilot systems for complete assembly of cooling water hoses;
- join examination with different hose materials;
- gripper systems with flexible diameter for hoses;
- FEM analyses of joining operations (two- and three-dimensional);
- linking of sensor computer to robot control.

O-ring assembly   Another typical problem in the joining of flexible parts is the assembly of O-rings. These parts, which are generally made of non-preformable materials, must be separated initially and then inserted into surface grooves which need not necessarily be circular.

With the aid of two robot-compatible tools it is possible to separate the O-rings from a heap, check their dimensions, position them and insert them into a variety of different basic parts (Fig. 2.24). A specially developed tool first scoops a random number of O-rings from a hopper. As the tool passes through a brush integrated in the hopper, all but one of the rings are separated. A downline sensor station checks whether a single ring has been taken and also verifies that it is not deformed. In the event of an error, a fault strategy is triggered. Finally the tested sealing ring is deposited in a prismatic positioning station on the workpiece carrier.

The gripper integrated in the joining tool then takes the ring and positions it on the base part, so that it coincides with the groove on one side (Fig. 2.24). At this point an assembly roller descends on the workpiece and presses the ring into the groove. The roller then moves over the workpiece again to slip it precisely into the groove. Through the input of part-specific slip parameters the O-ring can be inserted even into highly non-circular groove contours.

Batch sizes of one or more parts can be handled without changing the tool. The information on the workpiece groove contour can be transferred to an industrial robot controller either visually by a camera or via coded workpiece carriers.

Wire harness manufacture   Even today, most wire harness manufacturing processes in industrial batch production are carried out manually. For complex harness productions, for example in a motor vehicle, consumer electronics and household goods, automation has been hindered by the following factors:

- difficulty in the handling of flexible wires and wire strands;
- the large number of different contacts and wires per harness;

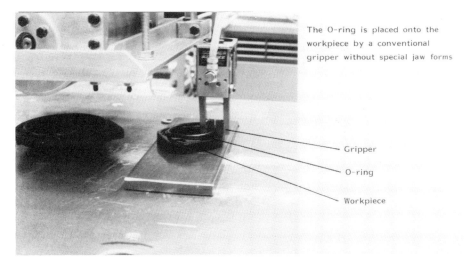

The O-ring is placed onto the workpiece by a conventional gripper without special jaw forms

Gripper

O-ring

Workpiece

Fig. 2.24 Assembly of O-rings with industrial robot.

- different connection techniques within a harness;
- absence of developments in equipment for automation, particularly for flexible pre-assembly, laying and sheathing of wire strands.

Flexible assembly of wire harnesses has been investigated. A fully automated off-line programmable assembly unit has been developed with insulation piercing connection devices. The operations performed by the unit include automated configuring of a layout board with the aid of a robot that positions different sockets for plug-in connectors and lays aids on a magnetic clamping plate. Special tools have been designed for laying and joining the wires. In view of the large amount of time required for laying wires and the demand for different types of wires, a multiple laying tool has been developed with which several wires (up to a maximum of three) can be laid simultaneously. The different wires are fitted in the tool by means of wire-change modules. Finished harnesses are tested and removed automatically.

Stitching cell for automated manufacture of trouser legs The clothing industry worldwide has been restructuring since the second half of the 1960s. Increasing unemployment and dwindling demand in the last 20 years have increased the pressure of competition, resulting in a greater drive toward rationalization and product innovation and a growing internationalization of production. With handling tasks accounting for 80% of production, the motivation for automation of these handling operations is evident.

A flexible stitching cell has been developed for trouser legs, comprising the following components:

- magazine unit for front and back parts;
- handling axis with high-speed changeover unit for fabric gripper;

- positioning system for large textile workpieces;
- standard contour stitcher.

The system has the following features:

- positioning system, handling system and magazine unit can be used universally;
- positioning system is flexible with respect to material and self-adjusting to different parts sizes;
- positioning accuracy of ±1 mm;
- programming via convenient user interface.

This system demonstrates that it is possible to automate the handling of flexible materials in both pre-production and assembly tasks in the clothing industry. In addition, the central component, i.e. the positioning unit, can also be used for other jobs and with other materials, for example awning cloth, bonded fibre fabrics, canvas, leather, filter material and carpets.

The technological feasibility of automated assembly for non-rigid parts has been demonstrated and proven. It is expected that in the future such automation will become more economic and hence more common in industry.

2.5.5   Assembly of small components

Assembly of small components has already been automated in several areas. Most applications involve 'hard automation' (mass production) in classical small-component assembly fields such as the watchmaking industry. More recently, the fitting of electronic components to printed circuit boards has been automated with impressive successes. Because the quantities are too small for rigid automation and in view of the large variety of parts involved, the assembly of many optoelectronic, precision electronic and high-grade precision mechanical appliances has been mostly manual to date. Flexible automation of the assembly of precision mechanical components has generally been thwarted because of a myriad of technical requirements:

- high accuracy and minimal joining tolerances;
- absence of assembly-optimized design;
- miniature size and sensitivity of connecting elements;
- integrated testing and adjustment operations; and
- cleanliness requirements in assembly.

The IPA has investigated the development of special assembly tools and joining procedures which would allow industrial robots to be used for the assembly of precision mechanical products and hence enable automation. The following problems are currently being investigated:

- development of grippers for handling miniature joining elements for a large variety of parts;
- development of assembly tools and joining strategies for tolerance com-

pensation in assembly tasks with low joining tolerances and no chamfer for the insertion of elements;
- development of instruments for the assembly of sensitive joining elements.

A test bench with a high-accuracy six-axis gantry robot is being used to establish the technical feasibility of the assembly of precision mechanical components and to analyse the problems which occur. It is necessary, for example, to determine the required joining forces as a function of the joining tolerances, parts tolerances and directions of join. For this purpose a universal gripper with programmable gripper opening is used, to which different gripper jaws can be flange-mounted for the assembly of small parts. With the six degrees of freedom of the industrial robot, assembly tests can be carried out in any joining direction. The strength and direction of the joining forces generated in the assembly tests are determined with strain gauges mounted on all sides of the gripper.

Figure 2.25 shows a test installation to determine the joining forces generated with the assembly of small parts. These parts are mounted as latching connections on the housing of model locomotives.

One component in the development of robot tools for precision mechanical joining tasks is a gripper system in which the position of the jaws and the gripping force can be programmed with high precision. The gripper opening is programmed by means of a stepper motor with a resolution of less than 0.001 mm. This fine resolution is achievable by a high step-down gearing consisting of a worm gear and rack and pinion pair. The design of the gripper system permits tolerance-free positioning of the gripper jaws. The gripping force can also be programmed indirectly in increments of 0.005 N through the controlled resilience of the pressure string-operated gripper jaws as they open

Fig. 2.25 Joining test with small mounted parts.

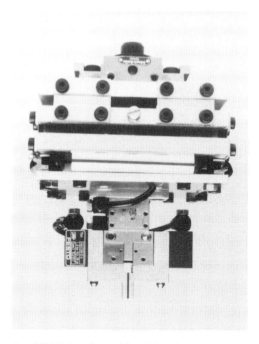

Fig. 2.26   Gripper for SCARA robot with minimal gripping force for the joining of tiny parts.

and close. This gripper system enables the tolerances of joining elements to be compensated when handling fragile small parts with minimum diameters of 0.5 mm, thereby enabling assembly of sensitive items such as optical lenses. Figure 2.26 illustrates the test assembly of hybrid sockets by a SCARA robot involving the gripping of fragile glass beads with a diameter of 1.45 mm and diameter tolerances of ±0.04 mm. The gripper tool permits non-destructive handling and assembly of sensitive joining elements with industrial robots.

With the use of these recently developed tools further rationalization potential can be achieved in the precision mechanics, electronics and optical assembly industries. Assembly tasks with minimal joining tolerances and sensitive parts such as optical lenses, galvanized parts or components without chambers can be flexibly automated with the aid of such tool developments in conjunction with industrial robots.

2.6   Summary

This chapter describes the main tasks of assembly and disassembly, and their technological aspects. Task details may differ, depending on whether the task is executed manually, by an assembly machine or by a robot. Tasks may also differ by their purpose, e.g. to produce a new product, to repair or maintain, or to recycle. The selection of automation level for a given industrial operation is therefore an important decision for designers.

Tasks are described and illustrated for electrical and mechanical products. The key requirements and tools applied for the various tasks are reviewed, as well as the special requirements for handling unique components such as pliable cables and hoses or tiny watch parts.

Understanding the technology and requirements is a key to task selection. Furthermore, there is a strong relationship between the task and the design of components and facilities **for** assembly. The latter topic is the focus of Chapter 3.

2.7 Review questions

1. Consider the design of a multi-model television assembly facility, and explain the specific implications of:

   (a) adaptation flexibility;
   (b) application flexibility.

   Can the facility be flexible with regard to one but not to the other?

2. Rank the degree of automation (using the Automation Grade Scale) in the following:

   (a) totally manual assembly line of upholstered furniture;
   (b) clean-room assembly of semiconductors;
   (c) printed circuit board SMT machine;
   (d) fast-food restaurant;
   (e) jet-engine disassembly line for periodic maintenance.

   (Note: Make the necessary assumptions following the example in section 2.2.2.)

3. Repeat question 2, this time **calculating** the degree of automation.

4. Prepare a comparison table displaying the key arguments for and against task automation in assembly of:

   (a) holiday decorations;
   (b) mattresses;
   (c) safety air-bags for cars.

5. Explain several reasons why:

   (a) SMT is replacing through-hole technology;
   (b) through-hole technology remains useful.

6. Describe how certain assembly tasks have been transformed from manual to automated in:

   (a) electrical products;
   (b) mechanical products.

7. What are the main reasons to disassemble products?

8. When is it justified to design flexible disassembly?

9. Identify the main components of a flexible disassembly cell, and explain the differences from a flexible assembly cell.
10. Explain the considerations for and against automating tasks using:
    - (a) screws;
    - (b) rivets;
    - (c) clinching.
11. Identify the three main types of rivets.
12. What factors influence the increase in use of rivets?
13. Explain the relative advantages of clinching.
14. Explain how human hand–eye coordination can be duplicated by machines in handling pliable components such as rubber hoses and electric cables.
15. Describe the specific challenges of small-part assembly.

References

Abele, E. (1984) 'Einsatzmöglichkeiten von flexibel automatisierten Montagesystemen in der industriellen Produktion', *Montagestudie* (Vol. 61: *Humanisierung des Arbeitslebens*), VDI-Verlag, Düsseldorf.

Badalamenti, J. and Bao, H. (1986) 'Simulation modeling of robotic assembly cell incorporating carousel', *Robotics and Material Flow* (ed. S.Y. Nof), Elsevier Science, New York, pp. 121–30.

Bralower, P.M. (1989) 'Flexible systems aid electronics assembly', *Assembly Engineering*, Vol. 32, No. 12, pp. 19–23.

Bright, J.R. (1958) *Automation and Management*, Harvard Business School, Boston, MA.

Dobrikow, M. (1992) *Erfahrungen beim Druckfügen von Aluminiumblech-formteilen in der Serie*, Seminarvortrag, Aluminium-Zentrale e.V.

Erhardt, K. (1981) *Zeitgemäße Schraubenpraxis*, uta groebel-infotip, Limeshain.

Erhardt, K.F. (1982) 'Moderne Montage von Schrauben und Muttern', *Verbindungstechnik*, Vol. 14, No. 6, pp. 17–22.

Gießner, F. (1975) *Gesetzmäßigkeiten und Konstruktionskataloge elastischer Verbindungen*, Braunschweig, Techn. Univ. Diss.

Huber, B. (1991) *Hochautomatisierte Recyclinganlagen für Metallfässer – ein Jahrzehnte bewährtes Vorbild künftiger Kreislaufarten?* 23rd IPA Forschung und Praxis T21, Umweltbewußtes Produzieren, Springer Verlag, Berlin.

Junker, G. (1974a) *Grundlagen der Berechnung hochbeanspruchter Schraubverbindungen*, VDI-Berichte No. 220, VDI-Verlag, Düsseldorf.

Junker, G. (1974b) 'Reibung – Störfaktor bei der Schraubmontage', *Verbindungstechnik*, Vol. 6, No. 11/12, pp. 25–36.

Kobrow, H. *et al.* (1990) 'Teilautomatisierte Demontage von Verbrennungs-motoren', *Fertigungstechnik und Betrieb*, Berlin.

Liebig, H.P. (1985) 'Neue Möglichkeiten des Verbindens von Blechteilen durch Druckfügen', in *Tagungsunterlagen: Zukunftssicherung in der Blechverarbeitung*, 7.2.1985, Stuttgart; 13.11.1985, Hagen/Deutsche Forschungs-gesellschaft für Blechverarbeitung e.V., Hannover.

Lotter, B. (1986) 'Automated assembly in the electrical industry', *Automated Assembly*, 2nd edn, Society of Manufacturing Engineers, pp. 49–56.

Muck, G. (1985) *Elektronische Messung und Streuung von Drehmomenten in Montagemaschinen*, VDI-Berichte No. 323, VDI-Verlag, Düsseldorf.

Nof, S.Y. and Drezner, Z. (1986) 'Part flow in the robotic assembly plan problem', Robotics and Material Flow (ed. S.Y. Nof), Elsevier Science, New York, pp. 197–205.

Pfaff, H. and Thomala, W. (1982) 'Streuung der Vorspannkraft beim Anziehen von Schraubenverbindungen', *VDI-Zeitschrift*, Vol. 124, No. 18, pp. 76–84.

Schwartz, W.H. (1988) 'Electronic assembly systems', *Assembly Engineering*, Vol. 31, No. 9, pp. 49–52.

Schweizer, M. (1978) *Taktile Sensoren für programmierbare Handhabungsgeräte*, Springer, Berlin, 1979; Zugl. Stuttgart Univ. Diss.

Spur, G. and Stöfele, Th. (1986) 'Fügen, Handhaben, Montieren', *Handbuch der Fertigungstechnik*, Vol. 5, Carl Hanser Verlag, Munich.

Steinhilper, R. (1988) *Produktrecycling im Maschinenbau*, Springer, Berlin, 1998; Zugl. Stuttgart Univ. Diss.

Strelow, D. (1981) 'Reibungszahl und Werkstoffpaarung in der Schraubmontage', *Verbindungstechnik*, Vol. 13, No. 6, pp. 19–24.

Truman, R. (1988) 'Component assembly onto printed circuit board', *International Encyclopedia of Robotics: Applications and Automation* (eds R.C. Dorf and S.Y. Nof), Wiley, New York, pp. 202–21.

VDI 2243 E (1990) *Recyclinggerechte Gestaltung technischer Produkte*, December 1984 (Vorentwurf zur Neuauflage, May).

VDI 2860 (1990) *Montage- und Handhabungstechnik; Handhabungsfunktionen, Handhabungseinrichtungen; Begriffe; Definitionen, Symbole*, May.

Warnecke, H.J. and Walther, J. (1984) 'Automatisches Schrauben mit Industrierobotern', *Zeitschrift für industrielle Fertigung*, Vol. 74, No. 3, pp. 137–40.

Westkämper, E. (1977) *Automatisierung in der Einzel- und Serienfertigung – Ein Beitrag zur Planung, Entwicklung und Realisierung neuer Fertigungskonzepte*, Aachen, RWTH. Diss.

Weule, H. (1983a) *Schrauben in der automatisierten Montage*, 5th Deutscher Montagekongress, Munich.

Weule, H. (1983b) *Schrauben in der automatisierten Montage*, VDI-Berichte No. 479, VDI-Verlag, Düsseldorf.

Zimmermann, J. and Hartung, R. (1986) 'Robotergestützte Instandsetzung von Verbrennungsmotoren', *Fertigungstechnik und Betrieb*, Berlin.

3

Design for assembly

3.1 Introduction

In the 'old days' assembly methods were selected by engineers only after the product design had been completed, approved and authorized. As long as all the assembly work was manual, human assemblers could be expected to learn how to assemble even complicated products. There were some guidelines on how to plan the assembly method for effective manual assembly. When parts had to be designed for the non-forgiving automatic assembly machines, and later for robotic assembly (whose cost increases exponentially the more forgiving it is expected to be), designers realized the need to 'design ahead'.

Analyses of the cause and emergence of costs in various production areas reveal that while a predominant portion of the product cost arises during its assembly, these costs are determined at the design phase (Solhenius, 1992). Consequently, thorough assembly-oriented design of the product and the method of assembly are warranted as part of the lifecycle engineering approach (Eversheim and Muller, 1984; Nevins and Whitney, 1989; Alting, 1991). Furthermore, the objectives of design for assembly (DFA) have changed over the last two decades not only in the recognition of need for concurrent design, but also because of increasing demand for quality, variety and timely delivery (Fig. 3.1).

Techniques for assembly-oriented product design must therefore consider the handling of product components and the joining/fastening of these components. Similarly, the system and organization of assembly must also be designed for cost-effective, reliable operation. Objectives and issues of design for assembly are summarized in Table 3.1 for companies' strategic and operational objectives.

DFA can be broken down into product-oriented DFA and system-oriented DFA; ideally, both should be integrated. This chapter focuses mainly on the product DFA; the integration of system DFA is covered in Chapters 4 and 5.

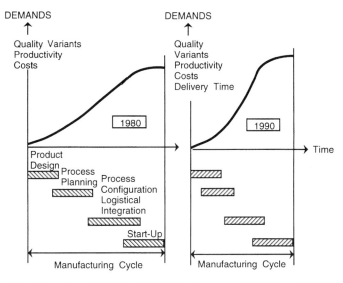

Fig. 3.1 Changing objectives in design for assembly (Wiendahl, 1989).

Table 3.1 General objectives and issues in design for assembly (DFA)

	Strategic	Operational
Product:	• Customer needs	• Subassemblies, components
	• Market structure	• Part form and functions
	• Volume trends	• Part features, characteristics
	• Quality, reliability, safety	• Assembly sequences
	• Economics – make or buy	• Assembly operations
	• Assemble-to-order or stock	• Tolerances, clearances
	• Lifecycle issues	• Tests, inspections
Assembly facility:	• Cost, quality, productivity	• Equipment selection
	• Organization, labor needs	• Layout
	• Location, space needs	• Task allocation
	• Degree of automation	• Material handling
	• Flow, control interfaces	• Part feeding, fixturing

The general subject of product DFA is actually part of the more general DFM/T or design for manufacturability and testability (Warnecke and Bassler, 1988; O'Grady and Oh, 1991). More recent approaches also address DFS, design for service of products after their purchase or installation, and even the recycling considerations, DFR. As a result of the lifecycle design trend and concurrent engineering emphasis, the whole array of these design methodologies is sometimes termed DFX (e.g. Gatenby, 1988).

Systems for comprehensive assembly-oriented product design are still being developed. However, major advancements have already occurred over the last two decades, and there are well-documented success cases of DFA applications, as discussed later in this chapter.

Four essential technologies for DFA engineering include:

- product design for assembly;
- group technology for assembly;
- computer-aided design systems;
- artificial intelligence methods for design.

These technologies are explained in the following sections.

3.2   Product design for assembly

When a product is designed or redesigned, consideration is generally given to the ease of manufacture and cost-effectiveness of its individual components, its function and the appearance of the final product. Several simple principles are illustrated in Fig. 3.2. One of the first steps in the design of any product is to consider if it is possible to produce **and** assemble it more cost-effectively and reliably without compromising the intended quality.

A product design directly affects the number of components that must be assembled and the difficulty of the assembly method. It can determine whether assembly must be manual or if it could be automated. For instance, there is a design trade-off between inexpensive, 'throw-away' products, for example computer diskettes, and products that are designed for maintenance and future repair, for instance household appliances. In the first case, the recommended design may include complete sealing of the product, applying force-fitting and snap-on components that reduce material cost and production time. A different design approach is required for products that need to be disassembled for periodic adjustments, preventive maintenance or replacement of worn modules. A more complicated example of trade-off is the design of electric motors. In certain cases, it may be justified to maintain the motor, while in other situations it would be preferred to discard the motor.

Product DFA can be divided into three areas: (1) manual, automated or flexible assembly; (2) feasibility and ease of assembly; and (3) component feeding and orienting. These areas share the general objectives of DFA.

3.2.1   *General objectives of product DFA*

Objectives of product DFA include the analysis of product design alternatives to:

1.   **minimize the number of components and subassemblies**, hence reduce material and production costs;
2.   **minimize the time and cost and maximize the reliability of assembly tasks**, e.g. eliminate component tangling;

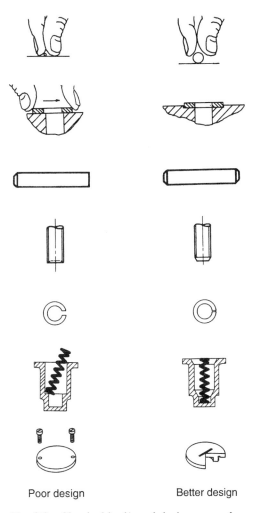

Poor design   Better design

Fig. 3.2 Classical bad/good design examples.

3. **maximize assembly stability**, e.g. adjust tolerance to part-feeders' dimensions;
4. **maximize process yields** at each workstation and for the whole assembly system;
5. **eliminate 'hidden' assembly faults and defects** that are initially within tolerance but surface after the product is used (for example, because of vibration or fatigue);
6. **standardize by common components, processes and methods** to reduce costs across the whole system.

There are three general types of approach to product DFA, listed below in the order of DFA evolution and refinement over the years:

1. **rule-based approaches**, following a list of guidelines developed and es-
    tablished as 'best practice' (e.g. Tipping, 1969);
2. **procedural methods**, providing systematic and sometimes computer-
    assisted analysis of products, including calculation of comparative
    metrics (e.g. Hitachi's AEM method, Ohashi *et al.*, 1984; Boothroyd and
    Dewhurst, 1986; various commercial systems);
3. **artificial intelligence-based approaches** that are still emerging, applying
    various techniques such as case-based reasoning, feature-based analysis,
    and knowledge-based design (e.g. Delchambre, 1992).

Design for assembly guidelines have evolved for manual, automated and
flexible (robotic) operations. They comprise a multidisciplinary combination
of experiential, analytical and theory-based recommendations, serving as
checklists and advice (Karger and Bayha, 1966; Tipping, 1969; Warnecke and
Schraft, 1982; Nof, 1985; Warnecke, Schraft and Wanner, 1985; Heginbotham,
1986; Arabian, 1989; Boothroyd, 1992; Boothroyd et al., 1994; Edan and Nof,
1995).

3.2.2   General DFA guidelines

Various guidelines are summarized in Tables 3.2 to 3.4.

3.2.3   DFA guidelines specific to manual assembly

Manual assembly has the following advantages.

- It applies simpler and less costly hand-tools.
- A greater variation in part dimensions can be tolerated.

Table 3.2   DFA: product considerations

Issue	Guidelines
1. Number of parts	Minimize the number of parts and levels of assembly, and simplify product complexity.
2. Modularity	Design products from modular subassemblies so that modules can be scheduled, built and tested independently.
3. Base part	Ensure the product has a suitable base part on which the rest of the assembly can be built; usually, it is the largest, heaviest part.
4. Locating features	Ensure the base part has features to locate it readily in a stable position, preferably in a horizontal plane.
5. Layers	Design a product to be built up in layers, so each component can be added from above and located positively, without a tendency to move during subsequent motions or steps.

Table 3.3   DFA: component considerations

Issue	Guidelines
1. Symmetry	Attempt to design symmetrical parts to avoid need for extra orienting devices or motions.
2. Asymmetry	If symmetry cannot be achieved, exaggerate asymmetrical features to facilitate orienting; alternatively, provide asymmetrical features that can be used to easily orient the parts.
3. No tangling	Avoid projections, holes or slots that will cause tangling with other parts (parts being 'hooked' into each other) when placed in bulk, bin or feeder. Example: design holes and slots that are smaller than any projection.
4. No jamming	Provide features to prevent jamming, such as nesting.
5. Fastening	Select lower cost mechanical fasteners, from the least costly snap-fit (includes use of Velcro) to the progressively more costly plastic bending, rivets, screws, bolts, special-purpose fasteners.

Table 3.4   DFA: operation considerations

Issue	Guidelines
1. Chamfers	Facilitate assembly operations by providing chamfers or tapers to help guide and position fasteners.
2. No repositioning	Eliminate or minimize the need for repositioning an assembly once it is fixtured.
3. Shortest distance	Minimize motion distance, within practical limits, to reduce motion time and improve accuracy.
4. No adjustment	Eliminate or minimize the number of electrical and mechanical adjustments.
5. One-way assembly	Design 'foolproof' operations: parts can be assembled only one way; if misassembled, subsequent parts cannot be added.

- A greater flexibility within the normal human motion abilities can be assumed.
- Mechanically defective components are usually detected quickly.
- It is generally faster to switch between designs and between processes, although some learning may be required.
- Judgment can be assumed for solving unexpected problems.
- Costs are relatively constant.

Beyond the general guidelines listed in Tables 3.2 to 3.4, guidelines that pertain specifically to manual assembly are given in Table 3.5.

Table 3.5 DFA: manual assembly

Issue	Guidelines
1. Hazardous parts	Avoid parts that are difficult or hazardous for human hands: that are either too small, too large or too thin to pick up; that stick together or are slippery; that are too hot or too cold, sharp or chemically unsafe.
2. Clearance	Eliminate or minimize resistance to manual insertion; allow sufficient clearance for insertion, but not too much, to prevent jamming.
3. Vertical motions	Design for vertical, progressive motions about one axis of reference, preferably working from above and placing larger components first.
4. No holding	Avoid holding parts to maintain their position during assembly, but if necessary, secure their position as soon as they are inserted.
5. Locating	When holding a part, locate it before it is released. For instance, avoid having to position a short peg inside a longer cylinder.
6. Motion economy	Apply manual motion economy principles, including:
(a) use both hands	Both hands can and should be used for operations, if possible.
(b) vision	Tasks requiring use of eyes should be limited to normal human field of vision.
(c) less motions	The number of motions should be minimized by eliminating or combining operations.
(d) multiple tasks	Tasks on more than one product and/or with more than one component at a time are preferred.

A DFA procedure estimates alternative designs for time, cost or difficulty. Designs are ranked by how closely they follow DFA guidelines or deviate from them. For instance, time estimation for manual assembly is illustrated in Fig. 3.3.

3.2.4 DFA guidelines specific to robotic, flexible assembly

Robotic assembly is considered flexible because, unlike traditional assembly machines, it is not designed for a limited, rigid set of assembly operations. Where robotic assembly can be justified economically (as discussed in Chapter 4) its advantages are:

- ability to reprogram equipment to accommodate design changes and product varieties, thus affording automation even for small-to-medium production lot sizes;

- consistency and accuracy that improve quality and reduce scrap and material use;
- programming of new assembly tasks can be done off-line without disrupting production;
- robots can be integrated with other production, material handling and test equipment under computer control;
- process and production monitoring can be provided by computer control.

DFA guidelines which are specific to robotic, flexible assembly are given in Table 3.6.

Information to support these guidelines is developed by experimental and analytic methods. For instance, Fig. 3.4 illustrates an analysis to identify the best work location for a given assembly robot. Figure 3.5 illustrates analysis based on evaluation of robot reachability in executing tasks such as part placement and kitting. In terms of reachable work points, the medium-size

Table 3.6 DFA: flexible assembly

Issue	Guidelines
1. Part handling	Provide component features, such as tapers and chamfers, lips, leads that make parts self-guide, self-align and self-locate readily, so that less accuracy is required during part handling.
2. Self-locating	Ensure that parts are self-locating if they are not secured immediately after insertion.
3. Motion economy	Apply robot motion economy principles, including:
(a) grippers	Minimize the number of grippers needed, and consider adding a simple feature for easier, more secure gripping.
(b) simpler robots	Minimize the complexity of required robots. **Reduce** (1) number of arms and joints (determined by the number of necessary orientations); (2) dimensions (determined by where robots must reach); (3) loads to be carried. **Result:** simpler robots, less energy use, simpler maintenance, smaller workspace.
(c) motion path	Simplify required motion paths; point-to-point motions require simpler position and velocity control compared with continuous path control.
(d) sensors	Do not apply sensors, but if needed, simpler touch and force sensors are preferred to vision systems; minimize the number and complexity of sensors to reduce hardware and operation cost.
(e) given abilities	Utilize robot abilities already determined as justified. Example: a sensor justified for one function can improve other tasks.

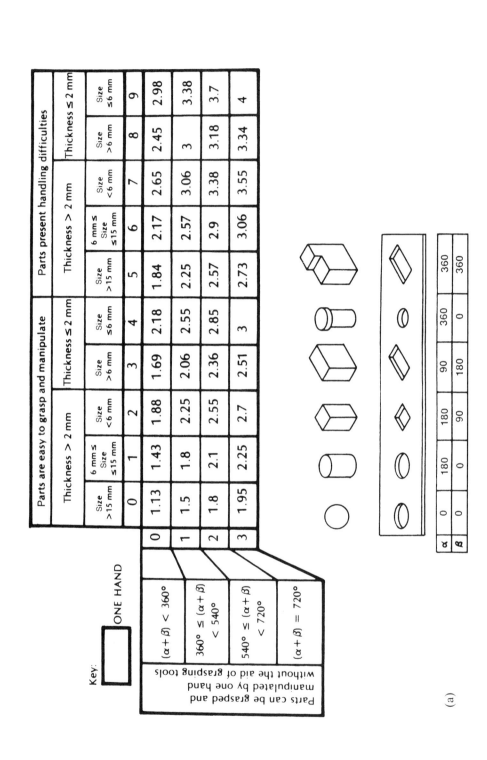

Key: ☐ ONE HAND

	Parts are easy to grasp and manipulate					Parts present handling difficulties				
	Thickness > 2 mm			Thickness ≤ 2 mm		Thickness > 2 mm			Thickness ≤ 2 mm	
	Size >15 mm	6 mm ≤ Size ≤15 mm	Size <6 mm	Size >6 mm	Size ≤6 mm	Size >15 mm	6 mm ≤ Size ≤15 mm	Size <6 mm	Size >6 mm	Size ≤2 mm / Size ≤6 mm
	0	1	2	3	4	5	6	7	8	9
0 $(\alpha + \beta) < 360°$	1.13	1.43	1.88	1.69	2.18	1.84	2.17	2.65	2.45	2.98
1 $360° \le (\alpha + \beta) < 540°$	1.5	1.8	2.25	2.06	2.55	2.25	2.57	3.06	3	3.38
2 $540° \le (\alpha + \beta) < 720°$	1.8	2.1	2.55	2.36	2.85	2.57	2.9	3.38	3.18	3.7
3 $(\alpha + \beta) = 720°$	1.95	2.25	2.7	2.51	3	2.73	3.06	3.55	3.34	4

Parts can be grasped and manipulated by one hand without the aid of grasping tools

| α | 0 | 180 | 180 | 0 | 90 | 90 | 0 | 360 | 360 | 360 |
| β | 0 | 0 | 90 | 180 | 180 | 180 | 360 | 360 | 0 | 360 |

(a)

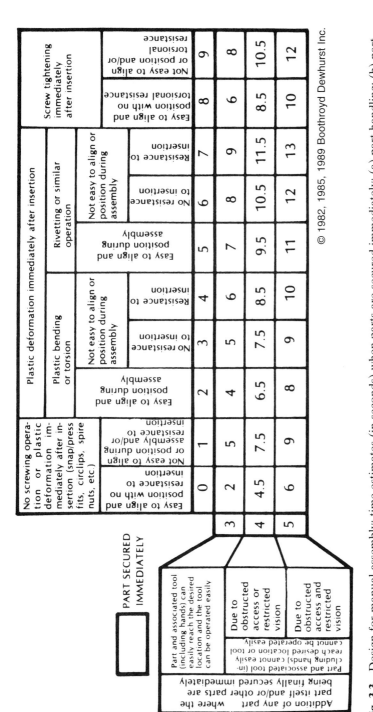

PART SECURED IMMEDIATELY

	No screwing operation or plastic deformation immediately after insertion (snap/press fits, circlips, spire nuts, etc.)		Plastic deformation immediately after insertion						Screw tightening immediately after insertion	
			Plastic bending or torsion			Rivetting or similar operation				
	Easy to align and position with no resistance to insertion	Not easy to align or position and/or resistance to insertion	Easy to align and position during assembly	Not easy to align or position during assembly, No resistance to insertion	Not easy to align or position during assembly, Resistance to insertion	Easy to align and position during assembly	Not easy to align or position during assembly, No resistance to insertion	Not easy to align or position during assembly, Resistance to insertion	Easy to align and position with no torsional resistance	Not easy to align or position and/or torsional resistance
	0	1	2	3	4	5	6	7	8	9
3 — Part and associated tool (including hands) can easily reach the desired location and the tool can be operated easily	2	5	4	5	6	5	6	7	6	8
4 — Due to obstructed access or restricted vision (Part and associated tool (including hands) cannot easily reach the desired location or tool cannot be operated easily)	4.5	7.5	6.5	7.5	8.5	7	8	9	8.5	10.5
5 — Due to obstructed access and restricted vision	6	9	8	9	10	9.5	10.5	11.5	10	12

Addition of any part where the part itself and/or other parts are being finally secured immediately

© 1982, 1985, 1989 Boothroyd Dewhurst Inc.

(b)

Fig. 3.3 Design for manual assembly: time estimates (in seconds) when parts are secured immediately: (a) part handling; (b) part insertion (courtesy of Boothroyd Dewhurst Inc.).

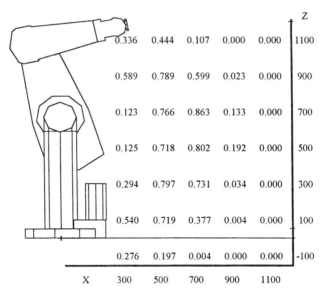

	0.336	0.444	0.107	0.000	0.000	z 1100
	0.589	0.789	0.599	0.023	0.000	900
	0.123	0.766	0.863	0.133	0.000	700
	0.125	0.718	0.802	0.192	0.000	500
	0.294	0.797	0.731	0.034	0.000	300
	0.540	0.719	0.377	0.004	0.000	100
	0.276	0.197	0.004	0.000	0.000	-100
X	300	500	700	900	1100	

Fig. 3.4 Composite ranking of reachable points in the workspace of a given assembly robot (PUMA 560) indicates the optimal work surface position when the ranking is 0.863 (Nof, Witzerman and Nof, 1996).

SCARA is significantly better, but in the other robot-size groups, the articulated type is preferred. Based on this study, it was also found that small-size and medium-size SCARA robots (a large one was not studied) perform relatively better than other robot types in terms of shorter cycle time and lower wear.

3.2.5 *DFA guidelines specific to automatic assembly*

The main advantages of automatic assembly machines are the relative advantages of automation in general: high speed, accuracy and consistency under predesigned conditions. A major goal of automation is fast-feeding of components, for example by prepackaged electronic components in feeder tubes or tape reels and by vibratory feeders. This topic is discussed further in Chapter 4.

   DFA guidelines for assembly machines and workheads generally follow the DFA guidelines listed above for robotic, flexible assembly. Table 3.7 summarizes considerations for automatic assembly. DFA data for automatic and robotic assembly are illustrated in Fig. 3.6.

3.2.6 *Integration of product DFA and facility DFA*

Assembly of well-designed components by an ineffective facility makes no sense. Therefore integration of product DFA and facility DFA is necessary.

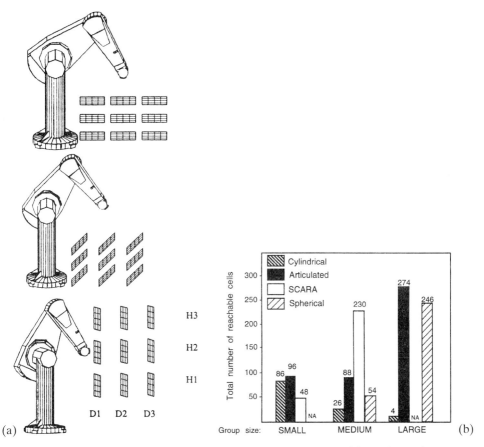

Fig. 3.5 Robot motion economy analysis for ten robot types: (a) 27 alternative positions for bin or work surface; (b) reachable cells (out of 432 possible cells) (Edan and Nof, 1995).

Table 3.7 Design features that determine cost-effectiveness of automatic assembly

1. Frequency of simultaneous operations
2. Orienting efficiency
3. Feeder(s) required
4. Maximum feed rate possible
5. Difficulty rating for automatic handling
6. Difficulty rating for insertions required
7. Assembly operations required – number and type
8. Total number of components per assembly

Source: Tanner (1985).

Addition of any part where final securing is taking place

		No screwing operation or plastic deformation immediately after insertion (snap or press fits, etc.)		Plastic deformation immediately after insertion						Screwing immediately after insertion	
		Easy to align and position no resistance to insertion	Not easy to align or position and/or resistance to insertion	Plastic bending			Rivetting or similar plastic deformation			Easy to align and position no resistance to screwing	Not easy to align or position and/or resistance to screwing
				Easy to align and position	Not easy to align or position (no features provided for the purpose)		Easy to align and position	Not easy to align or position (no features provided for the purpose)			
					No resistance to insertion	Resistance to insertion		No resistance to insertion	Resistance to insertion		
PART SECURED IMMEDIATELY		0	1	2	3	4	5	6	7	8	9
Straight line insertion	From vertically above — 3	1.2	1.9	1.6	2.4	3.6	0.9	1.4	2.1	0.8	1.8
	Not from vertically above — 4	1.3	2.1	2.1	3.2	4.8	1	1.5	2.3	1.3	2
Insertion not straight line motion — 5		2.4	3.8	3.2	4.8	7.2	1.8	2.8	4.2	1.6	3.6

(a)

		Force or torque levels within robot capability									
		Part can be gripped and inserted using standard gripper or gripper used for previous part						Part requires change to special gripper		Special workhead operation	
		Snap or push fit		Push and twist or other simple manipulation		Snap or push fit or simple manipulation		Screw fastening or nut running		Robot positions part	
		Self-aligning	Not easy to align	Self-aligning	Not easy to align	Self-aligning	Not easy to align	Self-aligning	Not easy to align		
		0	1	2	3	4	5	6	7	8	
Part added and secured immediately	Using motion along or about the vertical axis — 3	1.0 0.55	1.0 0.6	1.0 0.7	1.0 0.75	1.0 0.6	1.0 0.65	1.0 0.7	1.0 0.8	1.0 1.15	
		0	0	0	0	1.5 0.7	1.5 0.7	1.5 0.7	1.5 0.7	4.0 0.7	
	Using motion along or about a non-vertical axis — 4	1.5 0.55	1.5 0.6	1.5 0.7	1.5 0.75	1.5 0.6	1.5 0.65	1.5 0.7	1.5 0.8	1.5 1.15	
		0	0	0	0	1.5 0.7	1.5 0.7	1.5 0.7	1.5 0.7	4.0 0.7	
	Involving motion along or about more than one axis — 5	1.5 1.05	1.5 1.1	1.5 1.15	1.5 1.2	1.5 1.05	1.5 1.1			1.5 1.6	
		0	0	0	0	1.5 0.7	1.5 0.7			4.0 0.7	

© 1989 Boothroyd Dewhurst Inc.

(b)

Fig. 3.6 Design for assembly data samples: (a) relative workhead cost for automatic insertion; (b) time and cost (in seconds; cents) for a single-station, two-arm robot system (*table entries: top left* – relative robot cost; *top right* – relative effective basic operation time; *bottom left* – relative additional gripper or tool cost; *bottom right* – relative time penalty for gripper or tool change) (courtesy of Boothroyd Dewhurst Inc.).

Fig. 3.7 A cell designed for assembly of lighting fixtures (courtesy of Adept Technology, Inc.).

Such integration is illustrated in Fig. 3.7, showing mechanical assembly of a variety of fluorescent lighting fixtures with two robots and one operator. Tasks are allocated as follows: the left-side robot measures, cuts and strips wires, mounts a connector on some of them, and assembles them with other components to a base subassembly. The latter is prepared ahead by the right-side robot. This robot picks up fasteners and mechanical components from three bowl feeders and conveyors and assembles them to a base part. The human operator assembles other components, replenishes parts and supervises the cell. When a lighting fixture is ready, the left-side robot hangs it on an overhead hook-conveyor which carries it out of the cell.

The impact of DFA on such a cell can be described as follows.

1. **Product considerations.** By reducing the number of components, there are less operations, shorter cycle time, and smaller cell area, e.g. less feeders are required.
2. **Component considerations.** Components are designed for the simplest positioning and self-location, and easier feeding and gripping. The result influences the design of grippers, work methods and control.
3. **Operation considerations.** Operations are simplified by elimination of repositioning and adjustment at the assembly fixture, and minimizing motion distances for both robots and human operator.

4. **Operator considerations.** Design of adjustable work surface height, effective visibility and motion economy results in less fatigue, safety, more effective production and better product quality.
5. **Robot selection** matches the tasks, accuracy, reachability and workload requirements, resulting in better quality and productivity.
6. **Cell considerations.** Selection and placement of the assembly fixture, part feeders, accessories and conveyors determine the overall accuracy requirements, motion distances, access to setup and repair, and so on.

The hybrid (human–robot) cell in Fig. 3.7 produces 45 assemblies per hour and operates 2.5 shifts per day. Without careful DFA, such a cell will be wasteful, may not operate smoothly and may be difficult to justify economically.

Another example is DFA with robotic kitting (Fig. 3.8). Four alternative configurations, shown in Fig. 3.8(a), are compared to select the most effective for a particular company. Analysis of each robot operation is by RTM, the robot time and motion evaluation method (Lechtman and Nof, 1983; Nof, 1985; Nof and Rajan, 1994) as illustrated in Fig. 3.8(b). Then, each configuration is simulated under alternative kit flow control policies (Sellers and Nof, 1989). The results, shown in Fig. 3.8(c), indicate that in this case a gantry (overhead) robot at the staging area is recommended.

For DFA to be effective, each company must develop guidelines that are relevant for its particular practice. Another key element is training for consistent DFA application throughout the organization. Additional DFA illustrations are discussed in section 3.4 and in Chapter 4. DFA is part of design for manufacturability and of assembly process planning. Group technology, discussed in the next section, offers relevant techniques for these functions.

3.3 Group technology for assembly

Group technology (GT) is a rationalization approach based on commonalities among parts and processes (Ham, Hitomi and Yoshida, 1985). A classification and coding system for GT can provide users with a consistent method for part identification based on both design and process characteristics. Such a method can also be used as a key to integrating computer-aided design, process planning and manufacturing. GT codes allow for direct retrieval of similar parts, leading to benefits from previous experience.

Five motivating factors for companies that have implemented GT coding are (Tatikonda and Wemmerlov, 1992): (1) improved communications through a common terminology; (2) increased capacity of existing production facilities through use of cellular layout; (3) increased productivity of design change and process planning functions; (4) standardization in design and process planning procedures; (5) reduced cost of quality. A prominent advantage is the avoidance of design duplication. The main objection to GT imple-

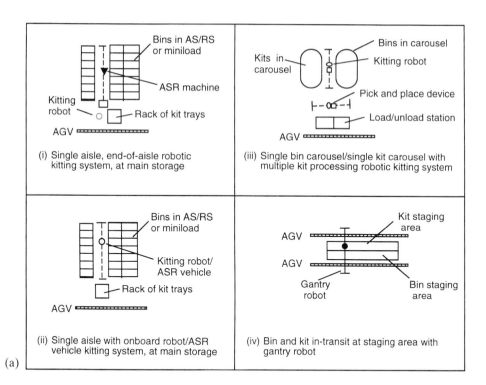

(a)

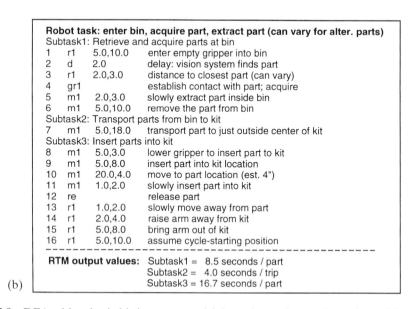

(b)

Fig. 3.8 DFA with robotic kitting systems: (a) four alternative configurations; (b) RTM analysis of on-board robot picking operation; (c) performance simulation results (EDD – earliest due date of kit orders; FIFO – first in, first out kit flow) (Tamaki and Nof, 1991).

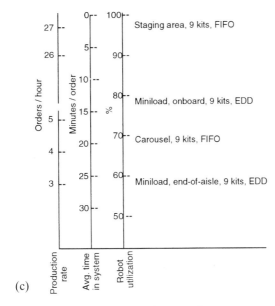

Fig. 3.8 *Continued*

mentation is that it may fail without good communication between design and manufacturing personnel. Nevertheless, GT is recognized as an effective ingredient for CAD–CAM integration and concurrent engineering. With maturation of database technology, it is easier to organize design and manufacturing information to reflect the key commonalities.

A technology database, built with classification and coding, enables a user to group components by common attributes (see example in Table 3.8) and take advantage of similarities. By grouping parts, the number of retrieval searches can be reduced dramatically. Groups of parts that are retrieved by class can be analysed for standardization and assemblability (Bao, 1988).

While general classification and coding schemes such as Optiz, CODE, MICLASS and DCLASS (Bedworth, Henderson and Wolfe, 1991) contain descriptive attributes as in Table 3.8, focus on assembly attributes is needed for DFA. For instance, the following additional attributes are listed as affecting part handling, insertion and fastening (Boothroyd, 1992):

- thickness;
- nesting;
- tangling;
- fragility;
- slipperiness;
- stickiness;
- necessity for using two hands, or grasping tools, or optical magnification, or mechanical assistance;

Table 3.8   Classified part/product attributes for various applications

Classification attribute	Typical application objective			
	Design retrieval	Process planning	Layout design	DFA
Shape	√	√		√
Size envelope	√	√	√	√
Form features	√	√		√
Dimensions	√	√		√
Tolerances		√		√
Surface finish		√		√
Functions	√	√		
Material	√	√		√
Raw material form		√		
Quantity		√	√	√
Next assembly			√	√
Routing		√	√	√

- accessibility of assembly location;
- ease of operating an assembly tool;
- visibility of assembly location;
- ease of alignment and positioning during assembly;
- depth of insertion.

For electronic assembly, the following additional attributes have been listed as relevant for DFA (Styslinger and Melkanoff, 1985):

- electrical type;
- maximum component height;
- lead length;
- hand assembly requirements;
- % auto-SIP (single in-line package) or auto-DIP (dual in-line package) insertable;
- % auto-axial insertable;
- terminal/connector types;
- special features.

The BMCODE for electronic assembly (Bao, 1988) follows the code structure of DCLASS, and addresses four classification areas: (1) part family; (2) part assembly; (3) board layout; (4) process and equipment. The part family classification is by a variable length code:

X(type)–DDDD(descriptions)–RRRRR(ratings)–PP(packaging type)–
YYY(temperature range, certification)

For example, a particular resistor is classified and coded by a 15-digit code: **1st digit:** R (type, for resistor); **digits 2–5:** descriptions of application purpose, voltage type and construction (e.g. wire, film, composite, ceramic, etc.); **digits 6–10:** five rating ranges of resistance, tolerance, wattage, voltage and temperature coefficient; **digits 11–12:** package style – for fixed applications, axial lead, tubular, cement embedded, flat strip, DIP or SIP, etc.; for variable applications, shape or terminal; **digits 13–15:** temperature range and certification source.

The part assembly code includes 14 digits: **digits 1–2:** M (type, for mounting method), through-hole, surface mount or other; **digits 3–6:** specific mounting attributes such as axial, radial, SIP, DIP, etc.; **digits 7–8:** body dimensions; **digits 9–11:** generic attributes such as polarity, heat, moisture and static sensitivity; **digits 12–14:** coating requirements, use of heat-sink and circuit connection. For non-lead mounting, digits 3–6 designate how the component is secured, positioned and fastened.

Group technology techniques help to improve the design and quality of assembled products. How easy or difficult (and costly) a given design would be to assemble is the subject of evaluation procedures that are described in the next section.

3.4   Procedural systems for assemblability evaluation

Procedural assemblability evaluation is applied by designers for quantitatively estimating the degree of difficulty and associated cost of assembly. While DFA guidelines are general, a quantitative ranking enables designers to systematically compare and analyse trade-offs. Hitachi developed an original method of assemblability evaluation (AEM) in 1975 (Ohashi *et al.*, 1984) which was later extended as the General Electric-Hitachi method. Figure 3.9 shows the structure of AEM and its application to VTR (videotape recorder) design. Most of the parts, such as pressed parts, rubber belts and coil springs, are now assembled from above with a simple insertion and joining movement, so they can be attached by pick-and-place workheads or cheaper robot arms with only three degrees of freedom.

The basic steps of AEM are:

1.   evaluation of assembly difficulty by means of a 100-point index;
2.   simple analysis and calculation of product assemblability during early design;
3.   evaluation indices are correlated to assembly cost. A reasonable cost estimation can be made as shown in Fig. 3.9(b).

By iteratively applying AEM, designers can improve the design in its early development. Figure 3.9(c) depicts improvement examples in the VTR mechanism. In the illustration, the total improvement achieved was a reduction from 460 to 379 components to assemble, and increased rating from 63 to 73.

Other companies, the University of Massachusetts, the Institute of Technology in Denmark (Andreasen, Kahler and Lund, 1983) and others have devel-

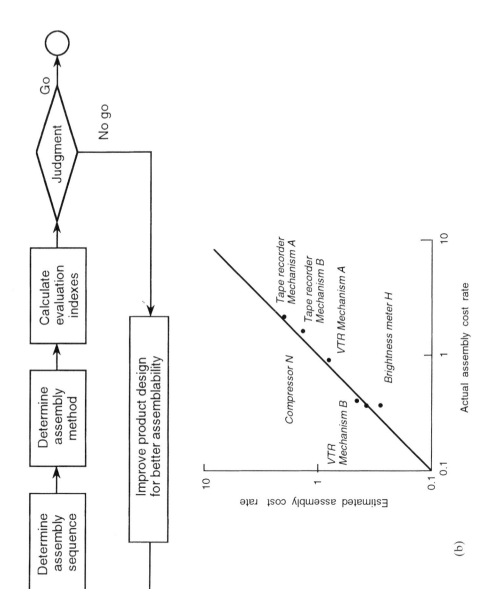

Go

Judgment

No go

Calculate
evaluation
indexes

Determine
assembly
method

Determine
assembly
sequence

Improve product design
for better assemblability

(a)

10

1

0.1

Estimated assembly cost rate

0.1

1

10

Actual assembly cost rate

Tape recorder
Mechanism A

Tape recorder
Mechanism B

VTR Mechanism A

Brightness meter H

Compressor N

VTR
Mechanism B

(b)

Purpose	Reduction of part number	Stabilize parts positioning	Easy handling	Easy insertion	Prevent entanglement	Prevent jamming
Means	Combine parts	Prepare positioning guide	Prepare parallel segment for gripping	Prepare chamfering	Use connector to eliminate lead wires	Increase parts thickness
Part name	Tape guide	Loading motor assembly	Capstan motor assembly	Bolt	Loading motor assembly	Washer
Before design improvement			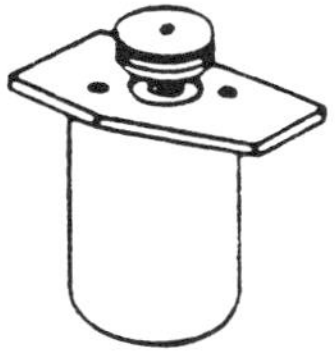			
After design improvement			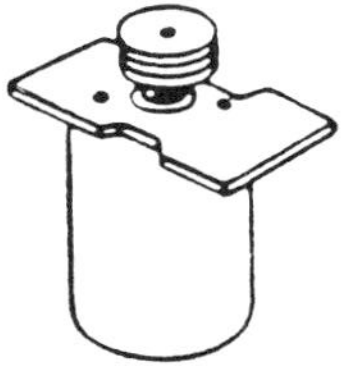			

Fig. 3.9 Assemblability evaluation method (AEM): (a) AEM procedure; (b) accuracy of AEM cost estimation; (c) AEM table (from Ohashi *et al.* 1984; courtesy of Hitachi Ltd.).

(c)

oped similar methods. Corporations such as Sony, Toshiba, NEC and Nippon-denso followed Hitachi in developing their own methods for product designs (see Fig. 3.10). The method developed originally at the University of Massachusetts (Boothroyd and Dewhurst, 1986) and later at the University of Rhode Island has been computerized and implemented widely in the United States with company-specific adaptations, for example by XEROX, Motorola, NCR, Digital Equipment, Texas Instruments and Hewlett Packard (Coleman, 1988).

Significant benefits have been reported by companies applying product DFA. For instance, Westinghouse (Funk, 1989) reports that improvements in electrical products by DFA have lowered costs, on average, of PCB assembly (–32%), wire assembly (–48%), mechanical assembly (–73%), and mechanical part cost (–76%). A 1994 survey by Boothroyd Dewhurst Inc. of 500 customers found the following average improvements: part count (–51%); part cost (–37%); time to market (–50%); quality and reliability problems (–68%); assembly time (–62%).

Assemblability evaluation methods have initially focused on conceptual design, with a goal to limit the number of parts and achieve greater savings from the outset. However, the accuracy of evaluation at this early stage cannot be as high as more detailed analysis which is integrated with design for manufacturability and testability. The system of DFA evaluation developed by Boothroyd and Dewhurst and later expanded to DFMA evaluation (Boothroyd, Dewhurst and Knight, 1994) addresses details as illustrated in Figs 3.3 and 3.6. To increase the usefulness of detailed approaches, computer-aided packages have been developed (section 3.6).

3.5   Design for disassembly and recycling

Disassembly is defined as the dismantling of a complete assembly down to its individual components. Disassembly is important in three areas: (1) as a logical step in design for assembly; (2) maintenance and repair; (3) recycling. These areas are discussed next.

3.5.1   Disassembly for DFA

Disassembly is considered a logical step in DFA. The purpose is to assess the relationships and constraints among components in building up an assembly. Kinematic disassembly models are used as part of assembly CAD (Khosla and Mattikali, 1989; Woo and Dutta, 1991; Zussman, Shoham and Lenz, 1992). The idea is that starting from a goal assembly and working backward, it is easier to find the constraints on related components.

Since assembly and disassembly are usually reversible for rigid parts, then for a given assembly it is possible to generate automatically all the opposite

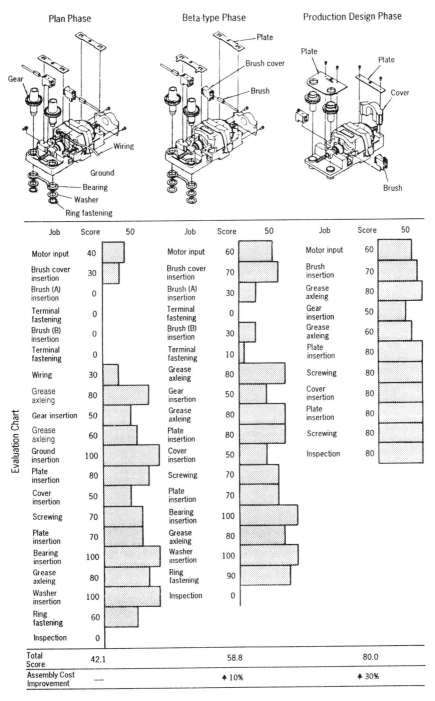

Fig. 3.10 A three-phase product redesign by the Sony Assemblability Evaluation Method (courtesy of Sony Corp.).

operations. Assemblies are represented in such analyses either as geometric 3-D models or as object graphs. Limitations are that the approach is restricted to rigid objects and applies only to local motions. In addition, a complete assembly must be designed first and its geometric model computerized before any disassembly analysis can begin. The advantage is that disassembly analysis provides a mathematical tool for automatic assessment of assembly/disassembly. By generating the kinematic characteristics, a designer can automatically select robots or assembly machines for a planned assembly. An automatic plan for partial disassembly can also help a machine to correct errors during assembly.

Disassembly analysis has been used to determine precedence constraints and assembly sequences. However, the number of ways to assemble a product is larger than the number of ways in which it can be disassembled, hence only a subset of all possible sequences can be provided. Similarly, the set of directions along which a component can be located in its final position, called the **location direction set**, is smaller than the full set generated by considering build-up assembly. This limitation impacts on task planning because sequencing based only on disassembly can cause problems during automatic assembly execution.

As a logical step in assembly planning, disassembly has certain advantages and certain limitations. However, it is a highly logical approach when maintenance and repair are considered, as discussed next.

3.5.2   *Disassembly for maintenance and repair*

This type of disassembly is considered an integral part of product functionality. Regular DFA approaches can be applied with emphasis on good access, visibility and ease of replacing worn modules. The areas of maintainability, repairability and serviceability are considered important functions in the design for reliability. The objective is to consider these performance characteristics for the useful life of a product. In recent years, the lifecycle of products has been extended (Fig. 3.11) and a consideration of recycling economics has become highly relevant.

3.5.3   *Disassembly for recycling*

Lack of natural resources and energy, shortage of landfill, and environmental hazards involved with the disposal of products are forcing industry to seek ways to recycle. Recycling means recovering materials or components of a used product in order that they may be used for new products. Studies (e.g. Jovane *et al.*, 1993) identify the main product types associated with recycling concerns in order of annual scrap rates: cars, household appliances, consumer electronics, control equipment, lighting systems and computers.

A used product can be recycled by one of two approaches (Seliger, Zussman and Kriwet, 1994): by shredding, followed by sorting the shredded materials,

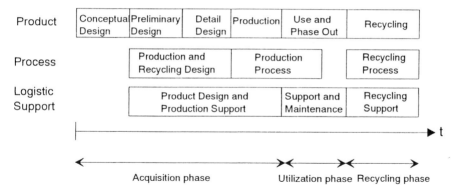

Fig. 3.11   The extended lifecycle concept (Seliger, Zussman and Kriwet, 1994).

or by disassembly. Shredding is presently more common, but disassembly has two advantages. It is non-destructive, and it allows complete material recycling with potential reuse of some parts and subassemblies.

Proper disassembly for recycling includes four steps: draining all fluids; dismantling all parts and components that should not be destroyed by shredding; shredding and sorting the remaining materials; disposing of non-recyclable materials. While new products can be designed for disassembly, older products pose three types of difficulties (Seliger, Zussman and Kriwet, 1994):

- certain parts of the product may have been modified during repairs;
- worn joints cause some disassembly to be difficult or impossible;
- traditionally, products have been optimized for functional and assembly requirements and are not easy to disassemble.

As a result, disassembly requires highly skilled workers and it is difficult, although highly desirable, to develop flexible disassembly equipment. Companies have to consider the extended lifecycle (Fig. 3.11) including the environmental impact and logistic support for the recycling of their products. Recommended guidelines for design for recycling by disassembly are given in Table 3.9.

A useful tool in planning disassembly for recycling and for product redesign for recycling is the **recovery graph** (Fig. 3.12). The graph (essentially an AND/OR graph) represents the dependencies and consequences of disassembly options for a particular product. In the example, the analysis indicates that the total recycling benefit per unit is greater than the cost of complete disassembly up to the level shown. In the original design, labels with adhesives, a metal contact and a leaf-spring (that are eliminated in the new design) made the device more costly to assemble, hence its recycling could not be justified. Design considerations relevant for manual, automated and flexible disassembly have been discussed earlier, in section 2.2.

Table 3.9 Design for disassembly guidelines (following Jovane *et al.*, 1993)

Objective	Guidelines
Less disassembly work	• Combine elements • Limit material variability • Use compatible materials • Group harmful materials in separate, accessible modules • Provide easy access to valuable and reusable parts
Predictable product configuration	• Avoid combining ageing and corrosive materials • Protect subassemblies against soiling and corrosion • Provide clear identification of replacement/repair modules
Easy disassembly	• Provide access to drainage points • Use fasteners that are easy to remove or destroy • Minimize number of fasteners • Use the same standard fasteners for all parts • Provide easy access to disjoining or breaking points • Avoid multiple disassembly directions, complex movements • Set center elements on base part • Avoid metal inserts in plastic parts
Easy handling	• Provide a surface for grasping • Avoid non-rigid parts • Enclose harmful materials in sealed modules
Easy separation	• Avoid secondary finishing, e.g. painting, coating, plating • Color-mark different material types differently for simpler sorting • Avoid parts, materials that will damage disassembly machinery
Reduced variability	• Use standard subassemblies and components • Minimize number and variety of fasteners • Avoid specialized fasteners

3.6 Computer-aided methods for DFA

Applications of computer-aided assembly engineering include cost estimating (e.g. Ong, 1993), time estimates (e.g. Nof, Knight and Salvendy, 1980; Nof and Paul, 1980; Nof and Lechtman, 1982; Rembold and Nof, 1991), process planning (e.g. Chiu, Yih and Chang, 1991), computerized assemblability data systems (e.g. Arai, 1985; Feldman and Geyer, 1991), development of assembly workplaces (e.g. Zulch and Waldhier, 1992), cellular assembly planning (e.g. Ho and Moodie, 1994), assembly planning and control with CAD/CAM (e.g. Warnecke and Domm, 1989), computer-aided line balancing (e.g. Dar-El and Rubinovitch, 1979; Shtub, 1993) and assembly scheduling (Van Brussel, 1990). Computer-aided DFA issues are discussed in this section.

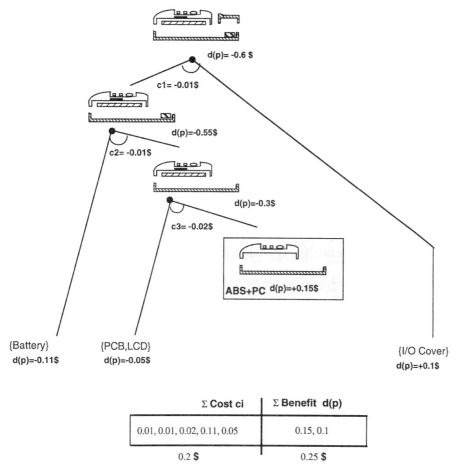

	Σ Cost ci	Σ Benefit d(p)
	0.01, 0.01, 0.02, 0.11, 0.05	0.15, 0.1
	0.2 $	0.25 $

Fig. 3.12 A recovery graph to assess design for recycling by disassembly of a remote control device (Zussman, Kriwet and Seliger, 1994).

3.6.1 Integration of CAE and assembly

Integration of CAE, CAD and CAM for assembly faces challenging problems, since assembly tasks are often less well defined compared with typical machine processing. Development of CAD/CAM systems for assembly requires that CAD data are directly linked to the control of assembly machines or robots. Presently, welding and painting robot controllers can be connected to CAD systems. But it is relatively more difficult for assembly robots to be integrated directly with CAD systems for the following reasons: (1) poor absolute accuracy of robots; (2) limited accuracy of geometric modeling techniques of a robot and its operational environment in CAD systems; (3) complex specification of assembly sequences, and trajectory planning of robot motions according to each assembly sequence. Because of the importance of flexible, robotic assembly, progress in all these areas is already occurring.

Six issues have been proposed by Kakazu (1987) to enable the integration of CAD/CAM systems and assembly automation:

1. development of new assembly task description languages;
2. development of computer-based visual robot simulator;
3. effective use of geometric modeling for assembly in CAD technology;
4. design of a good user interface for computer simulators of robotic assembly;
5. establishment of dynamic simulation environment based on accuracy and timing of variables and of assembly operations;
6. ability to communicate between virtual geometric modeling on a computer simulator and the real world.

Several researchers have already been working in these areas, including pioneering investigations of assembly languages (Popplestone, Ambler and Bellos, 1978), assembly database (Roy and Liu, 1989) and geometric models of assembly (Wesley *et al.*, 1980). In the area of assembly automation, practical implementations have been reported, for instance linking CAD with PCB assembly (Kiang, Gay and Chang, 1992). Recent implementation of robot calibration techniques (Duelen and Schroer, 1991) and interactive graphic simulators of robotic cells (Nof, 1994), including commercial robot CAD and simulation packages, also offer useful solutions to these problems.

3.6.2   CAD and DFA

Several computer-aided methods have been developed to assist design for manufacturability and assemblability. Commercial CAD systems already include, as an integral part, a DFA function. For instance, Fig. 3.13 shows an assembly planner with evaluation routines developed in the ICAD software environment. Assembly-oriented design processes have been developed to evaluate suitability for automated assembly, including assemblability measures (Hales, 1989).

A design process developed by Hales is divided into four stages, including planning, rough design, drafting and detailed design. Computer aids include: (1) guidelines and examples; (2) catalog of relative cost; (3) assembly priority graph; (4) value analysis and (5) evaluation of suitability for assembly, including measures for product structure, for subassemblies, for individual components and for joining techniques.

Product DFA methods were evaluated at Daisy Systems Corp. (Hales, 1989) including data entry and calculations, computer-aided drafting, solids modeling, color graphic results, concurrent design processes, networking, archiving, data management, reporting management and automatic initialization. Integration of tools was found to be essential. For example, as the number of CAD assembly drawings increases, simultaneous access to both

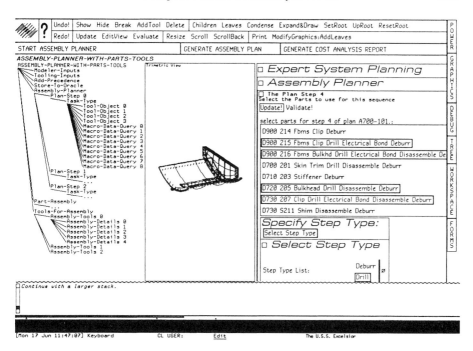

Fig. 3.13 The automated assembly planner for airframe subassemblies (includes CAD analysis, cost and time evaluation, assembly instruction generation and downloading capability to shop machines) (courtesy of Northrop Corp.).

CAD-generated drawings and data entry and calculation routines is necessary. Integrated support overcomes the inconvenience and delays caused by separate software which is working on separate systems and at remote sites.

An integrated CAD system for the design of an assembly facility and its control has been developed by Witzerman and Nof (1995a, 1995b). Its purpose is to utilize a graphic simulator/emulator workstation to specify parts, tasks, facility layout and material handling resources, and control hardware (e.g. sensors). The software extends the ROBCAD software TDL (task description language) functions and databases. An illustration is given in Fig. 3.14. The advantage is that the same platform is used for concurrent design and simulation/emulation of the assembly facility and the control system. Thus the implications of control for facility design, e.g. control field of view, hazardous collisions and near-collisions during execution, physical accessibility, etc., can be checked directly and improved in a virtual, simulated environment. For example, the black (bold) links in the figure (to the two robots and the rod conveyor) indicate active devices. Grey (thin) links indicate devices in ready-state. The color of each link changes according to the changing state of equipment, and serves to visualize how the control program operates during the kinematic simulation, permitting designers to identify logical errors in device task programs.

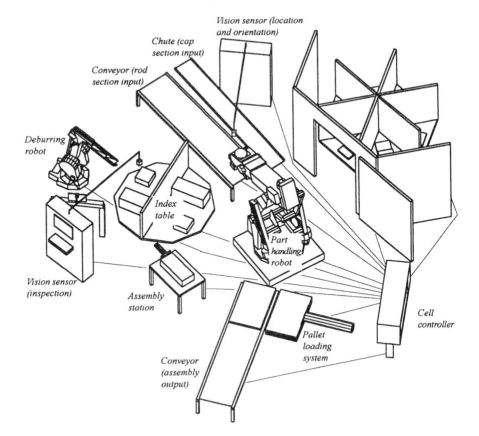

Fig. 3.14 An integrated CAD for an engine rod assembly facility (the top-right area is a shot-pinning cell) layout and control (Nof and Witzerman, 1994).

In recent years, several knowledge-based expert systems for DFA and assemblability evaluation have been developed, as described below in section 3.7. Computer-aided systems should provide alternative solutions to designers. It remains up to a designer to decide if a proposed solution is optimal (or at least satisfactory) and whether it violates or compromises other requirements, such as safety, functionality, reliability or ease of maintenance.

3.6.3 Computer-aided process planning (CAPP) and DFA

Assembly process planning translates assembly and component design data into assembly process information. Eight steps can be specified in the assembly process planning activity (following Alting and Zhang, 1989):

1. interpretation of design data;
2. selection of assembly processes;

3. selection of equipment/system;
4. determination of tooling;
5. assembly sequencing;
6. selection of inspection devices;
7. selection of assembly process parameters;
8. calculation of assembly times.

Some assembly process planning systems and approaches are discussed in this section. Assembly planning with artificial intelligence is discussed in section 3.7, while assembly planning with emphasis on system planning is discussed in Chapter 5, sections 5.4 and 5.5.

Relatively few systems have been implemented in practice so far to automatically generate process plans for assembly. Several methods and algorithms are suitable for planning assembly sequences (DeFazio and Whitney, 1987) and for dynamic and kinematic analysis of assemblies (Kim and Lee, 1989). A theory of part-mating in assembly has also been developed (Whitney, 1985). Emerging methods for assembly planning are discussed below in section 3.7.

Proliferation of product types and models and competitive constraints are increasing pressure on production planners with regard to deadlines, qualities and costs. Because the technical requirements of an assembly process influence the product design, the technical problems of assembly have to be integrated in the computer-aided design process. Also, evaluation procedures have to be applied during preliminary design to optimize it in terms of manufacture **and** assembly. Evaluation procedures consist of analysing (Ham and Lu, 1988): (1) flexibility, (2) functions, (3) feeding, (4) gripping, (5) insertion and (6) manufacturing.

Tasks of assembly process planning range from selection of suitable system principles, development of a preliminary process concept, development of layout variants, and determination of performance data parameters, costs and yield to simulation of system behavior and evaluation of alternatives (Warnecke *et al.*, 1987). Concepts from knowledge-based systems and AI-based planning techniques are found to be particularly suitable for automatic process plan generation, and are discussed in the next section.

3.7 Artificial intelligence approaches to DFA

Artificial intelligence (AI) techniques add to computer-aided DFA a variety of reasoning and decision-support capabilities. Once AI techniques are implemented effectively, three advantages can be gained. First, designers may need less training and expertise for utilizing CAD tools. Designers can obtain advice on how to improve their work. In addition, better quality designs with less errors can be expected, resulting in lower assembly cost. These advantages are important for wider proliferation of DFA practices in industry. Therefore, significant efforts have been invested in developing AI-based DFA. Examples of AI approaches for DFA include:

- rule-based knowledge (expert) systems;
- constraint net knowledge systems;
- neural networks for knowledge systems;
- fuzzy logic knowledge systems;
- search techniques for assembly planning;
- feature-based assembly design.

3.7.1 Knowledge-based approaches to DFA

Several systems have been developed with the relatively mature technology of rule-based knowledge. Such systems are programmed in LISP, PROLOG or expert system shells, and have been applied in industry (O'Grady and Oh, 1991).

For mechanical assembly, DACON or Design for Assembly Consultation (Swift, 1987) provides a CAD interface for drawing assembly components after they are designed with expert analysis; Hernani and Scarr (1987) developed an expert system interfaced with CAD to recommend assembly design rules; FADES or Facility Design Expert System (Fisher and Nof, 1987) provides economic analysis and selection of assembly technology; ADAM or Assisted Design for Assembly and Manufacture (Sackett and Holbrook, 1988) generates advice on reducing the number of components, rationalizing the assembly and insertion guidelines. Chen and Pao (1993) combine neural network and rule-based systems for the design and planning of mechanical assemblies.

For electronic assembly, Randhawa, Barton and Faruqui (1986) developed Wavesolder Assistant, an expert system for the troubleshooting of soldering assembly design; Chang and Terwilliger (1987) developed a rule-based system for planning printed wire board assembly; Bao (1988) developed an expert system for SMT printed circuit board design for assembly; a process planning expert system for electronics PCB assembly was developed by Sanii and Liau (1993).

In a constraint network approach for DFA (Oh, O'Grady and Young, 1995) design knowledge is represented not as a collection of rules, but as a collection of interconnected assembly constraint objects. An efficient search can be performed over these networks to evaluate the propagation of design changes. Other reasoning improvements are offered by a neural network approach (Naft, 1989) providing pattern recognition to address the variability of assembled components. An integrated CAD and expert system for the design of manual assembly workplaces, CARLA (Zulch, Braun and Schiller, 1995), is depicted in Fig. 3.15.

A knowledge-based approach to DFA is illustrated next by an example of assembly technology selection. This DFA problem is typical of situations

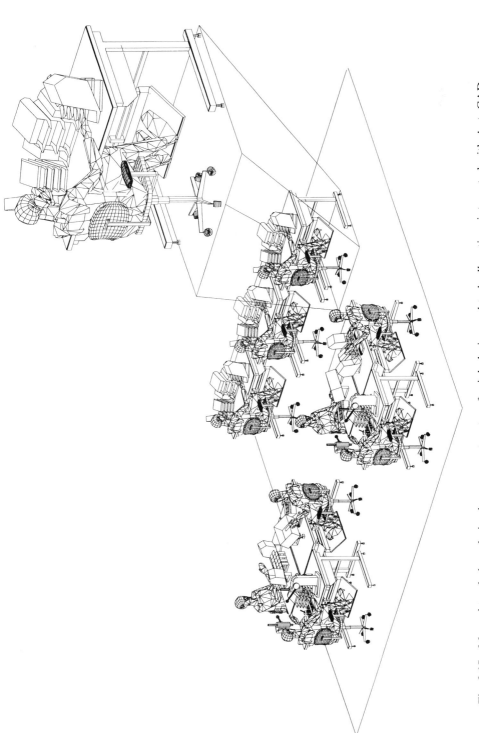

Fig. 3.15 Manual workplace design by an expert system for job design and task allocation, integrated with AutoCAD (Zulch, Braun and Schiller, 1995).

calling for a knowledge-based approach. The problem is time-consuming and has long-term impacts; it may occur under conflicting constraints; standard decision-models may prove inadequate because model assumptions may be violated; advice is needed on how to handle missing data and on the sensitivity to design changes; the required knowledge is available, but not necessarily where and when it is needed.

A knowledge-based system for DFA differs from conventional design and analysis aids in several ways. It can help by offering default values inferred from context knowledge, or by attempting to reason when some information is missing. Redundant data and fuzzy logic rules can help in this case. Certain systems provide learning capabilities that update the knowledge base automatically, and support for conflict resolution by trade-off comparisons is provided to handle conflicting recommendations that occur in reality. Evaluation of decision sensitivity to time or to design assumptions is included. Other support capabilities include interface with simulation models for determining production estimates of selected alternatives; pertinent databases, such as equipment catalogs, for retrieving necessary input parameters; and databases for sorting and storing results of decision analyses.

Knowledge-based economic analysis for DFA   An expert system prototype for facilities design, FADES (Fisher and Nof, 1984, 1987), was extended with a knowledge-based economic analysis for DFA. It includes the following elements:

1.   assembly technology selection module;
2.   rule-based logic (written in PROLOG) for economic evaluation of alternative assembly technologies, considered for acquisition or replacement;
3.   general facility design modules in FADES, including knowledge for individual workstation selection, capacity analysis and workstation site selection.

In addition to knowledge-based modules, FADES includes an inference engine, an external database containing economic, technology and facility data, and an interactive interface with designers. Facts in the knowledge base are developed and accumulated during the course of solving a particular design problem for a given product or group of products. As a general procedure for technology selection, a required assembly technology level is first examined relative to problem conditions. Next, 'good' technology candidates for given conditions are identified. A more detailed economic analysis is performed by the replacement analysis module to determine if and when a presently available system should be replaced with one of the other recommended candidates.

A simple example of a rule in FADES is Rule 21 below (converted from PROLOG to English). Its purpose is to evaluate the total cost of assembly personnel that use an automatic assembly machine.

Rule 21

IF   the number of parts per assembly, N1 and N1 ≤ 6
AND   the number of operators additional to machine super-
        vision on a rotary indexing machine is N2,
AND   the annual cost of one assembly operator is A,
AND   the annual cost of one machine supervisor is S,
THEN   use a *rotary indexing machine* and calculate the total
        cost of personnel for this machine, T, as T = N2 * A + S.

$N1$, $N2$, A and S are provided as default values and can be changed by designers. The limit value 6 applied in Rule 21 is based on general and local company experience. The cost list calculated for several relevant technologies is sorted and the best economic candidates are recommended for detailed consideration. The comparison is by judgmental logic rules for calculating comparative cost values.

An important ability in a knowledge-based system is **dealing with uncertainty**. For example, certain conditions may suggest that more than one technology may be recommended, but at different relative confidence. This situation can be addressed with certainty factors, as illustrated in Rule 53:

Rule 53

IF   automation-possible = *yes*
AND   product-styles = *few*
AND   design-changes = *few*
AND   investment-potential = I
AND   I < 5 AND R1 > 2
AND   production-volume = large
AND   no.-of-parts-per-assembly = N1
AND   N1 ≤ 15 AND N1 ≥ 7
THEN   assembly-technology = *af* cf 100
AND   assembly-technology = *ai* cf 80

where: *af* is automatic, assisted part feeding; *ai* is indexed, mechanical part feeding; and cf is a certainty factor at levels of 100% and 80%, indicating the confidence level in each recommendation.

Other examples of knowledge-based assembly design systems include the assembly planner for airframe assemblies (Fig. 3.13); an expert system to select design rules for automated assembly (Hernani and Scarr, 1987); a knowledge-based system for assembly line design (Suer and Dagli, 1994); and a collaborative assembly planning and problem-solving system developed at IBM (Balakrishnan *et al.*, 1994).

A knowledge-based approach to DFA is limited by currently available design expertise and models, and by a knowledge representation to completely represent all practical design nuances. Additional aspects of knowledge-based planning and DFA are discussed in the next section.

3.7.2   *Assembly planning*

Assembly planning is an active area of research attempting to develop a theoretical foundation for DFA (as discussed earlier in the section on CAPP). A review of assembly planning research (Rajan and Nof, 1996) describes five main subjects that have been studied:

1.   assembly knowledge representation;
2.   assembly sequence generation and planning;
3.   integrated assembly and task planning;
4.   multi-robot assembly planning and control;
5.   multi-machine cooperation.

These subjects are discussed next.

Assembly knowledge representation   This involves the interpretation of design data for generating assembly sequences. The three representation types are **language-based**, **data structures** and **graph-based** (see also section 1.7). Graph-based representation has been most popular, because it is more suitable for the automatic generation of valid assembly sequences (Homen de Mello and Sanderson, 1991).

Assembly sequence generation   Graph-theoretic methods and geometric models of components are applied with some form of disassembly (e.g. Zussman, Shoham and Lenz, 1992) to generate mating directions and sequences of assembly. DeFazio and Whitney (1987) describe a query-based approach to generate all possible valid sequences, given a liaison diagram of mating relationships between components. Because query-based methods become difficult to use when assembly size and complexity increase, Rajan and Nof (1996) provide algorithms to automate the sequence generation by the CRP method, which is described below. Huang and Lee (1990) describe a knowledge-based assembly planning system based on disassembly reasoning. For a given CAD description, it generates a plan subject to the resource constraints of a given cell.

After assembly sequences are generated, assembly tasks can be planned for execution. Preferably, sequence generation and task planning are integrated.

Integrated assembly and task planning   An assembly-planning system, RALPH (Nnaji, Chu and Akrep, 1988), generates robot commands (assemble, re-orient, hold and release) from the configuration space-graph of given components. Tamaki, Hasegawa and Ishidate (1988) have developed RMT, a method to design a desired assembly robot from given assembly motion trajectories. With workspace specification, robot joints and links can be determined. Some theoretical developments in task planning have already been implemented in commercial CAD systems as TDLs (task description languages).

Multi-robot assembly planning, control and cooperation Multi-robot planning and control addresses the specification of workcells, the execution of tasks assigned to each robot and the resolution of conflicts during execution. Potential benefits of multi-robot assembly are an improved reliability by mutual backup, reduced workstation space and cost reduction by shorter cycle time and use of shared resources (such as fixtures and feeders). Disadvantages are the need for more complex control to prevent collisions and to 'orchestrate' the operation.

Chang, Goldman and Yeralan (1986) plan work for a robot team with wait conditions to ensure coordination. Nagata, Honda and Teramoto (1988) plan multi-robot stacking operations and eliminate conflicts by delays or by changing arm movements. Nof and Hanna (1989) and Rajan and Nof (1990, 1992) define and analyse three types and measures of task cooperation in a multi-robot cell: mandatory, optional and concurrent.

- **Mandatory task cooperation.** Successful completion of a task involves the mandatory participation of two or more robots (or machines.) Two subclasses are: **sequential mandatory** where the participation is in a certain specified sequence, and **parallel mandatory** where robots participate simultaneously in performing a task. This mode captures the limitations of individual machines but exploits their ability to cooperate. **Example:** Three robots must hold and balance a large, pliable sheet during its positioning in a fixture. **An advantage of this mode:** It can save cell resources, e.g. when a robot also serves as a fixture. **Disadvantage:** Mandatory task cooperation usually reduces the productivity per robot, therefore, it should be minimized.
- **Optional task cooperation.** Tasks can be completed independently by any one of several robots in the cell. Two subclasses are: **redundancy within** a robot type, and **redundancy between** robot types. This mode captures the overlap in capabilities of robots in a cell. **Example:** Any one of five robots in a cell can assemble the cover of a printer. **Advantage:** Redundancy provides backup and reliability, and can improve overall performance. **Disadvantage:** Analyses show that inconsistent improvement in performance may occur due to imbalance in work allocation. Therefore, the lowest level of such cooperation (and redundancy) should be designed if there is no significant difference in performance for higher levels of optional cooperation (i.e. more optional robots).
- **Concurrent task cooperation.** Use of additional, concurrently working robots leads to better efficiency (assuming the quality remains the same). This mode captures the flexibility available in a facility and the benefits due to cooperation. **Example:** In a certain workstation one robot can rivet a panel in four minutes, while three robots (when they are available) can complete the task in one minute by dividing it among themselves. **Advantage:** This mode always results in shorter lead time and is highly desirable. **Disadvantage:** Too many cooperating robots may cause

physical collisions and logical conflicts, therefore collision avoidance and conflict resolution are needed. Additional discussion can be found in Nof and Hanna (1989).

Nof (1989) and Rajan and Nof (1990) apply game-theoretic logic with search techniques for task assignment and resource allocation in multi-machine workstations. Game theory provides useful models for cost/benefit analysis of cooperation and negotiation in multi-machine assembly.

Automatic assembly plan generation is followed logically by execution planning and control. An activity controller for multi-robot assembly is described by Maimon and Nof (1985). It was implemented with four modules: a process scheduler; a robot scheduler that assigns tasks to each individual robot; an activity controller that translates a specific assignment into detailed execution; and a machine controller. The activity controller receives a given task assignment information, then coordinates and synchronizes the multiple robot operations and motions. It also resolves conflict situations that arise due to sharing of resources.

Cooperation requirement planning   The cooperation requirement planning (CRP) methodology (Rajan and Nof, 1992; Nof, 1992; Rajan, 1993; Nof and Rajan, 1993) does not assume a given task plan. It is an integrated assembly, task planning and execution system that seeks to exploit the benefits of cooperation among robots. An overview of CRP is shown in Fig. 3.16. For a given product, defined in a CAD file, the CRP-I procedure applies assembly planning techniques to generate the cooperation measures for given sets of assembly tasks. The CRP-II procedure generates the optimal global assembly plan by using a best-first search method, or heuristics can be used for task assignment, and interaction conflicts can then be resolved by time delays, communication methods or path modifications.

CRP was developed and implemented on a ROBCAD graphic simulation/ emulation workstation as part of the PRISM (Production Robotics and Integration Software for Manufacturing) program at Purdue University.

In the CRP methodology, cooperation is considered throughout the planning. To maintain maximum flexibility, it is necessary to minimally constrain the planning. On the other hand, it is important to generate plans that can be completed most effectively with the available equipment. These issues are studied in detail by Rajan (1993) and Rajan and Nof (1996), and include:

1.   **minimal precedence constraints**: identify a set of constraints and associated location direction sets that will be correct and complete, and will provide the maximum possible flexibility throughout assembly task planning;

2.   existence of a **global execution plan**: consider the minimal precedence constraints along with the constraints of a given assembly cell, and prove whether a global cell-level execution plan exists.

Planning with minimal precedence constraints is important for the following reason. A plan selected by traditional process planning as optimal (e.g. because the sequence is defined as 'best' by the lowest cost) is likely to be suboptimal when the current, dynamically changing state of assembly facility is considered. In the worst case, it may not be possible to execute the plan. This observation is true of both process plans (Ham and Lu, 1988) and assembly plans.

Integration of **assembly sequence planning** with **execution task planning** is also an important objective. Assembly constraints must be reconsidered throughout the task execution planning, rather than first selecting a particular ('best') assembly plan and then attempting to generate a feasible execution plan for it. Another important consideration is non-mating constraints that may restrict the assembly of a component to a subset of its own mating direction set. The integrated planning process follows three steps.

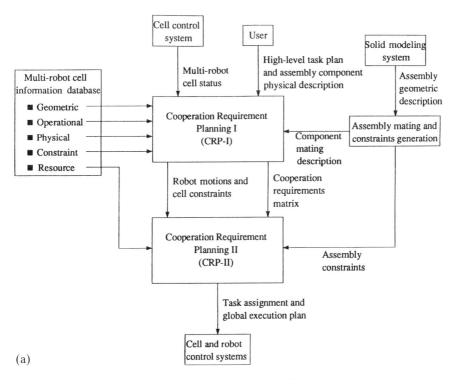

(a)

Fig. 3.16 Cooperation requirement planning (CRP) for multi-robot assembly: (a) overview of the planning methodology; (b) parts of a sliding mechanism to be assembled; (c) the generated assembly cell; (d) task allocation among three cooperating robots in the cell. The optimal global execution plan generated by CRP-II is shown with the actions and action times in seconds (Rajan, 1993; Nof and Rajan, 1993; Rajan and Nof, 1994).

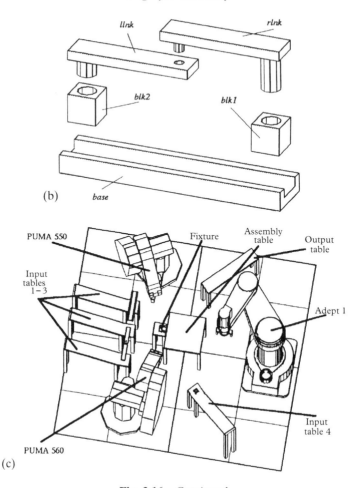

Fig. 3.16 *Continued*

- **Step 1: Generation of minimal precedence constraints for components.**
  From the geometric model of a given assembly, three logical sets are
  generated automatically for each pair of assembled components i, j:
  mating directions, MD_{ij}; constraint directions, CD_{ij}; location directions,
  $LD_{i,S}$ where S is the current state of components already assembled (see
  example in Fig. 3.17). It has been proved that for any assembly, precedence constraints exist only for those components that are fully constrained in the final assembly state. The minimal precedence constraints
  for a completely constrained component can be defined as follows:

  The minimal precedence constraints for a component i represent the alternate combinations of other assembly components, whose presence in the
  current assembly state will preclude the assembly of component i.

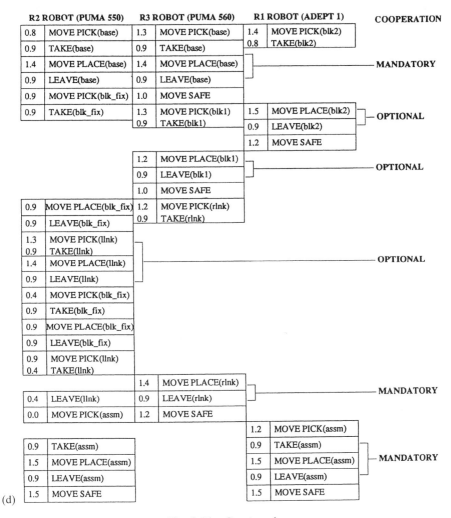

(d)

Fig. 3.16 *Continued*

Three algorithms have been developed by Rajan (1993; also in Rajan and Nof, 1996) to compute and determine mating direction sets, constraint direction sets and component precedence relations, leading to the minimum precedence constraints.

- **Step 2: Assembly representation and sequence generation.** Generate the feasible sequences of component mating that **do not** violate the precedence constraints from Step 1. Mating sequences can be represented by a mating graph, which is a graph consisting of the set of vertices V representing each one of the components in the assembly.
- **Step 3: Global execution plan existence and generation.** Integrate the planned sequences from Step 2 with the task execution planning. Prior to

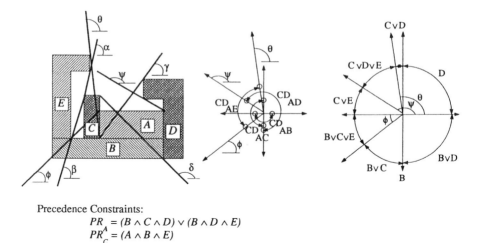

Precedence Constraints:
$$PR_A = (B \wedge C \wedge D) \vee (B \wedge D \wedge E)$$
$$PR_C = (A \wedge B \wedge E)$$

Fig. 3.17 Example demonstrating the assembly analysis process (*left to right*: the planar assembly; constraint direction sets; partitioning of the location direction set for component A; *below*: the minimal precedence relations for the two completely constrained components, A and C) (Rajan and Nof, 1996).

converting the assembly motion requirements to robot trajectory and grasp plans, it is important to determine whether the given cell equipment is capable of completing all the assembly tasks. If so, then the best possible task assignment and execution plans can be generated, as illustrated in Fig. 3.16(d). The existence of a global execution plan is established by considering the interactions between the geometric assembly constraints and equipment limitations. If these interactions can be resolved, a global execution plan exists, otherwise it does not. An algorithm for determining the existence of a global execution plan was developed by Rajan (1993).

Assembly planning theory and methods can provide a powerful computerized basis for DFA. Implemented in a CAD system, assembly planning will enable designers to apply automatic, comprehensive evaluations of assembly design alternatives in a truly virtual environment.

3.8 Summary

In an increasingly competitive world market, companies realize that a well designed and well run assembly operation provides a competitive weapon in their strategic arsenal. How can they achieve such a competitive operation? Rationalization methods and approaches that can be applied in the development and improvement of assembly have been introduced in Chapter 1. The most promising of them, and those that have proved most effective for a variety of industrial sectors, focus directly on design for assembly. In this chapter, we have covered design for assembly with a comprehensive approach: design for product assembly, design for assembly facility and design at different product lifecycle stages.

The role of computer-aided design methods is essential. Among the many benefits of computers, CAD systems have enabled us to design with more depth, complexity and analytical power. They have provided fast and accurate computation, thus allowing us to create and consider many more alternatives **before** committing to a decision. The design guidelines of yesteryears have advanced to the computer age refined and extended. Now anyone that has access to a computer can utilize them profitably. The rewarding result: higher quality at lower cost of the designed product **and** operation.

The chapter concludes with a review of emerging developments in design and planning for assembly. Research in this area continues to enhance design quality with attention to continuously improving design logic and artificial intelligence.

3.9   Review questions

1. From a market point of view, the lifecycle of a product begins with a start-up phase, followed by phases of rapid growth and maturity. Finally, the product either becomes a commodity or its market share diminishes.
   (a) Can the same design for product assembly be reasonable throughout the product lifecycle phases, or should the design be different for each phase?
   (b) Illustrate your answer to (a) with specific reference to the following products: (i) a refrigerator; (ii) a remote-control device; (iii) a desk.
2. Repeat Question 1 with consideration of the extended product lifecycle, which includes the phase of recycling.
3. What are the goals of design for assembly, and how do they change during the lifecycle of a given product?
4. Concurrent engineering means that in order to reduce the total product lifecycle cost, early design considerations must include subsequent phases such as production, maintenance and field support, and possibly recycling. What is the implication for design for assembly?
5. Disassemble your pen and identify design for assembly guidelines that appear in its design. Distinguish between the cases of manual, automatic and robotic assembly of the pen.
6. Can you offer any improvements in the pen design (in question 5) from the point of view of assemblability?
7. What are some design for disassembly guidelines that might have been considered in the pen design (in question 5)?
8. What are the advantages that group technology offers a designer for assembly?

9. What are some ways by which group technology can help in the assemblability evaluation?
10. Use the Hitachi method procedure to evaluate the assemblability of:
    (a) a pen;
    (b) scissors;
    (c) a stapler;
    (d) a chair.
11. Design a workstation to assemble chairs and estimate the time it would take to assemble one chair (a) manually; (b) with one robot; (c) with multiple robots.
12. Which type of assembly robot(s) would be recommended for the workstation in question 11?
13. Illustrate how the 10 computer functions in CAD apply specifically to DFA.
14. What is the relationship between assembly process planning and assemblability evaluation?
15. Suggest how a knowledge-based system can benefit the assemblability evaluation.
16. Formulate, using pseudo-code, knowledge-based system rules for DFA guidelines in the areas of:
    (a) component design;
    (b) product design;
    (c) assembly operations.
17. What are the advantages of a knowledge-based system for DFA?
18. Explain how the objectives of DFA can be accomplished more effectively by use of particular:
    (a) CAD methods described in the chapter;
    (b) artificial intelligence methods described in the chapter.
19. Develop a recovery graph and analyse the cost/benefit of recycling the following products:
    (a) a lamp;
    (b) a calculator;
    (c) a sofa;
    (d) a pen.
20. Explain the potential advantages of emerging assembly planning approaches compared with
    (a) traditional assembly process planning;
    (b) design guidelines;
    (c) assemblability evaluation.
21. Suggest several types of assembly tasks that can benefit from cooperating robots.
22. Consider the lighting fixture assembly cell shown in Fig. 3.7 and describe:

(a) assembly tasks that can be planned for cooperation;
(b) non-mating constraints in planning operations in the cell;
(c) possible conflict states in the cell, and how to prevent them or recover from them.

23. Using the terminology of assembly planning (section 3.7.2) model the assembly of a pen, including the sets of:

(a) mating directions;
(b) constraint directions;
(c) location directions.

24. Following question 23, illustrate conflict states that would prevent the assembly process from progressing towards a successful completion.

25. Use a graph to define several assembly sequences for the assembly of your pen. Would the sequences be different if the pen is assembled by one or two robots?

References

Alting, L. (1991) 'Life cycle design', *Concurrent Engineering*, Vol. 1, No. 6, pp. 19–27.

Alting, L. and Zhang, H. (1989) 'Computer aided process planning: the state-of-the-art-survey', *International Journal of Production Research*, Vol. 27, No. 4, pp. 533–85.

Andreasen, M.M., Kahler, S. and Lund, T. (1983) *Design for Assembly*, IFS Publications, London.

Arabian, J. (1989) *Computer Integrated Electronics Manufacturing and Testing*, Marcel Dekker, New York.

Arai, T. (1985) 'Application of knowledge engineering on automatic assembly of parts with complicated shapes', *Proceedings of the International Conference on Assembly Automation*, May 1985, Birmingham, UK, pp. 67–76.

Balakrishnan, A., Kalakota, R., Whinston, A.B. and Ow, P.S. (1994) 'Designing collaborative systems to support reactive problem-solving in manufacturing', *Information and Collaboration Models of Integration* (ed. S.Y. Nof), Kluwer, Dordrecht, Netherlands, pp. 105–33.

Bao, H.P. (1988) 'Group technology classification and coding for electronic components, development and applications', *Productivity and Quality Improvement in Electronic Assembly*, McGraw-Hill, New York, pp. 785–836.

Bedworth, D.D., Henderson, M.H. and Wolfe, P.M. (1991) *Computer-Integrated Design and Manufacturing*, McGraw-Hill, New York.

Boothroyd, G. (1992) *Assembly Automation and Product Design*, Marcel Dekker, New York.

Boothroyd, G. and Dewhurst, P. (1986) *Product Design for Assembly*, Boothroyd Dewhurst Inc., Wakefield, RI.

Boothroyd, G., Dewhurst, P. and Knight, W. (1994) *Product Design for Manufacture and Assembly*, Marcel Dekker, New York.

Chang, C.A., Goldman, J. and Yeralan, S. (1986) 'A time buffer method for multi-arm operational planning', *Robotics and Material Flow* (ed. S.Y. Nof), Elsevier Science, Amsterdam, pp. 187–96.

Chang, T.C. and Terwilliger, J. Jr (1987) 'A rule based system for printed wiring assembly process planning', *International Journal of Production Research*, Vol. 25, No. 10, pp. 1465–76.

Chen, C.L. and Pao, Y.H. (1993) 'Integration of neural network and rule-based systems for design and planning of mechanical assemblies', *IEEE Transactions on Systems, Man and Cybernetics*, Vol. 23, No. 5, pp. 1359–71.

Chiu, C., Yih, Y. and Chang, T.C. (1991) 'A collision-free sequencing algorithm for PWB assembly', *Journal of Electronic Manufacturing*, Vol. 1, No. 1, pp. 1–12.

Coleman, J.R. (1988) 'Design for assembly – users speak out', *Assembly Engineering*, July, pp. 25–31.

Dar-El, E.M. and Rubinovitch, Y. (1979) 'MUST – a multiple solutions technique for balancing single model assembly lines', *Management Science*, Vol. 25, No. 11, p. 1105.

DeFazio, T. and Whitney, D. (1987) 'Simplified generation of all mechanical assembly sequences', *IEEE Robotics and Automation*, Vol. RA-3, No. 6, pp. 640–58.

Delchambre, A. (1992) *Computer-Aided Assembly Planning*, Chapman & Hall, London.

Duelen, G. and Schroer, K. (1991) 'Robot calibration – method and results', *Robotics & CIM*, Vol. 8, No. 4, pp. 223–31.

Edan, Y. and Nof, S.Y. (1995) 'Motion economy analysis for robotic kitting tasks', *International Journal of Production Research*, Vol. 33, No. 5, pp. 1213–27.

Eversheim, W. and Muller, W. (1984) 'Assembly oriented design', *Programmable Assembly* (ed. W.B. Heginbotham), IFS Publications, London.

Feldman, K. and Geyer, G. (1991) 'AI programming tools for planning assembly systems', *Journal of Design and Manufacturing*, Vol. 1, No. 1, pp. 1–5.

Fisher, E.L. and Nof, S.Y. (1984) 'FADES: knowledge-based facility design', *Proceedings of the IIE Spring Annual Conference*, May 1984, Chicago, IL.

Fisher, E.L. and Nof, S.Y. (1987) 'Knowledge-based economic analysis of manufacturing systems', *Journal of Manufacturing Systems*, Vol. 6, No. 2, pp. 138–50.

Funk, J.L. (1989) 'Design for assembly of electrical products', *Manufacturing Review*, Vol. 2, No. 1, pp. 53–9.

Gatenby, D.A. (1988) 'Design for "X" (DFX): key to efficient profitable product realization', *Productivity and Quality Improvement in Electronics Assembly* (eds J.A. Edosomwan and A. Ballakur), McGraw-Hill, New York, pp. 639–52.

Hales, H.L. (1989) 'Designing for assembly: a computer-based approach', *Production and Quality Improvement in Electronics Assembly*, McGraw-Hill, New York, pp. 751–66.

Ham, I. and Lu, S.C.-Y. (1988) 'Computer-aided process planning: the present and the future', *Annals of CIRP*, Vol. 37, No. 2, pp. 591–601.

Ham, I., Hitomi, K. and Yoshida, T. (1985) *Group Technology*, Kluwer-Nijhoff, Norwell, MA.

Heginbotham, W.B. (1986) 'Programmable and robotic assembly', *Robotics and Material Flow* (ed. S.Y. Nof), Elsevier Science, Amsterdam, pp. 153–74.

Hernani, J.T. and Scarr, A.J. (1987) 'An expert system approach to the choice of design rules for automated assembly', *Assembly Automation* (ed. M.M. Andreasen), IFS Publications, London, pp. 129–36.

Ho, Y.-C. and Moodie, C.L. (1994) 'A heuristic operation sequence-pattern identification method and its application in the design of a cellular flexible assembly system', *International Journal of CIM*, Vol. 7, No. 3, pp. 163–74.

Homem de Mello, L. and Sanderson, A. (1991) 'Representations of mechanical assembly sequences', *IEEE Transactions on Robotics and Automation*, Vol. 7, No. 2, pp. 211–27.

Huang, Y.F. and Lee, C.S.G. (1990) 'An automatic assembly planning system', *Proceedings of the IEEE International Conference on Robotics and Automation*, Cincinnati, OH, pp. 1594–9.

Jovane, F., Alting, L., Armillotta, A., Eversheim, W., Feldmann, K., Seliger, G. and Roth, N. (1993) 'A key issue in product life-cycle: disassembly', *Annals of CIRP*, Vol. 42, No. 1, pp. 651–8.

Kakazu, Y. (1987) 'Integration of CAD/CAM and Robot' (in Japanese), *Journal of the Robotic Society of Japan*, Vol. 5, No. 3, pp. 64–9.

Karger, W. and Bayha, F.H. (1966) *Engineered Work Measurement*, Industrial Press, New York.

Khosla, P.K. and Mattikali, R. (1989) 'Determining the assembly sequence from 3-D model', *Journal of Mechanical Working Technology*, Vol. 20, pp. 153–62.

Kiang, A.M., Gay, R.K.L. and Chang, C.L. (1992) 'Integration of PCB CAD with automatic insertion machines', *Proceedings of ASME Japan–USA Symposium on Flexible Automation*, San Franciso, CA, pp. 1647–54.

Kim, S.H. and Lee, K. (1989) 'An assembly modeling system for dynamic and kinematic analysis', *Computer-Aided Design*, Vol. 21, No. 1, pp. 2–12.

Lechtman, H. and Nof, S.Y. (1983) 'Performance time models for robot point operations', *International Journal of Production Research*, Vol. 21, No. 5, pp. 659–73.

Maimon, O.Z. and Nof, S.Y. (1985) 'Coordination of robots sharing assembly tasks', *ASME Journal of Dynamic Systems Measurement and Control*, Vol. 107, No. 4, pp. 299–307.

Naft, J. (1989) 'Neurocomputing for multi-objective design optimization for printed circuit board component placement', *Proceedings of the International Joint Conference on Neural Networks*, Washington DC, IEEE, Vol. 1, pp. 503–6.

Nagata, T., Honda, K. and Teramoto, Y. (1988) 'Multirobot plan generation in a continuous domain: planning by use of plan graph and avoiding collisions among robots', *IEEE Journal of Robotics and Automation*, Vol. 4, No. 1, pp. 2–13.

Nevins, J.L. and Whitney, D.E. (1989) *Concurrent Design of Products and Processes*, McGraw-Hill, New York.

Nnaji, B.O., Chu, J.-Y. and Akrep, M. (1988) 'A schema for CAD-based robot assembly task planning for CSG-modeled objects', *Journal of Manufacturing Systems*, Vol. 7, No. 2, pp. 131–45.

Nof, J., Witzerman, J.P. and Nof, S.Y. (1996) *Motion Economy Investigation with a Robot Simulator/Emulator Workstation*, Research Memo 96-J, School of Industrial Engineering, Purdue University, West Lafayette, IN.

Nof, S.Y. (ed.) (1985) 'Robot ergonomics: optimizing robot work', *Handbook of Industrial Robotics*, Wiley, New York, pp. 549–604.

Nof, S.Y. (1989) 'The impact of advances in information technology on interactive robotics', *Advanced Information Technology for Industrial Material Flow Systems* (eds S.Y. Nof and C.L. Moodie), Springer Verlag, Berlin.

Nof, S.Y. (1992) 'Collaborative coordination control (CCC) in distributed multi-machine manufacturing', *Annals of CIRP*, Vol. 41, No. 1, pp. 441–5.

Nof, S.Y. (1994) 'Recent developments in simulation of industrial engineering environments', *Proceedings of the International Symposium on Computer Simulation and Artificial Intelligence*, February 1994, Mexico City.

Nof, S.Y. and Hanna, D. (1989) 'Operational characteristics of multi-robot systems with cooperation', *International Journal of Production Research*, Vol. 27, No. 3, pp. 477–92.

Nof, S.Y. and Lechtman, H. (1982) 'Analysis of industrial robot work by the RTM method', *Industrial Engineering*, April, pp. 38–48.

Nof, S.Y. and Paul, R.L. (1980) 'A method for advanced planning of assembly by robots', *Proceedings of SME Autofact-West*, pp. 425–35.

Nof, S.Y. and Rajan, V.N. (1993) 'Automatic generation of assembly constraints and cooperation task planning', *Annals of CIRP*, Vol. 42, No. 1, pp. 13–16.

Nof, S.Y. and Rajan, V.N. (1994) 'Robotics', *Handbook of Design, Manufacturing and Automation* (eds R.C. Dorf and A. Kusiak), Wiley, New York, pp. 259–95.

Nof, S.Y. and Witzerman, J.P. (1994) *The Production Robotics and Integration Software for Manufacturing (PRISM): An Overview*, Research Memo 94-22, School of Industrial Engineering, Purdue University, West Lafayette, IN.

Nof, S.Y., Knight, J.L. and Salvendy, G. (1980) 'Effective utilization of industrial robotics – a job and skill analysis approach', *AIIE Transactions*, Vol. 12, No. 3, pp. 216–25.

O'Grady, P. and Oh, J.S. (1991) 'A review of approaches to design for assembly', *Concurrent Engineering*, Vol. 1, No. 3, pp. 5–11.

Oh, J.S., O'Grady, P. and Young, R.E. (1995) 'A constraint network approach to design for assembly', *IIE Transactions*, Vol. 27, pp. 72–80.

Ohashi, S., Miyakawa, Y., Arai, S., Inoshita, S. and Yamada, A. (1984) 'The development of an automatic assembly line for VTR mechanisms', *Programmable Assembly* (ed. W.B. Heginbotham), IFS Publications, London, pp. 53–62.

Ong, N.S. (1993) 'Activity-based cost tables to support wire harness design', *International Journal of Production Research*, Vol. 29, pp. 271–89.

Popplestone, R.J., Ambler, A.P. and Bellos, I. (1978) 'RAPT: a language for describing assemblies', *Industrial Robot*, Vol. 5, No. 3, pp. 131–7.

Rajan, V.N. (1993) *Cooperation Requirement Planning for Multi-Robot Assembly Cells*, PhD dissertation, Purdue University, West Lafayette, IN.

Rajan, V.N. and Nof, S.Y. (1990) 'A game theoretic approach for cooperation control', *International Journal of Computer Integrated Manufacturing*, Vol. 3, No. 1, pp. 47–59.

Rajan, V.N. and Nof, S.Y. (1992) 'Logic and communication issues in cooperation planning for multi-machine workstations', *International Journal of Systems Automation: Research and Applications (SARA)*, Vol. 2, pp. 193–212.

Rajan, V.N. and Nof, S.Y. (1994) 'Cooperation requirement planning for multiprocessors', *Information and Collaboration Models of Integration* (ed. S.Y. Nof), Kluwer Academic, Dordrecht, Netherlands, pp. 179–200.

Rajan, V.N. and Nof, S.Y. (1996) 'Minimal precedence constraints for integrated assembly and execution planning', *IEEE Transactions on Robotics and Automation*, Vol. 12, No. 2, pp. 175–86.

Randhawa, S.U., Barton, W.J. Jr and Faruqui, S. (1986) 'Wavesolder Assistant: an expert system to aid trouble shooting of the wave soldering process', *Computers & Industrial Engineering*, Vol. 10, No. 4, pp. 325–34.

Rembold, B. and Nof, S.Y. (1991) 'Modeling the performance of a mobile robot with RTM', *International Journal of Production Research*, Vol. 29, No. 5, pp. 967–78.

Roy, U. and Liu, C.R. (1989) 'Establishment of functional relationships between product components in assembly database', *Computer Aided Design*, Vol. 20, No. 10, pp. 570–80.

Sackett, P.J. and Holbrook, A.F. (1988) 'DFA as a primary process decreases design deficiencies', *Assembly Automation*, August, pp. 137–40.

Sanii, E.T. and Liau, J.-S. (1993) 'An expert process planning system for electronics PCB assembly', *Computers & Electrical Engineering*, Vol. 19, No. 2, pp. 113–28.

Seliger, G., Zussman, E. and Kriwet, A. (1994) 'Integration of recycling considerations into product design – a system approach', *Information and Collaboration Models of Integration* (ed. S.Y. Nof), Kluwer Academic, Dordrecht, Netherlands, pp. 27–41.

Sellers, C.J. and Nof, S.Y. (1989) Performance analysis of robotic kitting systems', *Robotics and CIM*, Vol. 6, No. 1, pp. 15–24.

Shtub, A. (1993) 'Increasing efficiency of assembly lines through computer integration, dynamic balancing and wage incentive', *CIM Systems*, Vol. 6, No. 4, pp. 273–7.

Solhenius, G. (1992) 'Concurrent engineering', *Annals of CIRP*, Vol. 41, No. 2, pp. 645–55.

Styslinger, T.P. and Melkanoff, M.A. (1985) 'Group technology for electronic assembly', *Proceedings FMS for Electronics*, Cambridge, MA, SME Paper EE85-132.

Suer, G.A. and Dagli, C.H. (1994) 'A knowledge-based system for selection of resource allocation rules and algorithms', *Handbook of Expert Systems Applications in Manufacturing, Structures and Rules* (eds A. Mital and S. Anand), Chapman & Hall, London, pp. 109–29.

Swift, K. (1987) *Knowledge-Based Design for Manufacturing*, Kogan Page, London.

Tamaki, K. and Nof, S.Y. (1991) 'Design method of robot kitting system for flexible assembly', *Robotics and Autonomous Systems*, Vol. 8, No. 4, pp. 255–73.

Tamaki, K., Hasegawa, Y. and Ishidate, T. (1988) 'Development of task description method for robotized large scale structure assembly system', *Journal of the Japan Industrial Management Association*, Vol. 39, No. 3.

Tanner, W.R. (1985) 'Product design and production planning', *Handbook of Industrial Robotics* (ed. S.Y. Nof), John Wiley & Sons, New York, pp. 537–48.

Tatikonda, M.V. and Wemmerlov, U. (1992) 'Adoption and implementation of group technology classification and coding systems: insights from seven case studies', *International Journal of Production Research*, Vol. 30, pp. 2087–110.

Tipping, W.V. (1969) *An Introduction to Mechanical Assembly*, Business Books, London.

Van Brussel, H. (1990) 'Planning and scheduling of assembly systems', *Annals of CIRP*, Vol. 39, No. 2, pp. 637–44.

Warnecke, H.-J. and Bassler, R. (1988) 'Design for assembly – part of design process', *Annals of CIRP*, Vol. 37, No. 1, pp. 1–4.

Warnecke, H.-J. and Domm, M.E. (1989) 'Assembly by industrial robots with CAD/CAM and vision control periphery', *Annals of CIRP*, Vol. 38, No. 1, pp. 41–4.

Warnecke, H.-J. and Schraft, R.D. (1982) *Industrial Robot Application Experience*, IFS Publications, London.

Warnecke, H.-J., Schraft, R.D. and Wanner, M.C. (1985) 'Mechanical design of the robot system', *Handbook of Industrial Robotics* (ed. S.Y. Nof), Wiley, New York, pp. 44–79.

Warnecke, H.-J. *et al.* (1987) 'Computer-aided planning of assembly systems', *Proceedings of the 8th International Conference on Assembly Automation*, Copenhagen, pp. 53–65.

Wesley, M.A., Lozano-Perez, I., Lieberman, L.I. *et al.* (1980) 'A geometric modeling system for automated mechanical assembly', *IBM Journal of Research and Development*, Vol. 24, No. 1, pp. 64–74.

134 Design for assembly

Whitney, D.E. (1985) 'Part mating in assembly', Handbook of Industrial Robotics (ed.
S.Y. Nof), Wiley, New York, pp. 1084–116.

Wiendahl, H.P. (1989) 'Shortening the manufacturing cycle of products with many variants by simultaneous assembly engineering', *Proceedings of the 10th International Conference on Assembly Automation*, Kanazawa, Japan, pp. 53–65.

Witzerman, J.P. and Nof, S.Y. (1995a) 'Integration of cellular control modeling with a graphic simulator/emulator workstation', *International Journal of Production Research*, Vol. 33, No. 11. pp. 3193–206.

Witzerman, J.P. and Nof S.Y. (1995b) 'Tool integration for collaborative design of manufacturing cells', *International Journal of Production Economics*, Vol. 38, pp. 23–30.

Woo, T.C. and Dutta, D. (1991) 'Automatic disassembly and total ordering in three dimensions', *Journal of Engineering for Industry*, Vol. 113, pp. 207–13.

Zulch, G. and Waldhier, T. (1992) 'Integrated computer aided planning of manual assembly systems', *Computer Applications in Ergonomics, Occupational Safety and Health* (eds M. Mattila and W. Karwowski), North-Holland, Amsterdam, pp. 159–66.

Zulch, G., Braun, W.J. and Schiller, E.F. (1995) 'Analytical determination of job division in manual assembly systems', *Proceedings of the 13th International Conference on Production Research*, Jerusalem, Israel.

Zussman, E., Kriwet, G. and Seliger, G. (1994) 'Disassembly-oriented assessment methodology to support design for recycling', *Annals of CIRP*, Vol. 43, No. 1, pp. 9–14.

Zussman, E., Shoham, M. and Lenz, E. (1992), 'A kinematic approach to automatic assembly planning', *Manufacturing Review*, Vol. 5, No. 4, pp. 293–304.

4

Design of assembly systems

4.1 Introduction

The success of any assembly system depends on the translation of its design into an implemented working facility. The purpose of this chapter is to describe and explain the system elements and how they are combined into an effective working system.

The chapter begins with the planning objectives and risks faced by the multidisciplinary design project team, the planning methodology and the tasks to be carried out throughout the design process. The orientation is mostly towards mechanized and automated system elements since they present more complex and challenging engineering problems. Many of the design principles and system elements, however, are also useful for manual or hybrid facilities.

The main elements include handling equipment and machines that are needed for a variety of assembly conditions. A section on economic justification considerations that are specific to assembly is included (section 4.6). Finally, four case examples at the end of the chapter illustrate the design and planning considerations of particular industrial assembly systems.

4.2 The planning of an assembly system

4.2.1 Planning risk

The planning of automated assembly installations represents a complex assignment. In order to consider all possible ways of rationalizing, the most imaginative aspects must be considered at the planning of the installation. Requirements which are generally not recognized at first can have a significant effect on the concept and layout of an automated assembly system. Often, requirements beyond the concrete assignment (scope of assembly, number of parts, required output, retooling costs, etc.). Considered here, for example, are:

- product configuration;
- quality of individual parts;
- condition of delivered parts to be assembled (in bulk, in magazines, cleaned up, etc.);
- joining into the flow of existing material;
- integration into the flow of information;
- integration of manual systems;
- training of operating and maintenance personnel;
- the effects of preceding and subsequent manufacturing areas;
- questions of future development of types and numbers of derivatives, customer requirements, new product technologies, etc.

The many aspects to be considered during the installation planning stage of assembly systems can be efficiently covered by the following actions:

1. Formation of a project team comprising several representatives of the future user of the installation, and of its supplier(s).
2. Close cooperation between installation builder and operator, from conception of the assembly system right up to the commissioning and optimization phase.
3. Agreement on a systematic procedure for planning and realizing the assembly installation.

The planning risk can be minimized by these actions, both for the system builder and its user. The planning risk is significantly higher with automation of assembly tasks relative to the automation of components manufacturing tasks, for which fewer fringe conditions need to be satisfied.

4.2.2 Planning team

The project team in which both installation user's and builder's employees collaborate on processing the assignments must be decided as the first step in the execution of the project. A typical composition of a project team is shown in Fig. 4.1.

The members of the project team must have technical knowledge in various areas in order to be able to solve the different problems in planning a rationalization project in the field of assembly. At different phases of the project various activities will be involved to a greater or lesser extent. At the beginning of the project, for example, it will be up to the user's sales or marketing department to determine future volumes and assembly models or variety distributions (derivative chart). When analysing the assembly assignment, design representatives from both user and builder are both required to investigate the range of products with a view to their potential for automatic assembly. They will also consider useful changes to the product structure in

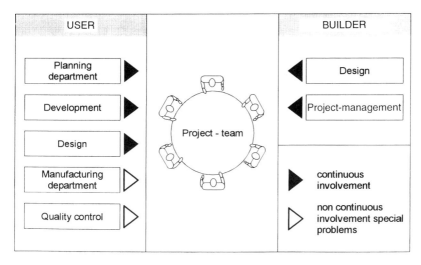

Fig. 4.1 Combined project team of installation builder and user for planning an assembly system.

agreement with the manufacturing department. The manufacturing department has, in addition, a great influence on the feasibility of automation in the assembly area. Since identical parts of a product have to be interchangeable in automatic assembly, regardless of time of manufacture, size of manufacturing batch and so on, their quality characteristics must be uniform and consistent. Manufacturing tolerances may, therefore, frequently be tightened in order to raise the functional reliability of assembly operations.

4.2.3 Planning methodology

Figure 4.2 shows a step-by-step systematic progress of planning. The progress shown is divided into six steps. These steps can also, contrary to the illustration, be carried out concurrently in part. Depending on the problem posed, repeated iteration loops may also be necessary.

The methodology shown here should be seen as an abstract form of the actual procedures followed when assembly installations are being planned. Transitions between planning sections are fluid. The flow chart given here is intended as a guideline and aid for a systematic approach.

Analysis of assembly task   The point of departure for planning is the assembly task, the analysis of which supplies information about assembly processes, workpiece data, etc., and also concrete demands on the assembly system to be developed, such as product volumes, volume distribution, degree of flexibility, quality requirements, etc. Analysis of the assembly task (Fig. 4.3) results in the definition of the conditions prevailing in the planned assembly facility.

Assembly processes, the range of workpieces and peripheral conditions are

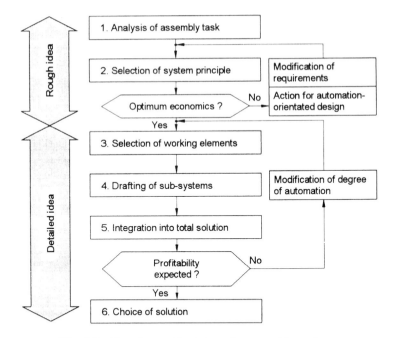

Fig. 4.2 Planning methodology for assembly systems.

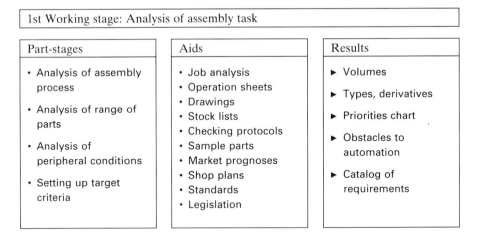

Fig. 4.3 Planning methodology for assembly systems: Stage 1.

registered thereby and target criteria for the appraisal of alternative solutions
are set up. When assembly procedures are examined, the main processes of
'manipulate', 'join' and 'inspect' will be subdivided into sub-functions such as
'make ready', 'feed in', 'isolate', 'position', 'spring into place', 'check for
torque', etc. Thereafter, the individual sub-functions will be assessed for automation. In many cases it is inefficient subsequently to automate an operation

which is likely to be manual. In order to be able to determine which assembly operation is most suitable for automation, a priority chart can be developed. The objective and structure of a priority chart for assembly planning are shown in Fig. 4.4. The automation cost for each complete assembly process can then be estimated with the aid of analysis of assembly sub-functions and the priority chart.

Selection of system principle   After analysis of assembly tasks, the characteristics are available by which assembly systems principles of structure can be distinguished. These characteristics are mainly the flexibility, workpiece total weight and required output. Figure 4.5 shows this stage. The type of assembly system which is optimally suited to the area of application investigated in the analysis can then be chosen.

After the system principle has been selected, its economic effectiveness must be determined. Often it will become apparent that with the given product structure and given configuration of parts, economic automation is not possible. For this reason, measures to design the product to suit assembly should be taken before the next planning steps are carried out for a rational assembly system.

Selection of working elements   To accomplish economic application of the selected system principle the working physical elements should be determined in this stage (Fig. 4.6). When a product is to be assembled, the individual parts must be positioned, fed in and joined by these elements.

Working elements for the positioning function are, for example, bins, palettes, magazines, etc.; for the function of feeding, for example, angled conveyors, vibrator chutes, grippers with manipulation systems; for the function of joining, for example, presses, riveting stations, grippers or nut-runners. The

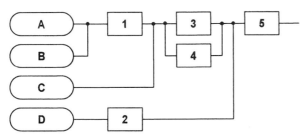

The priority chart indicates:

• which assembly activities ① have priority over other activities ③ ④ ⑤ and which parts are needed Ⓐ Ⓑ

• the earliest point when an assembly activity can be carried on ②

• which assembly activities can be performed in parallel or in any desired order ③ ④

• the latest assembly step ⑤ at which preassembly ② of a part Ⓓ must be completed.

Fig. 4.4   Explanation of an assembly priority chart.

2nd Working stage: Selection of system principle		
Part-stages	**Aids**	**Results**
• Determining scope of solutions • Estimating profitability • Selection of system principle and rough structure	• Priority chart • Types/derivatives framework • Workpiece data • Profitability requirements	▶ Suitable system principle ▶ Profitability framework ▶ Need for automation-oriented design

Fig. 4.5 Planning methodology for assembly systems: Stage 2.

3rd Working stage: Selection of working elements		
Part-stages	**Aids**	**Results**
• Allocation of joining operations to working elements • Allocation of manipulation operations to working elements • Optimization of product design	• Joining trials • Manipulation trials • Databases for working elements • Guidelines for automation-oriented design	▶ Segments of assembly that can be automated/manual ▶ Specifications for function-bearers ▶ Recommendations for automation-oriented design

Fig. 4.6 Planning methodology for assembly systems: Stage 3.

appropriate working elements can be matched with the various assembly sub-functions based on the analysis of the assembly task. At this point, joining or manipulating experiments can be simulated to facilitate the design process.

Experience can be gained about the potential application of various working elements. It may lead to certain product re-design. For example, manipulation surfaces, identifying features, positioning points, feeding aids, etc. should be finalized, thus facilitating automation.

As a rule only some of the necessary assembly steps can be carried out automatically. Analysis of the potential working elements thus leads to the determination of how far assembly may be automatic and which segments must be manual. Once working elements suitable for automation have been identified, the requirements for the component units which carry the working elements (such as insertion mechanisms for the grippers, travel slide units for nut-runners, industrial robots for soldering guns) can also be confirmed. These component units are designated as function-bearers in the following section.

Drafting of subsystems　As shown in Fig. 4.7 the selected working elements are assigned to appropriate function-bearers. Thus several working elements may be carried by one function-bearer, e.g. several grippers with different grasping tasks can be manipulated by several locating mechanisms, all according to the design of the subsystem. An alternative design is for several grippers to be carried by a single, common industrial robot which is equipped with a tool-turret or a system of interchangeable grippers.

There are extensive data on the tools, manipulation systems and special equipment that are suitable to serve as function-bearers. Work contents are combined according to the nature of the selected function-bearers and thus capacities are distributed over individual stations. In addition, the mode of operation of the station is determined by the individual function-bearer content in that station, so that the station construction and the decision of which components to make or buy can be planned.

Integration into total solutions　This planning stage is described in Fig. 4.8. Selection of interlinkage equipment, the layout of buffers and materials interfaces, the setting up of structure options and finally the layout of the control system leads to the conceptual construction of the whole system. The necessary information can, in turn, be taken from the appropriate component groups. In addition, materials flow, assembly process and fault conditions must be simulated. When the interlinkage equipment is being selected, a suitable

4th Working stage: Idea of subsystems		
Part-stages	Aids	Results
• Allocation of working elements to function-bearers • Selection of suitable function-bearers • Selection of interlinkage equipment	• Databases for: 　– tools 　– manipulation system 　– special equipment 　– interlinkage equipment • Outline of completed installations	► Distribution of work-content among the stations ► Mode of operation and construction of stations

Fig. 4.7　Planning methodology for assembly systems: Stage 4.

potential solution must be investigated to match the interlinkage structure of the selected system principle.

The layout of buffers and materials interfaces represents a problem in optimization, especially when buffer capacity is being planned. Here, it is a matter of registering the effects of various cost factors and selecting those buffer capacities at which piece costs reach a minimum.

The assembly structure is determined based on the priority charts and the derivatives 'tree' for the assembly product. The derivatives can be divided into main derivatives, special derivatives and individual special cases.

Choice of solution   Investment costs, labor costs, operating costs and economic values must first be investigated in order to select the optimal solution from among several system options (Fig. 4.9). In addition to the consideration

5th Working stage: Integration into total solutions		
Part-stages	**Aids**	**Results**

• Laying out buffers and materials interfaces
 	
• Setting up structural alternatives
 	
• Laying out control system
	• Databases for
	  – storage systems
	  – controls
	
	• Simulation of
	  – materials flow
	  – assembly procedure
	  – behavior under fault condition
		▶ Layout options for complete systems
		
		▶ Interlinkage structures
		
		▶ Ideas of control
		
		▶ Mode of functioning and construction of overall systems

Fig. 4.8   Planning methodology for assembly systems: Stage 5.

6th Working stage: Selection of solutions		
Part-stages	**Aids**	**Results**

• Determining mechanical values
 	
• Determining investment costs
 	
• Determining supply conditions
 	
• Choice of solution
	• Catalog of requirements
	
	• Mechanical values
	
	• Catalog of relative costs
	
	• Manufacturers bids
		▶ Optimal solution based on:
		  ▷ profitability
		  ▷ mechanical values
		  ▷ supply conditions
		
		▶ Amortization period

Fig. 4.9   Planning methodology for assembly systems: Stage 6.

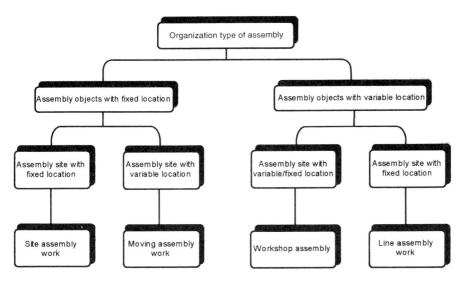

Fig. 4.10 Organization of assembly system by type.

of economic values and expected costs, supply conditions must also be investigated. Thus, a solution may be rejected if its development lead-time would significantly delay installation launch.

In determining the period of amortization, it should be noted that for different degrees of subsystems flexibility, a varying capacity for utilization will result. Thus, a longer amortization period should often be allowed than the two-year period which is typical in many cases.

4.3   Assembly organization

The assembly object (Fig. 4.10) is the most important criterion in defining the organization of the assembly system. It is necessary to differentiate between:

- objects with fixed location;
- objects with variable location.

Objects with fixed location are assembled without transport to another station; objects with variable location are assembled and are transported from one station to another.

4.3.1   Assembly organization for objects with fixed location

The organization for objects with fixed location is divided into site assembly work and moving assembly work (Fig. 4.11). According to the work division

Assembly objects with fixed location		
Site assembly work		Moving assembly work
Single-site assembly work	On-site assembly work	Non-orientated assembly site
Assembly site with fixed location	Assembly site with fixed location	Assembly site with variable location
No sort division	No definite sort division	Definite sort division
Quantity division possible	Quantity division possible	Quantity division possible

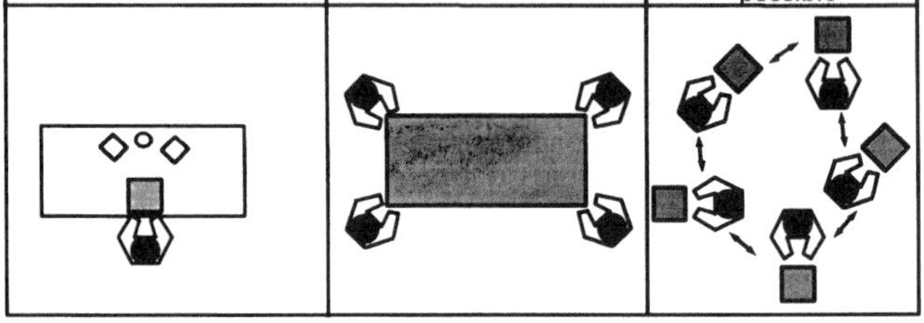

Fig. 4.11   Organization of assembly for objects with fixed location.

the field of site assembly is divided into single-site assembly and on-site assembly work:

- **site assembly:** the complete assembly of one product or one component is carried out in one assembly site with fixed location without any division;
- **on-site assembly:** step-by-step assembly of a product is carried out at assembly locations according to the local application. The object is handled from the beginning by one or more assemblers without any sharing of the jobs (Dolezalek and Warnecke, 1981).

With moving assembly, contrary to site assembly, the assemblers 'move' from one object with fixed location to another and the assembly order is exactly defined. The progressive assembly of the object defines the special location of assembly; the assembly locations are not lined up.

4.3.2   Assembly organization for objects with variable location

Workshop assembly and line assembly work are the two different ways of organizing the assembly of objects with variable location (Fig. 4.12).

- **Workshop assembly:** a rarely used way of organizing assembly. The configuration is a step-by-step, application-oriented assembly of products in combined assembly sites with fixed or variable location. The process of assembly is interrupted by the transport of the object to the next assembly site in the workshop.
- **Line assembly work:** the assembly sites are organized according to the

sequence of assembly tasks that have to be fulfilled in a range of assembly stations (task-oriented assembly sites). Due to this order there is a temporary dependency between arriving parts, parts to be assembled and parts to be distributed. This organization form is divided into line assembly work with and without defined sequence (Fig. 4.13) (Adamczyk, 1969).

The cycle time of stations along the line is the time interval in which every operation has to be executed. The longest operation determines the timed

Assembly objects with variable location	
Workshop assembly	**Line assembly work**
Process oriented assembly site	Task oriented assembly site
Assembly site with fixed/variable location	**Assembly site with fixed location**
Definite sort division	**Definite sort division**
Quantity division possible 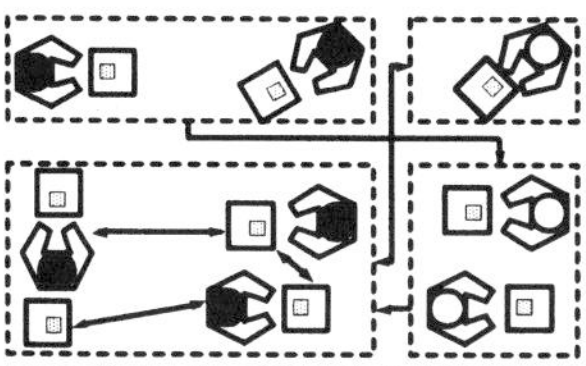	Limited quantity division possible 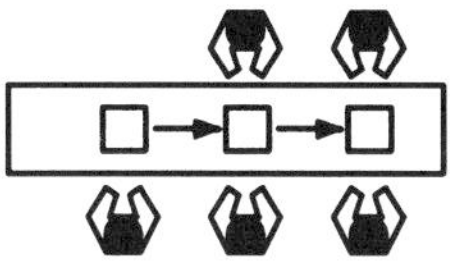

Fig. 4.12 Organization of assembly for objects with variable location.

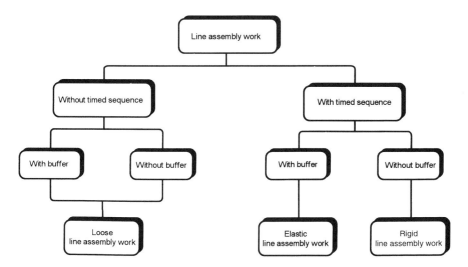

Fig. 4.13 Types of line assembly work.

sequence. Optical or acoustic signals as well as the handling device of the assembled product maintain the timed sequence.

Line assembly work without given time sequence is often called loose line assembly work. The time dependency of the stations due to a given assembly sequence can be decreased by buffers.

Buffers are technical units between the different sections of an assembly line, designed for picking up, storing and distributing parts from an earlier station to maintain the rhythm of the assembly line. The distribution of the objects is achieved in lots.

In elastic line assembly work it is possible for the operators to keep to a limited individual working rhythm. In this case, there is a possibility of building time limited buffers between the assembly sites or at the assembly site itself (e.g. by keeping a few workpieces at each working site).

In rigid line assembly work the operators are not able to change the working rhythm caused by the time sequence. There is no possibility of buffering between the assembly sites.

Another feature of line assembly work is the method of conveying the assembled objects. The conveying methods can be:

- **manual handling:** parts are conveyed manually – horizontal unmotorized handling devices (for example, roller conveyors) may facilitate the movement;
- **manual–mechanical handling:** after a part is loaded on the handling device (for example, a flat belt conveyor) it is moved without intervention to the next assembly site;
- **mechanical handling:** there is no manual handling at all and parts are processed on an automatic handling device (continuous or intermittent).

Due to product-specific characteristics (dimensions, number of parts to be assembled) it may be necessary to apply a combination of assembly organization types. For instance, a subassembly may be produced by a single-site assembly work and the final assembly by line assembly work.

4.4   Handling facilities

The flow of material, tools and workpieces to, from and between the active positions is the function of the handling facilities. The handling tasks incorporate all the functions necessary to deliver the correct number of components in alignment at the active position at the right time. After the assembly process the handling equipment removes the workpiece, stores it or passes it to the next destination.

Handling facilities can be divided into four main groups according to their main function (Fig. 4.14). The first main group contains equipment for storing workpieces before and after the assembly process. The workpieces could be delivered ordered in magazines or disordered in bins. Regardless of the stor-

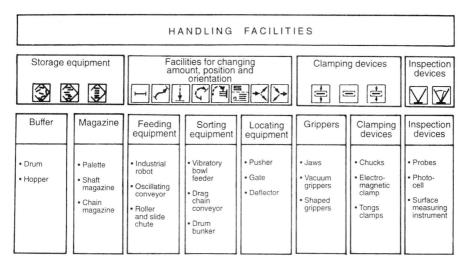

Fig. 4.14 Classification of handling facilities.

age function the equipment may have adaitional handling functions to enable the flow of workpieces.

The second group represents a wide variety of equipment to change the amount, position and orientation of workpieces. The orientation is achieved by rotating the workpiece around one to three axes and changing the position by translator movements in space. This equipment is used to execute the flow of workpieces in, from and between the active assembly positions.

The third main group contains gripper tools and clamping devices, which ensure the alignment of workpieces. Important factors are the prestressing force and the means of establishing an assembly connection. Gripper tools are generally integrated into the handling device and react with the prestressing forces in the machine station.

4.4.1 Storage equipment

Correct storage of workpieces is necessary for optimal operation of assembly installations. Waiting for delayed workpieces and too frequent charging or discharging of stores is not economic. Storing assembly output is necessary for timely flow and delivery at different positions along the assembly process.

The following kinds of storage may be found:

- **disposal store** for supply before the beginning of a handling process;
- **storage battery** for the supply of prefabricated units before removal;
- **emergency store** for overcoming delays at subsequent and preceding stations;
- **surge store** for compensating temporary differences in flow timing.

Workpiece stores can be constructed as bins or magazines. Factors affecting layout design are as follows:

- size, form, weight and space requirements of the workpieces;
- pattern of delivery and collection;
- processing speed (cycle time);
- value and sensitivity of the workpieces;
- cost and space requirements of the store.

Bins   Bins are installations for the storage of workpieces in disordered state. An overview of the different kinds of bin is given in Fig. 4.15. The simplest kind of bins are boxes, pile cases, box pallets, etc. Discharging by hand can be improved with grab tongues, sliding dosages and tipping devices (Fig. 4.16).

In terms of function there are movable bins and stationary bins as well as stationary bins with movable discharging elements. Movable bins (i.e. the whole supply of workpieces is moving) are used for unbreakable small components and relatively small amounts of storage (light weight).

Movable bins and bins with discharging elements are connected in chained installations preferably with a control. The discharging elements of the bins have to be adapted to the workpieces.

Stationary bins are closed containers with charge and discharge openings. The amount of storage is in general not functionally limited.

Magazines   Magazining is the storage of workpieces in a certain order to form stock ahead of and behind a manufacturing unit. Magazines for ordered storage of workpieces in their simple form comprise containers with grid inserts,

Type of bin		Stationary bins			Movable bins	
		With fall opening	With movable discharging elements		With stationary discharging elements	With movable discharging elements
Type of component discharging facility	Manual from top	By gravitation and tilting bracket	Circulate discharging elements	Dangle moved discharging elements	Gravitation trench	Power impulse or gravitation
Functional unit	Grid basket	Tank with sheer or conical walls	Bins with chain or belt	Funnel with lifting tube	Funnel	Cylinder or conical oscillating cask
	Drum	Funnel	Bins with discs or wheels	Bins with lifting slide valve	Drum	Drum with dangle Gravitation trench
					Shell	

Fig. 4.15   Overview of types of bin.

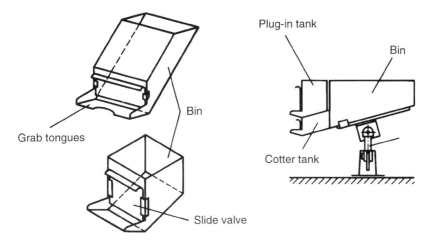

Fig. 4.16 Bin constructed as a grab container to enable removal of workpieces by hand.

plug-boards and others. On magazines for automatic manipulation the functions 'magazining' and 'transferring' mostly blend easily into one another.

Static magazines and moving workpieces are the most widespread application, especially in small part manufacture, while for easily damaged materials which have to be manipulated and for larger workpieces, driven or moving magazines are preferred, often by typical transfer or conveying devices.

Static magazines Static magazines (Fig. 4.17) are easy to construct at no great expense because they exploit the free movements of the workpieces (falling, rolling, sliding) and do not need a drive unit. However, the movement of the workpiece and the frictional behaviour have to be taken into account.

Palette magazines Palette magazines store workpieces mainly in flat arrangement and in a particular alignment (Fig. 4.18). Positioning of workpieces takes place usually by shape, less frequently by non-positive connection, so that the workpiece retains its attitude and position during transport. Transport is achieved with handling devices or by gravity. Palette magazines can be controlled by means of coding. They are mainly used in loose interlinkages.

4.4.2 Facilities for handling amount, position and orientation

Feed equipment As shown in the overview of handling facilities (Fig. 4.14) feed equipment can perform various manipulating functions on assembled components or workpieces, e.g. loading, unloading, selecting and filling hoppers and bins. In practice, feeding equipment takes many forms and is generally tailored to the assembly equipment to meet the requirements of the

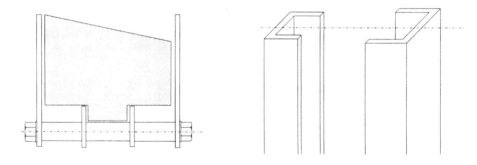

**flexible cable ducts
of spring steel strips**

**shaft magazine adapted
to the shape of the
workpiece**

Fig. 4.17 Types of static magazine.

Fig. 4.18 Flexible plastic magazine for different workpieces.

specific process. The following factors determine the feeding coordination features:

• workpiece input and output channels;
• workpiece acceptance at the assembly point;
• assembly time (work cycle).

In order to achieve short tool change times, the aim should be short insertion travel with positively controlled movements, especially as insertion normally has to meet the timing requirements of the production process.

Basically, a distinction should be made between two main groups of feed equipment:

- pusher devices;
- pick-up devices.

Pusher devices   For the selection of individual workpieces from hoppers and for loading, pusher devices (Fig. 4.19) only require one controlled movement. For this reason, they basically provide greater feed capacities than pick-up devices, which grip the item to be manipulated, lift, carry and then deposit it.

Pusher devices are suitable for workpieces which permit stacking and arranging in rows and, when moved on, maintain their orientation.

The simplest devices of this kind are tongue pushers with a straight-line alternating movement which is generated by a crank or cam mechanism, or a pneumatic or hydraulic cylinder. The pusher tongue is usually designed to match the workpiece.

Conveyor belts and conveyor chains work with a straight-line incremental movement, and in addition to selecting individual items and feeding them, also usually provide for the discharge or onward transfer by means of several operating stations.

Endless conveyor belts usually have cut-outs matched to the workpiece at uniform intervals for positive acceptance of the workpiece, while with conveyor chains this function is performed by 'dogs' (protruding pin-carriers).

Indexing tables operate in the same manner as conveyor belts, but with a circular incremental movement. Because of the limited number of dogs, the table is less suited for linking several operating stations compared to conveyor belts or conveyor chains.

Pick-up devices   Pick-up devices not only pick up workpieces but also load and unload machines and equipment, reposition workpieces, etc. They are also referred to as transfer devices, loading portals and pick-and-place units.

Pick-up devices are units of mechanical equipment which perform a specified movement cycle according to a fixed program and are usually equipped

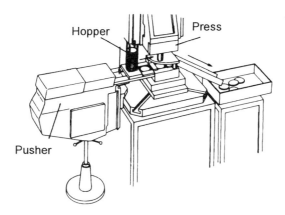

Fig. 4.19   Pusher feed equipment for a press (courtesy of VDI).

with grabs. With these devices, the movement cycle is specified, although it is usually still possible to adjust the stroke. They are frequently used for large production runs. For small runs or frequent repositioning, it is more appropriate to use easily programmable industrial robots.

In the main, pick-up devices are of simple construction. They consist of a drive, a control mechanism, kinematics and a grab.

Sorting equipment   The function of sorting equipment is to extract particular workpieces to be manipulated from a random storage, place them in sequence and select them, i.e. to provide the right quantities in a defined position and orientation at the correct time.

Sorting can be achieved by selection (Fig. 4.20) or according to an arbitrary principle (Fig. 4.21), but with both methods free movements are preferred. In the case of workpieces which have a complicated geometrical shape and which tend to cling together, however, power pulses or even forced movements are necessary.

In order to be able to place workpieces in sequence, it is necessary to know the specific orientation of the workpiece before and after this process. Orientation is determined with the manipulation function 'test'. During this function, the workpiece has to be guided precisely. Identifying the workpiece orientation can be done purely mechanically, electrically, electronically, pneumatically, hydraulically or fluidically using sensors.

The manipulation functions 'rotate', 'divert', 'pivot' change the workpiece's orientation by rotating it about a point or an axis within or outside the workpiece, and change the direction of movement of the manipulated items.

Whether the change in orientation or direction of movement can be brought about by appropriate guides and movement induced by gravity or only by driven equipment depends on the characteristics of the workpiece.

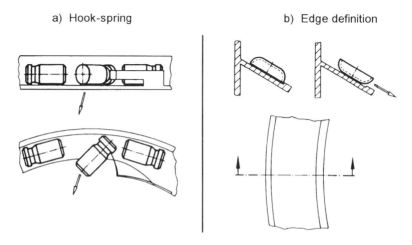

a) Hook-spring   b) Edge definition

Fig. 4.20 Sorting by selection.

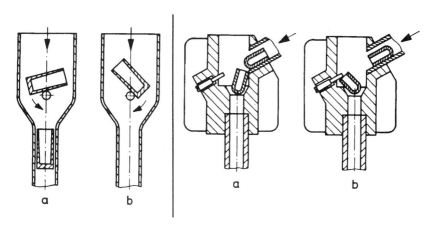

Fig. 4.21 Arbitrary sorting.

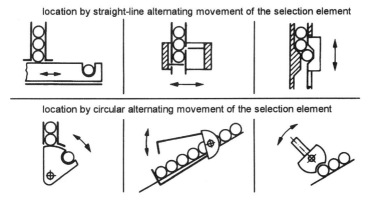

location by straight-line alternating movement of the selection element

location by circular alternating movement of the selection element

Fig. 4.22 Design of locating equipment.

Locating equipment Locating equipment (Fig. 4.22) usually has the task of picking up individual workpieces from hoppers (stored mainly in line), releasing a particular workpiece and blocking the movement of the remaining workpieces. Regardless of whether the movement of the workpiece in the hopper or feed equipment takes place by gravity or external forces, selection of the allocated workpiece has to be controlled.

Possible means of allocating workpieces are:

• by number, individually or by counting in the case of discrete items;
• according to dimension;
• according to volume, dosage in the case of fluid goods;
• according to weight.

The most frequent task in locating is the allocation of individual workpieces from hoppers. The location principle and the design arrangement of the locator are determined by the conditions before and after the location process.

Locators having an alternating movement (pushers, shut valves, rockers) are simple to design and suitable for allocating individual workpieces from hoppers with selection outputs of up to approximately 100 units/minute.

Locators with incremental movement (Fig. 4.23) (indexing tables, chain conveyors with speed-changing gears) achieve high allocation outputs (no backward movement of the allocation element).

Locators with continuous movement (Fig. 4.23) (chain conveyors and belts with continuous drive, location wheels and worms) achieve the highest location outputs. They can also be used to control the initial and final speeds of the allocated workpiece. An advantageous piece of equipment for allocating individual workpieces for both low-output and high-output feeder mechanisms is the allocation wheel. This device has an Archimedes screw which enables the release of workpieces to or from the hopper at uniform speed and permits very short pick-up times.

4.4.3   Clamping devices

Clamping devices are divided into:

- gripping tools (grabs);
- chucks.

Grabs are always used in conjunction with transfer equipment, pick-up devices and industrial robots. Chucks carry out the manipulation functions of positioning, clamping and releasing. Positioning means placing the workpiece or material in an exact position for a production process or for loading. In the

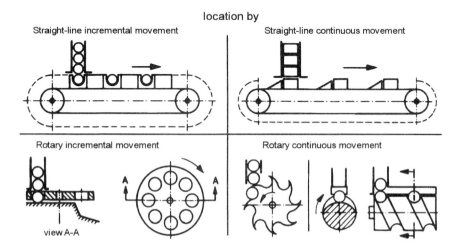

Fig. 4.23   The principles for the location of workpieces.

manipulation function of clamping, the workpiece or material is held in a specific position for a production process. Upon unclamping, the workpiece or material is freed by releasing the clamping pressure. Various positioning and clamping devices also handle release, i.e. the removal of the workpiece from the machining, assembly or measurement position.

A complete review of positioning and clamping equipment can be found in the appropriate literature. To solve the many different clamping problems there is a wide variety of clamping devices available. In addition to manually operated clamping devices, there are some operated by compressed air, hydraulics or electricity.

4.4.4 Chaining of manufacturing equipment

If production equipment is linked together with the aid of feed or transfer equipment and a control system, a transfer line results. Thus a transfer line is a production line automated with the aid of manipulating and data-processing equipment, comprising two or more units of production equipment with an automatic production cycle. A transfer line enables the timed completion of a production process based on the flow principle. According to the type of linkage, a distinction is made between rigid and loose linking or chaining.

Rigid linking   The characteristics of rigid linking are:

- manipulation of transfer equipment under common or coordinated control;
- common initiation of all processing and manipulation cycles (uniform timed sequence determined by the longest cycle, with central control);
- flexible transfer equipment, with location of production equipment at fixed intervals;
- shutting down one linking element results in the shut-down of the complete transfer line.

With rigid linking, precise timing and location of the production stations ensures the shortest lead-times. The timing of equipment for different production processes, however, often results in considerable problems. Rigid linking is therefore predominantly suitable for large workpieces and parts which are difficult to handle and which have long lead-times.

The disadvantage with rigid linking is that a fault in any unit causes the shut-down of the complete transfer line and shut-downs occurring one after the other have a cumulative negative effect on the output.

Loose linking   The characteristics of loose linking are (Fig. 4.24):

- manipulation and production devices are independent of one another and only linked at the junction points by controls (sequence control);

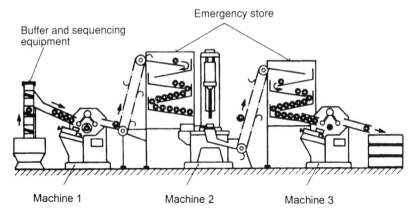

Fig. 4.24 Loose linking of production equipment.

- individual initiation of the machining and manipulation cycles (no uniform timing cycle);
- a hopper as an emergency buffer between consecutive production devices is used to compensate for short shut-down periods;
- flexibility for relocation and arrangement of the production equipment.

Transient faults on individual pieces of equipment have no effect on the equipment arranged before or after them because of the emergency buffer, so long as the buffer capacity is adequate. Shut-down times occurring one after the other do not accumulate with loose linking, which has a particularly favorable effect on output. Pieces of equipment for different production processes can be more easily linked together loosely because their timing cycles are independent. Furthermore, the erection in stages of such a transfer line, or the removal or introduction of production equipment due to a change in the production process, can be relatively easily carried out. Disadvantages are the longer workpiece lead-times, the cost of the emergency buffer and the greater space requirements of loose linkage compared to rigid linkage.

Combined linking   In addition to loose and rigid linking in pure form, one frequently finds a combination of the two systems in practice, i.e. sections of rigidly linked machines, loosely linked by interposing emergency hoppers. The advantages, first of the direct short workpiece lead-times and second of using a buffer hopper to compensate for shut-downs in some sections, are often combined in this way.

In particular, on long transfer lines with a large number of stations, combined linking may be designed for economic reasons. To improve the flow, pieces of equipment with identical technological processes and productivity can be rigidly connected to one another in series and in parallel.

4.4.5   Design of emergency buffers

The design of emergency buffers (stores) is an interesting problem, because the different demands on it are partially contradictory. The correct contingency storage capacity can have a significant influence on the economic efficiency of an entire assembly chain.

Relevant factors for the setup of storage capacity are:

- frequency and duration of the failures that have to be overcome (reliability statistic);
- the cost of production, operating and space of the emergency buffer;
- the capital that is tied up in the buffer in semi-finished products.

Because of the many factors that have to be considered, programs have been developed to find the optimal storage capacity. They compare the benefit of the installed stores (higher output of the assembly chain as a result of shorter stoppage times) with the expenses of installing and operating the store to find the lowest total cost per workpiece.

Emergency buffers must be designed to suffer minimum disturbances. They must ensure that workpieces can pass them singly and in groups with stable alignment. Emergency buffers should only be used to allow workpieces to catch up temporally and spatially with preceding workpieces (by self-propulsion of the workpieces or by a belt drive) (Fig. 4.25).

During standard operation of an assembly chain, a portion of the emergency store is filled, so that it is able to accommodate and supply workpieces. Passage magazines and back-flow magazines may be distinguished by their functional principle (Fig. 4.26).

Back-flow and shunt magazines are peripheral to the interlinkage installa-

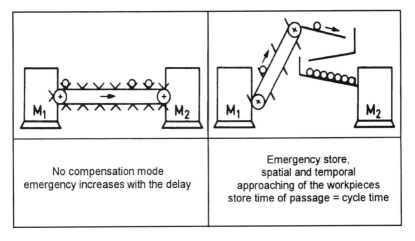

Fig. 4.25   Action mode of an emergency store.

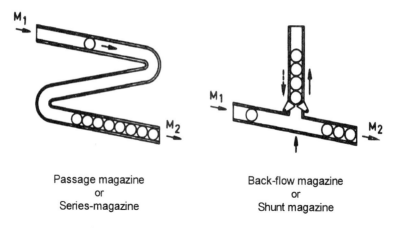

Passage magazine
or
Series-magazine

Back-flow magazine
or
Shunt magazine

Fig. 4.26 The principle types of emergency stores.

tions. Workpieces are fed into or taken from them only in case of emergency. The flow of workpieces bypasses them during standard operation.

4.5 Machines for assembly automation

Machines for automating assembly (automated assembly systems) comprise several assembly stations in which one or more assembly functions are performed automatically. These individual assembly stations are joined together by transfer equipment.

The layout of automated assembly systems is determined mainly by the workpieces to be assembled. The most important factors specific to the workpiece which have an effect upon the layout are (Warnecke, Löhr and Kiener, 1975; Spingler and Bäßler, 1983):

- total mass;
- dimensions;
- number of pieces (production volume);
- number of types and derivatives.

Automated assembly systems range from a plant for workpieces with a total mass of a few grams up to a plant for workpieces with a total mass of well over 100 kg. Cycle times of automated assembly systems vary between tenths of a second for small parts and several minutes for large parts and their flexibility, i.e. their levels of adaptability to different assembly tasks can also vary.

The main task in assembly is joining. All joining processes in assembly are applied in accordance with standards. Typically workpieces or materials have to be moved and the necessary equipment must have working elements such as grippers, nut-runners, etc. Automatic assembly stations are characterized by the various designs of such motion-equipment which are also used for inspection functions. They can be classified as follows:

- assembly stations with cam-operated motion-equipment, the movements of which in terms of sequence, travel or angle can be adjusted only by mechanical intervention (interchange of cams) (Fig. 4.27(a));
- assembly stations with pneumatically or hydraulically driven motion-equipment, the movements of which in terms of travel or angle can be adjusted only by mechanical intervention i.e. mechanical adjustment of stops (Fig. 4.27(b));
- assembly stations with modular-construction motion-equipment with pneumatically and electrically driven axes, the movements of which in terms of travel or angle can be adjusted and freely programmed (on electrical axes) partly without mechanical intervention (Fig. 4.27(c));
- assembly stations with industrial robots whose movements in terms of

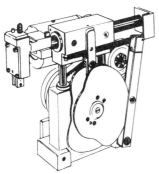

a: Cam-driven movement
   equipment

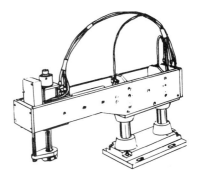

b: Pneumatically driven
   movement equipment

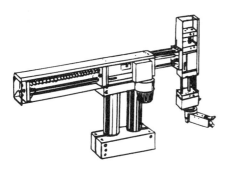

c: Modular-construction
   movement equipment

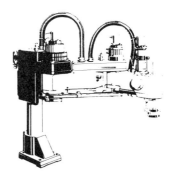

d: Industrial Robot

Fig. 4.27 Motion equipment for automatic assembly stations.

sequence, travel or angle are freely programmable and, if required, sensor-guided in several axes (Fig. 4.27(d)).

4.5.1 *Assembly automatics for short-cycle assembly processes*

The range of short-cycle assembly processes extends from cycle times of one second to approximately 10 seconds. Assembly automatics working in this range of cycle times are used for 'large-volume' assembly, i.e. annual volumes generally exceeding 1 million.

These short-cycle assembly automatics should be regarded as single-purpose automatics because they are specially designed for only one assembly task. Many assembly functions are repeated but in similar form on a wide variety of products. This repetition has enabled many manufacturers of these assembly automatics to standardize some engineering solutions. They have developed kit systems which contain standardized components not only for the interlinked equipment but also for the assembly stations (Fig. 4.28).

The kit system is a principle of arrangement, that represents the construction of a limited or unlimited number of different setups out of a collection of standardized building blocks, based on a program or prototype, in a certain area of application.

A component is a subassembly standardized in its interface-dimensions, structural features and characteristics, which can be variously combined with

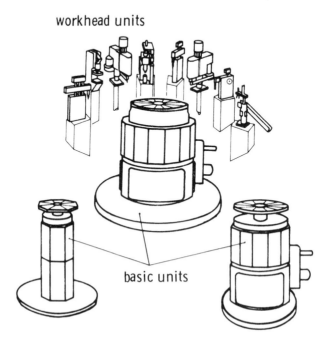

Fig. 4.28 Kit systems for rotary transfer automatics (courtesy of OKU-Automatik).

other components and, on dismantlement of production equipment, can be re-employed on new production requirements.

The manufacturer is approaching the objective of meeting the demands of many customers by the wide range of combinations possible with the kit, without having to create new assembly machines from the drawing-board stage. The kit system, however, also offers certain advantages. Because the kit system is flexible, a user can adapt existing assembly automatics to new situations by adding on, converting or retooling when the product to be assembled is redesigned.

Interlinkage of stations of short-cycle assembly automatics is carried out exclusively in rigid form, since the respective interlinking equipment permits very fast transfer motion. This interlinkage equipment can be designed for both rotary transfer and linear transfer systems.

Rotary transfer automatics   The relatively more cost-effective rotary transfer system is the most widespread of all assembly automatics employed for short cycle times, with a share of over 60% (Petri, 1982). Workpieces or partly assembled subassemblies pass through the rotary transfer automatic, indexing in step with the assembly process, in a circular path on workpiece areas that are mounted on a rotary table. The assembly stations are arranged in a radial pattern inside or outside the circular path.

Rotary transfer drive is achieved in accordance with one of three different principles (Lotter, 1982):

- ratchet drive (pneumatic);
- Geneva wheel drive (electrical);
- cam gears (electrical).

Rotary transfer systems with a Geneva wheel drive or cam drive have the advantage of working without jolting. Furthermore, the ratio of indexing time to pause time can be determined on cam gears by determining the cam contour, regardless of the number of index steps in the rotary transfer system.

Positioning of rotary transfer systems is usually central, so that indexing the workpiece carriers in the assembly stations is unnecessary. However, when workpiece carriers are positioned on a rotary table of larger diameter, tolerances increase, therefore the diameter of rotary transfer systems are generally smaller than 1000mm.

On the other hand, rotary transfer systems are distinguished by their highly compact construction. This advantage is, however, usually achieved at the cost of ease of inspection and accessibility. In addition, the base area of parts to be assembled is limited to about the size of a fist because of the relatively small space between workpiece carriers.

The number of assembly stations on a rotary transfer automatic is generally limited to a maximum of 24. Since the total plant susceptibility to faults rises with the increasing number of stations due to the principle of rigid interlinkage, some manufacturers have recently switched to design of rotary transfer automatics with a maximum of 16 assembly stations. This trend leads to better accessibility (Petri, 1982; Lotter, 1982).

Linear transfer automatics   On linear transfer automatics the assembly stations are arranged side-by-side or facing parallel to the direction of transfer. Workpieces or partly assembled subassemblies are indexed in linear motion from station to station. Chains, steel tape or plates are used as elements for carrying out the transfer movement and for securing workpiece carriers.

When linear transfer systems are being constructed, three different principles are applied (Fig. 4.29):

- merry-go-round transfer;
- above/below floor transfer;
- plate transfer.

In the merry-go-round transfer principle the workpiece carriers move around endlessly in a horizontal plane. They can thus be utilized across the whole extent of the transfer equipment, i.e. it is possible to arrange assembly stations distributed about the whole periphery.

Movements of the workpiece carriers take place in the vertical plane on the above/below floor transfer principle. On this construction principle only the upper part of the transfer equipment can be used for assembly purposes; the empty workpiece carriers return in the lower part.

In the plate transfer principle basic plates joined to one or more workpiece carriers are pushed in guide ways from one assembly station to the next station. Assembly stations may be arranged either outside or inside the circuit on plate rotation. The plate transfer principle offers the possibility of channel-

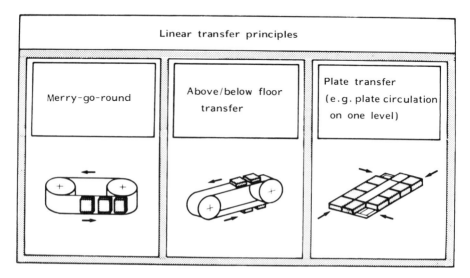

Fig. 4.29   Operating principles of linear transfer systems.

ling individual workpiece carriers out of cyclic rotation, in order to include, for example, manual rework stations in the material flow free of indexing.

In comparison with rotary transfer systems the manufacturing costs of linear transfer systems are generally 10 to 20% higher according to type. The main reasons for this difference are the higher number of workpiece carriers, particularly if the workpiece-carrier return cannot be used for assembly purposes, and that on some systems indexing of workpiece carriers is necessary in every assembly station. Space requirements are, according to type of construction, mostly greater than on comparable rotary transfer systems. The assembly stations are certainly more accessible for fault clearing and maintenance because of the side-by-side or front-facing arrangement, parallel to the direction of transfer. Also, larger products can be assembled compared to rotary transfer automatics (Petri, 1982; Lotter, 1982).

The possibility of adapting the length of the linear transfer system to the number of assembly stations needed can rarely, in practice, be fully exploited with rigid interlinkage. The reason is that with greater numbers of stations, there is a higher probability of failures. Therefore even linear transfer automatics are designed with a limited number of assembly stations, similar to rotary transfer automatics.

Interlinkage of several short-cycle assembly automatics   Manufacturers of assembly automatics for short cycle times guarantee as a rule an overall degree of availability of at least 80%. The remaining 20% is the result of factors which are partly dependent on the product, the plant or the peripherals (Petri, 1982).

For automatic assembly of complex products a high number of assembly stations will be required and the overall degree of availability could not be achieved with rigid interlinkage. If one considers the checks necessary with automatic assembly the capacity of these assembly automatics with rigid interlinkage is limited to about 10 individual parts per assembly. Therefore, assembly equipment with several loosely interlinked short-cycle assembly automatics has been preferred in practice for the assembly of complex products (Fig. 4.30).

Product subassemblies are assembled in these installations in rigid interlinkage within the cycle of an assembly automatic and fed in loose interlinkage to the next assembly automatic for the next level of assembly. Here, the interlinkage equipment has the function of a buffer against failures. The optimal storage capacity and engineering layout of this buffer depends on the failure characteristics of individual assembly stations, various cost factors and factors specific to the product. The engineering layout of this buffer/interlinkage equipment is determined mainly by the product. The types of linear transfer systems mostly employed are those in which subassemblies are conveyed directly or, if not possible, on workpiece carriers. The transfer motion of subassemblies or workpiece carriers is achieved by friction drive on belts, plastic conveyer chains or similar equipment and thus allows stock

Fig. 4.30 Assembly installation with two loosely interlinked short-cycle assembly automatics (courtesy of Mikron Haesler).

piling. Short-cycle assembly automatics or assembly installations can be interlinked on the same principle to assembly systems ahead of them such as final assembly systems (Warnecke, Löhr and Kiener, 1975; Abele *et al.*, 1984).

Cam-driven assembly automatics Besides the interlinkage equipment, the design of assembly stations is the second feature characterizing assembly automatics. On short-cycle assembly automatics assembly stations with either cam-driven or pneumatically-driven movement devices are used. They are generally designated cam-driven or pneumatically-driven assembly automatics according to this distinguishing feature (Spingler and Bäßler, 1983).

The lower range of cycle times (<1s to approximately 5s) is almost exclusively operated by cam-driven assembly automatics. The reasons are the high degree of accuracy of positioning and repetition and the almost vibration-free motion of cam-driven movement equipment, even with very fast movements. In addition, the movement process always runs, under varying loading, in a fixed position/time relationship. The simple and sturdy construction of cam-controlled movement equipment ensures reliable functioning with relatively low maintenance.

The movement process of the assembly station is achieved by means of hardened cams. Drive is achieved either by separate electric motors or (mostly) via the assembly automatics own centralized drive with control shafts. Subordinate movements frequently operate by means of pneumatic elements which are often controlled by the main drive.

Assembly stations are permanently programmed for a given task by the control and the predominant drive with cams (Fig. 4.31). When the assembly task changes (e.g. as a result of product redesign, new models etc.) movements or functional progress can only be changed by modifying the cams. Figure 4.32 shows the principle of a rotary transfer automatic with centralized drive and horizontal control shafts constructed from standardized component modules. However, even cam-driven linear transfer automatics are generally fitted with a mechanical central drive (Fig. 4.33).

The principle of centralized drive with predominantly mechanical command–variable control offers the advantage of good synchronization of motion actions which will continue to run without constant starting and stopping, thus making possible very rapid, functionally reliable cyclic sequences. Nevertheless, even in cam-driven assembly automatics, there has been of late an increase in the use of programmable controls by which the pneumatic and electrical elements are governed. The advantage lies in the programmability of these elements leading to simpler and quicker changes of process when product derivatives are to be assembled or when there are product design changes.

A range of different assembly processes is carried out by means of cam-driven assembly automatics, such as:

- screwing;
- inserting;
- pressing in;
- gluing;
- soldering;
- welding (electrically or ultrasonically);
- stamping;

Fig. 4.31 Example of drive and control by cams (courtesy of OKU-Automatik).

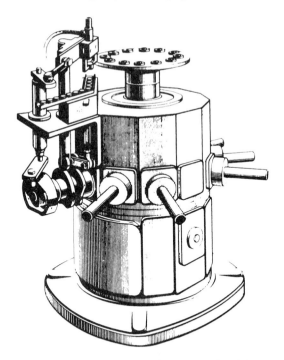

Fig. 4.32	Construction of a rotary transfer automatic with central drive and horizontal cam shafts (back shafts) (courtesy of OKU-Automatik).

Fig. 4.33	Linear transfer automatic with mechanical centralized drive (courtesy of OKU-Automatik).

- induction hardening;
- straightening;
- applying oils and greases.

Assembly processes are monitored by integrated inspection fixtures (mechanical feelers, inductive detectors, etc.) at assembly stations or sometimes at separate stations. This is an important function for quality assurance. Figure 4.34 shows a typical example of an application for a cam-driven rotary transfer automatic.

Pneumatically-driven assembly automatics The basic construction of the pneumatically-driven assembly automatic is very similar to that of the cam driven. They are also frequently built up from standardized modules (Fig. 4.35). The assembly processes automated by them are typically similar. The main area of application for pneumatically-driven assembly automatics is within a cycle range of approximately 3 to 10 seconds (Spingler and Bäßler, 1983).

Apart from the drive, the feature that distinguishes cam-driven from pneumatically-driven assembly automatics is the programmability of the latter and thus the flexibility of such assembly stations.

Control and monitoring of the assembly process are frequently undertaken by memory-programmable controls on more recent pneumatically-driven assembly automatics. The sequences of motion in the assembly stations thereby become directly programmable. Limit stops are commonly used to determine travel distances or angles for position scheduled control. The limit stops are either adjustable or the positions of which can be altered (partly also automatically), for example by swiveling in or out.

Fig. 4.34 Cam-driven rotary transfer automatic for assembly pressure regulating valves (courtesy of OKU-Automatik).

Fig. 4.35   Pneumatically-driven rotary transfer automatic for the assembly of two
types of projectile (courtesy of Microtechnik).

The programmability of the sequence of movements and the simple alteration of travel distances and angles which can be carried out automatically to a
limited extent enables the use of pneumatically-driven assembly automatics
for the assembly of product derivatives. In addition, assembly stations can also
be adapted to changes in the assembly processes which are frequently necessary when there are product design changes (Spingler and Bäßler, 1983).

Pneumatically-driven assembly automatics with time-schedule open-loop
control are also used for assembly tasks for which programmability of the
sequence of movements is not necessary. Pneumatic valves are directly and
mechanically controlled with the aid of cams built into what are called program rolls. The sequence of movements in the assembly stations is thereby
controlled by a fixed program. Alteration to this program can only be achieved
by adjusting or exchanging the cams. Therefore, this method of control is now
seldom applied. There are also pneumatically-driven assembly units on which
both the sequence of movements and the travel distances are determined by
interchangeable quadrant controls (command control). These units are, in
terms of regularity, similar in their action to the cam-driven type.

Standard units for pneumatically-driven assembly automatics are produced
by many manufacturers and often offered off the shelf. Individual units can be
combined almost as desired, usually with separate drive and control. This
separation enables the user to build up complex assembly automatics, even
from units made by different manufacturers.

In general, pneumatically-driven assembly automatics are slower as a
result of the longer response time of the control elements, but are somewhat
more flexible than the cam-driven types because of their better programmabil-

ity. Therefore, they are more suitable for the assembly of a family of product derivatives (models).

4.5.2   Flexible modular assembly systems

Modern assembly machines are designed predominantly as kit systems. The elements of the kit are designed as standardized units and capable of meeting requirements specific to each particular customer by adaptation. Reuse of such assembly equipment based on an equipment kit is limited because of the relatirely high cost of adaptation. Admittedly, for conventional assembly equipment adaptation to a different assembly task with a kit system is possible in principle but frequently not practical because of long retooling times and high cost of adaptation work. Application of a kit system therefore does not promise greater flexibility. The kit principle is, however, an important prerequisite for the adaptation of assembly systems to various assembly tasks, using rearrangeable or reconstructable versatile components. Flexible assembly systems are, therefore, mostly set up within the framework of a kit system, which is why they are frequently described as flexible kit systems.

The essence of kit systems is that all the parts in the system are constructed on a modular basis. Modularity applies not only to the actual assembly stations, including alignment devices, basic frames, etc., but also to the interlinkage equipment, control and closed-loop control systems, installation engineering, etc. (Spingler and Bäßler, 1983).

Flexible kit systems are designed to provide a wide range of application because of the constantly rising requirements for flexibility. A variety of assembly tasks can be automated by means of the multiple combinations possible with kit elements. The modular design of these systems ensures that:

- changes and additions can be made relatively easily;
- conversion is possible when design factors change;
- there will be a high degree of reusability;
- the control technology can be upgraded.

These factors also ensure that risk to investment is thereby reduced.

Flexible modular systems work mainly in the cycle range of 10 to 60 seconds and are frequently applied for the automatic assembly of product derivatives in medium to large volumes. The weight of workpieces assembled by these systems varies between a few grams and 10 kg (Spingler and Bäßler, 1983). A typical example of such a system is shown in Fig. 4.36 (Drexel, 1983).

These systems encompass all types of flexibility, not just flexible assembly stations and flexible interlinkage structures, but also forms of control that can be adapted quickly to new situations. Therefore, the equipment in flexible assembly systems is controlled by memory-programmable controls.

Flexible modular assembly stations   A distinguishing feature of flexible modular assembly systems is their construction from standardized modules. In particular, the configuration possibilities for motion equipment play an important role. Equipment can be adapted to a variety of requirements as a result of the compatibility of non-programmable and programmable modules. Sizes and travel distances can also be changed, if necessary, at a later stage.

Many combinations of individual modules, from exclusively pneumatically driven equipment (Fig. 4.37) to fully programmable industrial robots constructed from programmable axes (Fig. 4.38), are possible to meet the requirements of specific design solutions.

The layout and flexibility of the assembly station which meets the requirements of a particular application are possible with this type of station in spite of the high degree of standardization. Adaptation is possible when processes change, and there is no need to invest in 'excess' flexibility. In this

Fig. 4.36 Flexible modular assembly system for electric motors (courtesy of Bosch).

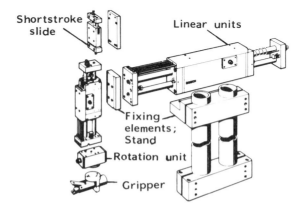

Fig. 4.37 Module of a pneumatically-driven manipulating mechanism (courtesy of Bosch).

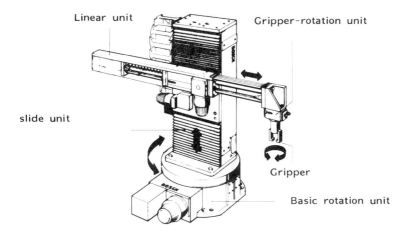

Linear unit   Gripper-rotation unit

slide unit

Gripper

Basic rotation unit

Fig. 4.38   Modular construction industrial robot (courtesy of Bosch).

design approach (particularly by combining pneumatic modules with electrically driven fully programmable modules) 'adaptive' flexibility of assembly stations can be achieved.

The flexibility of pneumatically-driven motion devices is related mainly to rebuilding, conversion and retooling based on modular construction. Numerically controlled (NC) modules offer, in addition, a potential for programming of positions and motions. It thereby becomes possible to accommodate as many different positions as desired, for example, when workpieces are to be removed from or loaded into flat magazines. An assembly station constructed from NC modules can also be used for carrying out:

- several (even different) assembly processes per cycle, for better loading when cycle times are relatively long; or
- different assembly processes, such as are often required when a variety of product models and derivatives are being assembled.

The number of required NC modules in an assembly station is determined by the number of motion devices in which several different positions have to be achieved for the assembly task. It is necessary to use NC modules in the horizontal plane when, for example, workpieces are being removed from flat magazines so that every workpiece position in the magazine can be approached. Vertical movements for the removal and joining of workpieces can, by contrast, be executed with a more economical pneumatically-driven module if removal and joining locations are on the same plane.

The example shown in Fig. 4.36 demonstrates potential combinations of the various types of modules. The left-hand illustration shows a station for transferring the motor casing from intermediate store to the workpiece carrier on the assembly belt. This task is achieved by two pneumatically-driven modules with fixed program travel. Since the casing has to be joined in different angular attitudes (according to type) a numerically controlled rotary unit is attached to the vertical module.

The right-hand illustration shows the stations for transferring the cases out of magazines into a temporary store. An electrically-driven programmable module combined with a pneumatic non-programmable module (vertical motion) removes six casings at a time from the magazine and loads them into locators in the temporary store. The temporary store is necessary in the present design to compensate for variations in cycle times (Drexel, 1983).

Interlinkage of assembly stations   Interlinkage installations in flexible modular assembly systems are also combined from single modules as shown in Fig. 4.39. The modularity of interlinkage allows one to customize a configuration specific to a given task just as with assembly stations of modular construction. The potential for adaptation of the interlinkage structure is retained for any change in processing. In flexible assembly systems the principle of loose interlinkage of assembly stations is applied, as flexibility in the interlinkage of individual assembly stations can only be achieved in this way (reversal flexibility). In linear transfer systems used as interlinkage installations the transfer motion is mostly produced by friction between the workpiece carrier and the transfer device (strap, belt, plastic-link chain, etc.). The power-and-free principle is also applied to some extent.

The transfer equipment of these interlinkage installations is usually moving continuously. Execution of assembly operations, however, occurs while the workpiece carrier is at a standstill. Therefore, there are devices on each assembly station to separate and time the workpiece carriers.

If sufficiently accurate positioning of workpiece carrier is not possible

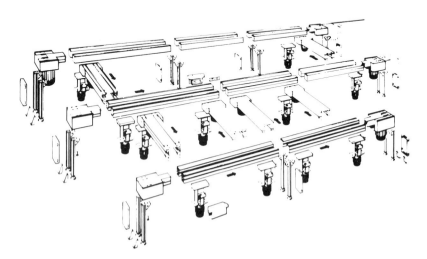

Fig. 4.39   Modules in a linear transfer system (friction principle) (courtesy of Bosch).

merely by means of timing, additional indexing devices are necessary by which workpiece carriers are positioned in the automatic assembly stations. In addition, the workpiece carriers must be lifted from the transfer equipment if high joining forces are to be applied, because the transfer equipment is constantly moving beneath the workpiece carrier which is stationary. This lifting is frequently achieved by indexing: vertical indexing pins are guided from below into indexing holes in the workpiece carrier, raising it a few millimeters and pressing it against a height stop.

In contrast to assembly automatics with rigid interlinkage, it is not essential that the period of time workpiece carriers stay in each assembly station is equal to the cycle time of the entire assembly system. Instead, the period of time depends on the assembly time of the individual station involved. Moreover, individual cycle times and any technical faults, capacity variations, etc. of each assembly station do not affect the stations ahead. Any delays are compensated by the buffering function of the transfer sections between assembly stations, in which several workpiece carriers can accumulate. The distance between assembly stations with loose interlinkage is thus determined not by the transfer equipment but by the optimal buffer capacity. This buffer capacity can differ from station to station. Therefore, not all assembly stations are equidistant.

The structure of linear installations may take the form of a straight line or a square or a combination of both. The line arrangement ensures good accessibility to assembly stations and interlinkage equipment from both sides and has advantages for maintenance and repair jobs. However, in contrast to the square structure, the workpiece carrier cannot easily recirculate for assembly purposes.

The significant advantage of interlinkage installations regardless of principle is the potential for branching of material flow. Branching can be utilized in order to:

- arrange identical stations in parallel with a shunt, to balance particularly large workloads or if the volume is highly variable;
- arrange alternate stations in shunt, which is frequently necessary in the assembly of sub-types and derivatives (reversal flexibility);
- divert individual products or subassemblies from the main stream (e.g. for any necessary rework);
- distribute pre-assembled subassemblies from a common assembly line onto several parallel assembly lines, e.g. to sub-type or derivative-specific assembly lines (reversal flexibility).

Coding of workpiece carriers is frequently necessary for controlling the material flow in order to distinguish particular product derivatives, or to identify a rework order. The code can either be stored directly on the workpiece carrier (mechanical, optical devices or magnetic strips) or indirectly via a workpiece carrier identifier linked to a central memory within the system control.

Which coding principle is applied depends on the number of different codes required and on the signal processing and control technology in use. Mechanical contact methods are simple to apply which is why they are most widely used. They are, however, generally limited in terms of the number of codes that are possible to apply.

The number of interlinked assembly stations can be relatively high with loose interlinkage, without adversely affecting the availability of the total system. Flexible modular assembly systems with well over 80 assembly stations for automatic assembly of complex products are thus not uncommon.

4.5.3   Flexible assembly systems

Flexible assembly systems with assembly robots are employed for automatic assembly of various products having cycle times of about 15 s up to several minutes (Spingler and Bäßler, 1983). The variety of products assembled by this type of system extends from small fine instruments or electrotechnical products to high-volume products in the automobile industry. The assembly tasks are carried out partly or completely by industrial robots (Warnecke and Haaf, 1982). Compared to the typical automatic assembly systems the application of assembly robots leads to the following essential advantages:

- programmable control and flexibility for different kinds of assembly tasks;
- because of their various possible applications and reuse, industrial robots can be depreciated over a longer period than the lifetime of a particular product;
- the possible application of sensors: the use of visual and contact sensors enables an industrial robot to perform auxilliary tasks associated with assembly such as testing, ordering of pieces, etc. and to control the assembly process. Therefore, the demand for peripheral equipment could be decreased and the quality of assembly could be increased.

The essential disadvantages of applying industrial robots in assembly are the following:

- the cycle times of assembly stations with robots are relatively longer than those of single-purpose stations;
- in general assembly robots are more expensive than simple pneumatically or cam-driven motion equipment.

The advantages of assembly robots are especially desirable if there are high flexibility demands. Flexible assembly stations with robots therefore are of most economic interest in the assembly of:

- various types and versions of products;
- distinct products;
- products with a small number of pieces;

• products with a relatively short production life (Spingler and Bäßler, 1983; Abele *et al.*, 1984).

Being highly programmable, these systems are often named *programmable assembly systems*.

Assembly robots   Feeding, joining, controlling and adjusting are the main functions of assembly robots. High demands are made on the repetitive accuracy, the process speed, the possibility of producing sensor signals and the ease of use of the programming language. The kinematic construction of the assembly robot, in terms of motion, must usually have from four to six degrees of freedom to carry out the various assembly tasks.

Assembly robots can be divided into five basic types (Fig. 4.40). Because of different assembly tasks, each has a special field of application in which the assembly robot is technically and economically used to the greatest advantage (Warnecke and Haaf, 1982). The useful load carried by the robots ranges from 100g in fine assembly work and up to 100kg as in automobile assembly.

The first robots applied in large numbers in Europe were constructed with a linear axis. By the middle of the 1970s the first of these devices were already installed. Versions of this type of assembly robot may be constructed with more than one arm (typically up to a maximum of four), each arm being independently programmable. Often these assembly robots are available for a particular size of work space which may be enlarged by lengthening the main axis (*x*-axis). These robots are distinguished by their kinematic structure of linear joints which provide consistently high process speed and accurate positioning and repetitive speed in all parts of their work space. The work space can also be relatively large. These advantages and the ability to provide parallel motions within the assembly work space enable optimal exploitation.

Linear axis manipulators are used mainly when assembly tasks require motions in the direction of the three main axes. Gantry robots are particularly

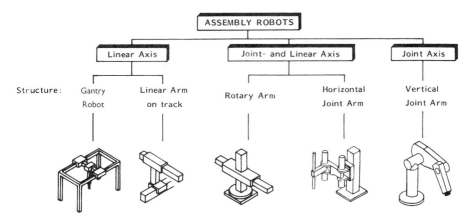

Fig. 4.40   Basic types of construction for assembly robots.

effective at vertical motions. In addition, such devices offer easy accessibility to the work space (from four directions) which is important for part feeding.

The widest variety of manipulators are combined rotation and linear axis robots. The so-called cylindrical robots, which are constructed out of two linear and one rotation axes, have been known for some time. They are applied mainly for vertical and somewhat less for horizontal (in radial direction relating to the rotation axis) motions.

The horizontal joint arm robot (Fig. 4.41), following the 'SCARA' principle (selective compliance assembly robot arm), is the type of robot most often applied for assembly. Its advantages are the acceleration and deceleration curve (cam curve) and the horizontal elastic foundation of the device during the joining process (selective compliance). These manipulators are equipped with three electrically-driven rotation axes, and one linear axis that is driven electrically or pneumatically. The sturdy mechanical design of the links enables relatively high positioning accuracy and repeatability.

The newest type of manipulator is the so-called directly driven robot. The specific feature of this device is that the motor of the drive axis is directly connected to the rotation axis of the robot. This enables not only very high acceleration and speed, but also very high accuracy of positioning that is achieved by the avoidance of play between motor and rotation axis. Until recently assembly robots with this type of drive have only been offered by US and Japanese manufacturers. The relatively higher cost of control and of motors imply that this device is significantly more expensive than the conventional SCARA type.

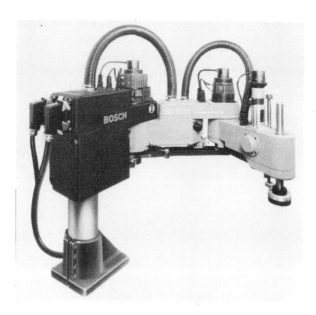

Fig. 4.41   Horizontal joint arm robot SR 800 (courtesy of Bosch).

Fig. 4.42   Vertical joint arm robot IR 163/19/30 (courtesy of KUKA).

As a result of their kinematics horizontal buckling arm horizontal joint robots may be used only for vertical motions from upside down. The motion range of the vertical linear axis is generally similar to that of larger manipulators (those with workspace diameter >1 m) and is about 250 mm. Therefore these robots can only be applied for particular assembly tasks, e.g. assembly of sandwich-like products or component insertion in printed circuit boards.

Another common assembly robot in use is the vertical joint arm (Fig. 4.42), constructed only with rotational joints. Their kinematics is similar to the human arm, enabling the approach to almost any direction of motion. To carry out cartesian motions in any direction i.e. motions along strictly straight lines while holding a workpiece or tool in alignment, six degrees of freedom and continuous-path control are required (Spingler and Bäßler, 1984).

Peripheral installations of assembly robots   The flexibility of assembly robots essentially results from their fully programmable sequences of motions. This programmability enables each robot to execute several distinct assembly tasks with different sequences of motions. Hence, assembly robots are able to carry out

- distinct assembly processes of one assembly task; or
- distinct assembly processes of several assembly tasks.

However, to execute different processes of assembly it is usually not sufficient just to carry out different sequences of motions, e.g. when welding or varnishing, but also different tools have to be handled. Therefore, a robot in principle requires the same peripheral equipment (handling devices and fixtures) as conventional automatic manipulation devices:

- installations for the disposal of workpieces;
- tools for handling and joining;
- fixtures.

The execution of different assembly processes with one robot has both quantitative and qualitative effects on the required peripheral equipment.

4.5.4   *Design of flexible assembly systems*

Despite the variety of flexible assembly systems in industrial assembly, in principle there are three basic types:

- flexible assembly lines (Fig. 4.43);
- flexible assembly cells (Fig. 4.44);
- flexible assembly systems (Fig. 4.45).

These basic types of programmable assembly systems have special system properties that make them suitable for particular areas of application.

Flexible assembly lines   The highest number of assembly robots in industrial assembly can be found in programmable assembly lines. Programmable assembly lines are, in terms of their construction, properties and field of application, comparable with flexible modular assembly systems.

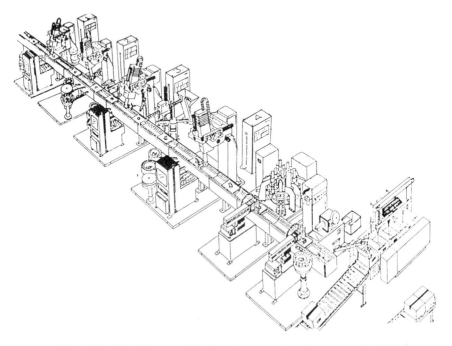

Fig. 4.43   Flexible assembly line for gas valves (courtesy of ASEA).

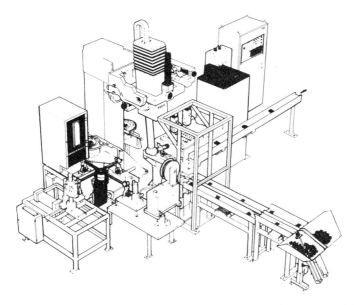

Fig. 4.44 Flexible assembly cell for the assembly of auxiliary contact blocks (courtesy of ASEA).

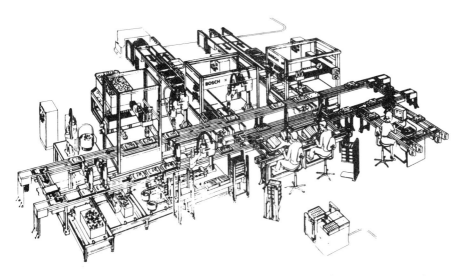

Fig. 4.45 Flexible automated assembly system (hybrid) (courtesy of Bosch).

Instead of modular-construction programmable manipulation installations, assembly robots are used in the automatic assembly station. The increasing application of assembly robots in this kind of assembly system is based mainly on the ratio of price to efficiency, which has clearly improved in recent years. SCARA devices in particular have become cost-effective. It is also this type of device that is mainly applied in programmable assembly lines. The cycle times of programmable assembly lines is predominantly between 15 and 30s. Therefore, the main field of application of these systems is the automatic assembly of products or product families with an annual batch quantity of 300000 up to 1 million. The application of a programmable assembly line, especially for the assembly of a product with several variations and/or short life time of each product variation, is economically sensible. Here the flexibility of assembly can be used for the execution of distinct assembly processes of future product redesign. Moreover, the robots can execute several assembly processes or assembly cycles, hence, especially in low batch quantities, the high rate of total utilization required for economic justification can be reached.

According to the cycle time range in programmable assembly lines a robot can manage a maximum of five or six assembly processes. The industrial application of assembly lines ranges from systems for component insertion in printed circuit boards to the assembly of automobile motors.

Flexible assembly cells   Flexible assembly cells are complex automatic assembly stations with one or two assembly robots for high operating capacity. This capacity is typically higher than in assembly stations with programmable assembly lines.

In assembly cells the peripheral devices are required to carry out the various assembly processes comprising complete or almost complete assembly tasks. Task-specific assembly installations such as presses, welding stations or the like are also integrated if necessary. Because of these peripheral devices the operating capacity of programmable assembly cells is limited – in general from eight to ten distinct workpieces are assembled per robot. There are several reasons for this limitation:

- only a limited number of peripheral devices can be accommodated in the work space of an assembly robot;
- susceptibility to interference increases with the number of subsystems in the cell;
- too many peripheral devices decreases accessibility for repair and maintenance work, and for failure clearance.

The majority of applications of industrial assembly cells are configured as island solutions, i.e. without linking to other assembly stations. They are often pilot applications to gain a first experience with assembly robots in production. This type of system is suitable because the required investment is relatively low.

Programmable assembly cells as assembly islands are designed for the automatic assembly of pre-assembly elements or for the assembly of simple products with less than 20 different components.

Based on the extensive programmability of the assembly process, and the programming of the sequence of motions of an assembly robot, this system type can be designed for several variants or types of a product. The cycle times for which this system design can be justified economically and technically ranges from 25s to 120s. This range allows its application for an annual assembly quantity of between 50000 and 500000 assembled units, depending on the volume of the assembly task and the number of shifts. Various studies of cells in practice indicate that there are assembly tasks for programmable assembly cells in several industrial sectors.

Linked flexible assembly systems   The effective applications of flexible assembly cells as assembly islands is limited to products or elements with no more than about 20 distinct components. But in industrial assembly there are many products or elements with an annual batch quantity below 500000 and far more than 20 components. For automatic assembly of these products it is possible to spread the assembly processes over several linked flexible assembly cells. There are two possible ways of linkage:

- with rigid sequence of linking;
- with flexible sequence of linking.

Linkage with rigid sequence means that the individual assembly cells are linked (e.g. by conveyor transfer systems) in a rigid given order. This kind of linkage is frequently applied, e.g. in assembly systems for automobile engines.

Because of the rigid sequence of linking this type of assembly system can only partially carry out different assembly tasks with distinct assembly processes. Hence, they are suitable for the automatic assembly of variants and sub-types or similar products or elements having a similar sequence of assembly.

For economic reasons, small annual batch quantities require several products or elements to be assembled on a flexible automated assembly system to reach a high level of utilization. As a consequence of the very different assembly processes which therefore have to be realized, a linkage structure (interlinking sequence) that is independent of any particular assembly process is required. This kind of assembly system design can satisfy the increasing demands for higher flexibility.

Flexible interlinking between single programmable assembly cells can be achieved by replacing bypass cells with a work-holding fixture or a pallet recirculation system or by absolutely flexible linkage with industrial trucks e.g. guided vehicles. Besides the linkage tasks inside the assembly system the industrial trucks can also link the assembly system with the preceding or the succeeding production system.

4.6   Economic justification of assembly systems

With an appropriate outlay of time, money and suitable professional personnel suitable technological solutions to automatic assembly may be designed. Automation of work operations merely for the sake of automation is uneco-

nomic, and consequently not viable for assembly work as in any other industrial applications. Automation of an assembly process becomes economic only
when the resulting benefits or savings are greater than the cost of automation.
In a factory, therefore, only those investment projects for automation should
be implemented for which the set functions can be achieved more rapidly, with
better quality and, above all, at more favorable costs. Existing or new assembly
processes still to be planned therefore have to be analyzed in detail to predict
the benefits of individual automation projects and the associated costs.

In practice, the estimated costs of two or more alternative designs for a
planned assembly process are compared with one another by means of a
detailed cost analysis. For this comparison, all the costs arising within the time
period to be considered are added up to determine the most economical
design based on the given planning document.

4.6.1   Changing costs with increasing degree of automation

The amount of total wages and its proportion of the total assembly costs, falls
as the degree of automation of the assembly system rises. If all assembly
operations are carried out automatically in one assembly process, it only
remains for operators to monitor, eliminate failures and maintain the technical
equipment. The wage costs then comprise only an insignificant proportion of
the assembly costs for a product. However, as the degree of automation
increases, capital costs rise in place of wage costs to cover depreciation and
interest on the production plant, and there are relatively higher maintenance,
power and design costs.

Capital costs   The extent of the capital costs is predominantly determined by
the level of invested capital. With a rising degree of automation the capital
requirements increase progressively. In contrast, the productivity of highly
automated means of assembly, following a saturation curve increases only
marginally.

The procurement costs include the purchase price of the assembly system
and the costs for transport, insurance and packing, as well as outlays for
installing the assembly system and preparing it for operation. In addition to
the procurement costs, with every new assembly system there are set up costs
which are the costs to cover the period from commissioning up to the reliable
operation of the equipment. In the case of manual operation, costs of this kind
occur substantially more frequently in the form of instruction or retraining
costs, due to the increasing fluctuations in assembly personnel, through illness,
holidays, etc.

Maintenance costs   Costs for maintaining assembly systems are approximately the same as for production equipment producing discrete parts and can
in general be set annually at 5–10% of procurement value. For instance, the
more complex are the control and drive systems of automated assembly systems, the higher is the cost for maintenance. It is therefore advisable to train

qualified factory personnel in the maintenance of the production plant in order that stoppage periods and the considerable costs arising therefrom can be kept as low as possible.

Design costs With increasing automation, the design costs for planning and preparation of an assembly process also rise. The cost for automated assembly systems is higher than for manual systems and the product to be assembled has in some circumstances to be redesigned if it is not suitable for automated assembly. In consequence, additional costs can arise for modifying individual production systems for the production of parts. This preliminary work considerably impairs the profitability of a unique automated assembly system. However, as soon as several similar machines are applied for a particular assembly process the proportion of these costs per machine falls considerably.

Power costs Power costs likewise rise with increasing mechanization and automation of operations. In comparison to the level of the remaining costs for automation, however, these costs are mainly insignificant when deciding for or against automation of the assembly process.

4.6.2 Assembly costs

Assembly costs are determined mainly by the costs mentioned above. Figure 4.46 illustrates these costs as a function of the degree of automation. According to this figure minimum assembly costs result from a certain optimum degree of automation. One can realize that minimum assembly costs can be achieved only when the automation of an assembly process does not mean that the services of the human operator are basically replaced by automation. The operator's abilities must be applied as appropriate when they can only be automated at great technical cost, and consequently also at great financial cost, especially in regard to highly complicated assembly work.

The optimum, and consequently most economic, degree of automation can be found by considering several design alternatives providing different levels of automation. The anticipated assembly costs are calculated and then compared as shown in Fig. 4.46.

4.6.3 Economic useful life and amortization period

The annual reduction in value of capital equipment is determined by the time period over which it is assumed the plant can be utilized. This time period is referred to as the useful life. Economic considerations lead to the determination of the most economic useful life of capital equipment. This time period extends up to the time when, by taking the equipment out of commission or replacing it, a more favorable return on investment is achieved overall than would be achieved with its continued use. In general the production equipment then has only scrap value.

The amortization calculation determines the period in which the net profit

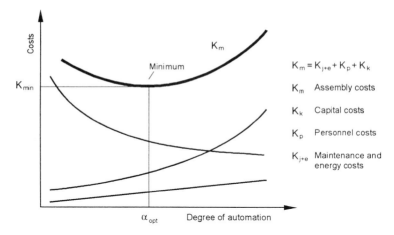

Fig. 4.46 Assembly costs as a function of the degree of automation.

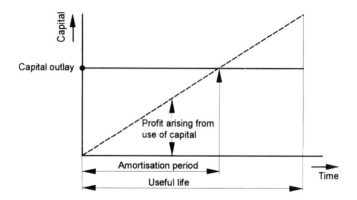

Fig. 4.47 Calculation of amortization.

return to the company exceeds the capital outlay for an investment (Fig. 4.47). The amortization period is calculated as follows:

$$\text{Amortization period} = \frac{\text{Capital outlay}}{\text{Annual cost savings plus depreciations}}$$

The annual cost savings due to the introduction of automatic assembly systems basically result from the savings in labor costs. On the other hand, the savings may be reduced by higher costs for maintenance, and by additional costs for stoppages and breakdowns, rejections and rework, etc.

The amortization period is an essential parameter for assessing the risk linked with an investment. The shorter the amortization period, the lower is the risk for the outlay in capital. In consequence, the following requirement is a prerequisite in planning an investment:

Amortization period < Economic useful life

Guide values for the economic useful life of automated assembly systems are different in individual companies. They are certainly dependent on individual experience.

4.7 Examples of assembly systems

4.7.1 *Flexible robot assembly cells for small oil pump*

The oil pumps to be considered here appear in many different forms and may consist of up to 11 individual parts. In designing the assembly cell (Fig. 4.48), the three alternative types of pump which are manufactured in the greatest numbers were taken as a reference.

For designing the assembly cell the scope of work to be considered ranges from preparation of the individual parts through the assembly operation up to the functional testing of the completed oil pumps for all three alternative types. The flexibility requirements for the assembly cell relate first to the processing of the selected oil pump alternatives in the model mix, and second to the need to change tools or extend the facilities when new alternative types of pump will become available.

On the basis of these design constraints, a multi-stage modular assembly cell concept was developed, in which the complex assembly operations are trans-

Fig. 4.48 Flexible assembly cell for small oil pumps.

ferred to peripheral equipment. An industrial robot is used as a central flexible manipulation device. The object was to use the inherent flexibility of the robot for various tasks:

- preparation of parts;
- manipulation of parts;
- assembly and changeover operations.

Flexibility is maintained by using peripheral components specific to the alternative types of pump. In order to achieve this design, while simultaneously maintaining the required flexibility for different types, the appropriate assembly and preparation of components had to be developed.

All the components relevant to the alternative types of pumps are of modular design. They have a basic construction for all alternatives and tool sets are specific to each alternative and can be exchanged to suit a change in pump type.

The idea of changing the tool with the industrial robot can also be extended to the assembly components which are fixed to the table in such a way that, using the robot, the assembly cell can be completely equipped for the individual oil pumps without manual intervention. After the plant has been started and programmed with instructions, the actual tool state of the assembly cell is determined and, if required, the robot can then equip it to meet those instructions.

The robot gripper system developed specially for this task consists of the modules:

- collision sensor;
- compliance element;
- gripper unit;
- gripper adaptor and tool changeover adaptor.

The interchangeable gripper adaptors serve to manipulate pump pistons and pump housing for the specific pump types. They represent the interfaces specific to the particular types between the robot and the various alternative components, and the uniform interface to the individual assembly equipment. Thus the interface between the individual items of assembly equipment can be considerably simplified. The tool changeover adaptor for manipulating tool modules is used in the tool changeover operation.

4.7.2 *Flexible system for automation of lamp wiring*

During the assembly of lamps, components such as chokes and sockets are fitted into lamp housings (Fig. 4.49) and then wired. From the point of view of assembly techniques, wires are difficult items to retain in position. Therefore they are very difficult to manipulate and assemble without the aid of methods and tools which have to be specially developed. In industrial production today, wire harnesses are predominantly assembled manually as pre-assembly groups and later fitted into the end product in a final assembly stage.

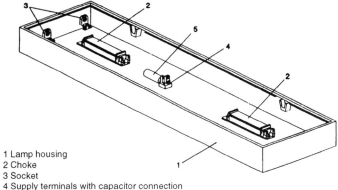

1 Lamp housing
2 Choke
3 Socket
4 Supply terminals with capacitor connection
5 Capacitor

Fig. 4.49   Unwired lamp housings.

The demands placed on a system for the fully automatic wiring of lamps are:

- fully automatic complete cycle;
- process-integrated quality control;
- program flexibility;
- CAD/CAM coupling;
- high level of availability and productivity.

Simplification, compared to the classic pre-assembly of wire harnesses, can be achieved by direct assembly of the individual wires in the lamp housings. This way it is possible to eliminate operations such as complete pre-cutting and terminating and order picking of the individual wires. In order to be able to carry out direct assembly economically and with appropriate component availability, it was unavoidable that the shapes of the components had be redesigned. Previously, individual lamp components were designed for the use of 'screws' or 'plugs' for connection purposes. It was necessary to redesign the connection geometry of sockets, chokes, etc. since the technique of making connections by piercing the insulation proves to be the most assembly-friendly form of connection for the given task (Fig. 4.50). Designing the individual parts and component assemblies in a form suitable for assembly can consequently be considered as the key to possible greater automation, subject to economic and technical constraints.

The requirements with regard to program flexibility and consequently minimum tool change and shut-down times with greater variety of types and variations necessitate the integration of tolerance compensating systems in a fully automatic cycle. Analysis of the product range shows that production tolerances (of housings, components, etc.) are always of the order of several millimeters. The systems to be applied must therefore be designed to handle the maximum tolerances which may occur in production. The system best suited to this purpose is a vision system, in which the exact position of the components before the assembly operation can be recorded (Fig. 4.51).

Before   **Afterwards**

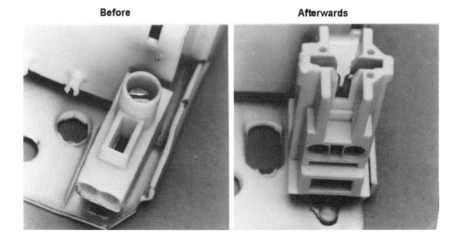

Fig. 4.50   Design for assembly of insulation piercing connection element.

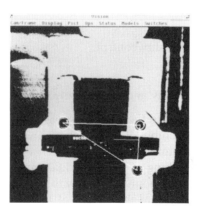

Fig. 4.51   Measurement of the connection zone with the aid of a vision system.

Knowledge of the exact position is obtained with the aid of markings provided especially for optical measurement on the components (three holes in each injection moulding in the region of the connection zone). After identification of the actual position, the deviation from the desired position can be entered in a sequential control program and the assembly cycle started.

The flange-mounted tool on the SCARA robot, with integrated slip control and synchronization of the tool path feed rate for feeding the wires, is designed to be multi-functional (Fig. 4.52). The vision system for determining the exact position of the components is fitted directly onto the tool, as are the two light sources required for measurement. The measurement of the components is carried out under incident light. With the aid of a pressing stamp and the robot moving in the z-direction the contact is formed. Cutting of the wire takes place in the tool, directly after the contact, with an integrated cutting system. The force transducers to measure the insertion forces are integrated directly into the *z*-axis of the industrial robot.

With the single-position system shown in Fig. 4.53, the lamp housings are fed in with the aid of transfer systems, each type is identified by the vision system

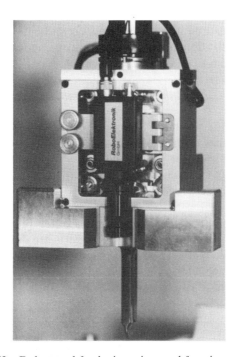

Fig. 4.52 Robot tool for laying wires and forming contacts.

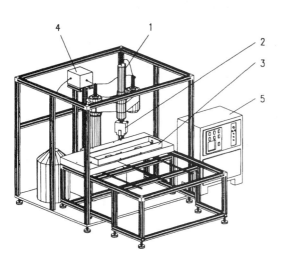

1 Robot
2 Tool with tolerance
    compensating system
3 Workpiece carrier with
    workpiece
4 Wire feed
5 Control cubicle

Fig. 4.53 Single-position system for wiring lamps.

and a bar code provided on the housings. After identification, the appropriate robot sequence program is loaded. In order to be able to compensate for tolerances, the industrial robot moves the vision system to the assumed desired position of the components and measures their exact position. For this operation the robot can refer back to CAD data and consequently drastically reduce not only the material flow but also the flow of information. After the components are measured, the wiring process can begin. Based on extensive test runs and optimization of the system, an average wire laying speed of approximately 530mm/s can now be achieved.

4.7.3 Linked flexible assembly system

The assembly line for printers shown in Fig. 4.54 is state of the art. The assembly of three types of printer takes place in a building with three floors. Material flow in/out is on the ground floor. On the middle floor there is the automatic assembly of the printers and on the top floor the assembly of printed circuit boards (Schweizer, 1988).

Ground floor Incoming parts and materials enter through five automatic racks divided into small and large parts. The parts are delivered on pallets by lorries; the piles are loaded on roller lines and are then transported automatically. Bar codes on the pallets can be read automatically. The pallets are made of molded or swagged plastics.

Middle floor The assembly system includes 127 stations in all (48 stations with industrial robots, 22 stations for screwing and about 50 test and special

Fig. 4.54 Assembly line for printers (courtesy of OKI DATA).

stations). The cycle time of the system is 18s. At present about 2000 printers a day are assembled. The lot sizes are about 1000 printers with a changeover time of 15 min. The assembly stations are linked by a double flat-belt. Because the system is winding, several bridges are necessary over the belt. Parts are delivered to the assembly stations by AGV, automatically passing the magazines to unloading systems. Each unloading system consists of different roller feeders. There are three lifting/lowering stations for automatic unloading of full pallets, moving the full and empty pallets, and piling the empty pallets. The unloading is identical at all stations. The area needed for the automatic assembly system is 2400 m²; 13 operators are employed.

The relatively small number of operators can be explained by good staff motivation and qualification. A robot breakdown is managed as follows: five operators approach the robot; four of them repair it while the other operator continues the assembly manually.

Top floor The component insertion into printed circuit boards is also highly automated. There are large numbers of different robots for the insertion of conventional parts, SMD, flat packs, etc. There are no special parts on the printed circuit boards. For the assembly of printed circuit boards only a few operators are necessary; each four to five robots are supervised by one operator (in contrast, in Germany one operator is assigned to each robot).

Costs The investment cost for the first and second floors is about $8 million for purchased parts only. The system was developed by OKI itself and the costs of development were borne by the company. The investment cost for the third floor was about $4 million for the assembly system.

Conclusion The automation was implemented completely, instead of adding automation modules to the previous facility. Several differences may be noted between the system in Japan and German practice:

- the relatively high number of assembled products with a cycle time of 18s each;
- the careful product design for assembly;
- assignment of operators;
- the simple construction of the automated assembly plant with simple installations and no backup jigs;
- willingness to invest the development costs of about US $13 million by the company.

4.7.4 *Flexible automation of automobile weather-strip assembly*

Introduction For a long time automotive weather-strip assembly has been a challenge to production engineers all over the world. The labor-intensive process of manual weather-strip assembly based on the use of rubber hammers typically involves eight to ten workers. Due to the heavy strain of working with the hammer – partially overhead – illness rates in this sector tend to be seriously above average. In addition, frequent problems with the quality of manual weather-strip assembly have led to an increasing demand for process automation.

Automation of weather-strip assembly has to overcome a variety of obstacles:

- assembly of non-rigid parts;
- inaccuracy of seals;
- car body tolerances;
- design deficiencies in the assembly process for both car flange and seal;
- restricted work space for robot end-effectors;
- high level of flexibility requirements.

State of the art   In recent years several car manufacturers, and weather-strip suppliers, have been working to find a feasible solution for the automation of weather-strip assembly. The German GM division Opel AG has developed an approach based on inflating a rubber hose to attach a previously mounted seal on the flange (Patent EP 0449044). German weather-strip supplier Draftex AG, Viersen has introduced a mechanized assembly device called roll-forming (Patent DE 40 34 212) to facilitate the manual assembly process. Both approaches lack the reliability required to be implemented on a production line.

Robotized assembly tools based on the methodology of driven pressure rolls being handled along the flange have been introduced by Kuka at the GM plant in Oshawa, Ontario (Patent EP 0253599), at Fiat in Cassino by Draftex (Patent DE 39 41 476) and at the Technical University of Munich (TUM). None of these systems deliver sufficient reliability due to problems with tolerances and structural instabilities of the seals. A recent publication of TUM concludes that the process of weather-strip assembly is too difficult to automate (Hoßman, 1992).

Design for assembly   At the Fraunhofer Institute for Manufacturing Engineering and Automation (IPA), the process of automotive weather-strip assembly has been studied by a scientific approach. Integrated product and process analysis has revealed a variety of design deficiencies in door seal structures. Due to a fairly flexible metal reinforcement made of wire or cut steel sheets, the seal tends to be too pliable and to drill itself. Moreover, inaccuracies of seal lengths can be up to 30 mm.

In close cooperation with Mesnel SA the structure of seals has been improved to facilitate a robotized assembly process. Figure 4.55 illustrates the weather-strip structure which has been developed, featuring stamped sheet material for reinforcement together with symmetrically arranged threads of fiber glass to ensure stability and inextensibility of the seal. The V-shape of the seal opening additionally cuts tolerance requirements for the end-effector along the body flange.

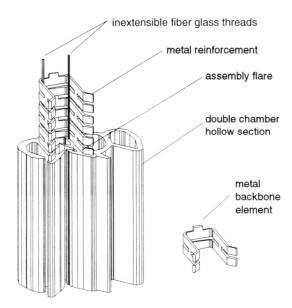

inextensible fiber glass threads

metal reinforcement

assembly flare

double chamber
hollow section

metal
backbone
element

Fig. 4.55   Door seal after design for assembly.

Concept development for flexible weather-strip assembly   Parallel to product design for assembly a variety of concepts for an automated assembly process were developed. Constraints for the process development included:

- 100% automation rate including parts feeding and quality control;
- 99% reliability;
- assembly of closed loop seals;
- assembly cycle time below 60 s for integration into final assembly line.

Using a morphological design approach a variety of solutions for the subsystems were investigated. Figure 4.56 gives an overview of the morphology including technical solutions for the subsystems, including:

- feeding of the seals;
- handling of the seals;
- joining process;
- compliance systems for tolerances.

Thorough analysis of alternative systems and an evaluation in close cooperation with Opel's production engineering systems division led to the selection of the robot tool system. Figure 4.57 illustrates the components of the developed end-effector.

The closed loop weather-strip is taken from a parts feeder and is handled by the robot using a gripper. The selected joining process based on a prototype development by Mesnel (Patent EP 0451023) closely copies the manual assembly process. A pneumatically driven hammer oscillating at 33 Hz drives the

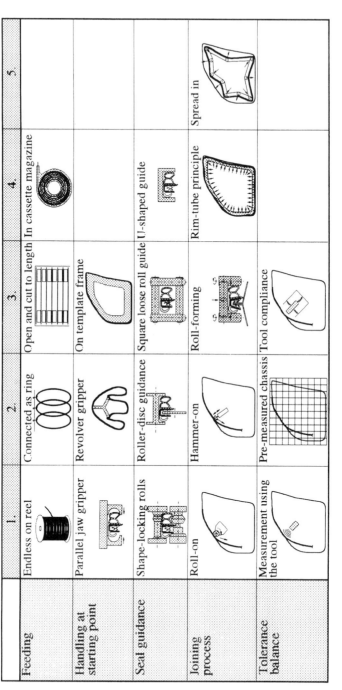

	1.	2.	3.	4.	5.
Feeding	Endless on reel	Connected as ring	Open and cut to length	In cassette magazine	
Handling at starting point	Parallel jaw gripper	Revolver gripper	On template frame		
Seal guidance	Shape-locking rolls	Roller-disc guidance	Square loose roll guide	U-shaped guide	
Joining process	Roll-on	Hammer-on	Roll-forming	Rim-tube principle	
Tolerance balance	Measurement using the tool	Pre-measured chassis	Tool compliance		Spread in

Fig. 4.56 Morphology.

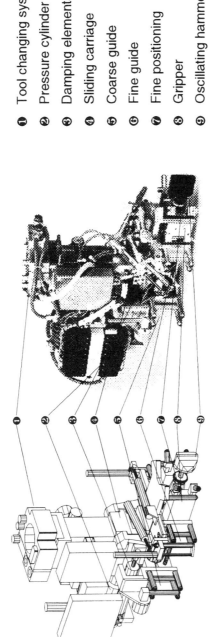

Fig. 4.57 Components of the developed end-effector.

① Tool changing system
② Pressure cylinder
③ Damping elements
④ Sliding carriage
⑤ Coarse guide
⑥ Fine guide
⑦ Fine positioning
⑧ Gripper
⑨ Oscillating hammer

seal onto the flange as it is guided by a robot along the door flange. Due to the design of the weather-strip opening, the process aligns itself along the path. Closed squares of non-driven guide rollers enable alignment of the weather strip.

Due to the hammer design the system accommodates to body tolerances in the tool y-direction of up to 6 mm. To adapt to body tolerances in the tool z-direction a compliance system has been integrated. The system consists of a hydro-pneumatic damped spring with a linear carriage and enables the tool to tolerate positioning errors of up to 15 mm.

Analysis of selected process parameters   In order to increase system availability and reliability on a long-term basis, critical factors of the assembly process were investigated by examining static and dynamic insertion tests. In a series of practical trials using various flange configurations, technical process parameters and their interrelation were established. Force requirements were determined in mounting and demounting processes in various conditions, e.g. different temperatures, various flange thicknesses, different tool velocities and with PVC contamination, and results were checked against the strict quality requirements.

Specific strategies were developed to guide the seal efficiently and to optimize the tool approach to the chassis in order to minimize excessive seal at the end of the operation. Experiments were carried out determining an optimal robot path and appropriate path velocities allowing for pre-known flange variations at given chassis positions. Motion adapting strategies have been investigated in critical parts of the door opening to ensure complete assembly without seal damage. The results of these experiments were essential for a successful cell integration.

Integration into robot cell   At IPA the developed tool for flexible weather-strip assembly has been integrated into a prototype robot cell. The assembly cell based on a Fanuc S-420F industrial robot includes a part feeder and a setup for the car body to evaluate the performance of the developed system. Figure 4.58 illustrates the weather-strip assembly process.

Intensive process analysis and improvement have driven the reliability of the tool to about 99% in the laboratory. Assembly time has been cut to approximately 33 seconds for a front door.

In a field test of the system with a Kuka industrial robot at Opel's Rüsselsheim plant 800 weather-strips have been mounted into the left front door of the Vectra. The test underlined the high performance of the developed system.

Currently, a four-robot assembly station for full automation of the weather-strip assembly is being manufactured by Thyssen Nothelfer, Hermeskeil and has been integrated into the final assembly line for the new Opel Omega in Rüsselsheim.

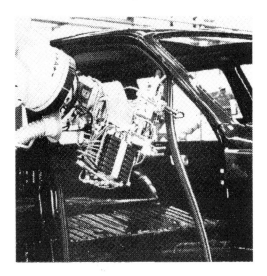

Fig. 4.58 Robot cell.

4.8 Summary

A system is only as good as its integrated building blocks, and assembly systems certainly depend on how they are built. This chapter has explained how to design effective assembly from a technological point of view: how to organize the whole operation and the object's location in the facility; how to store, transfer, position, locate, feed and sort the assembled objects; what equipment variety is available to perform these functions; and how to integrate and justify the selected system elements.

The case examples at the end of this chapter illustrate the material covered so far in the first four chapters. In assembly system design, there is a synthesis of management and labor considerations, economic issues, assembly tasks selection, design for product assembly and the assembly system configuration. Starting with Chapter 5 the rest of the book shifts attention to quantitative modeling and analysis of assembly design, operation and control.

4.9 Review questions

1. Discuss some of the planning risks that might have been faced by designers of the oil pump assembly cell (described in section 4.7).
2. Suggest which of the techniques described in Chapter 3 can be useful during the various stages of the system planning methodology.
3. Explain the role of the assembly priorities in the system design process.
4. Analyse the priorities that might have led to the design of the automated lamp wiring (section 4.7).
5. How does the system principle influence the selection of the system

elements? Illustrate your answer with the case of the printer assembly (section 4.7).

6. Select system organizations for the following products and justify your selection:

    (a) helicopters;
    (b) bicycles;
    (c) car radiators;
    (d) telephones;
    (e) book cabinets.

7. Give three examples of assembly systems that will most probably be implemented as assembly lines and classify them according to the types of assembly-line work.

8. What are the risks associated with a wrong choice of storage equipment?

9. Explain the role, advantages and disadvantages of using the following system elements:

    (a) storage magazine;
    (b) hopper;
    (c) pusher device ('ram');
    (d) chuck;
    (e) passage magazine;
    (f) continuous rotary positioning device.

10. Explain which device would be selected for the following requirements:

    (a) clamping and positioning, later releasing rigid parts;
    (b) overhead linear transfer of hooked parts;
    (c) rapid transfer of components at cycle time of 4 seconds;
    (d) individual workpiece delivery and positioning at a fast uniform rate;
    (e) storage and transfer of a set of different subassembly components to be assembled by a robot;
    (f) feeding flat components in a given alignment.

11. What are the main types of linking between assembly machines and under what conditions would they be preferred?

12. Describe the four main types of motion equipment for assembly workstations.

13. For automatic parts transfer, what is the advantage of pneumatically versus cam-driven devices?

14. Describe the main methods for interlinking assembly stations and their characteristics.

15. Explain the considerations in selecting each of the three main types of flexible assembly systems, referring specifically to the cases illustrated in section 4.7.

16. Explain a cost justification procedure for:

    (a) weather-strip assembly;
    (b) printer assembly.

In your analysis use the information provided in section 4.7 and add any necessary assumptions.

17. Imagine that Henry Ford had today's technology available in 1913. Would he prefer a flexible assembly system over the single-model assembly line? Explain the pros and cons (in the context of this chapter).

References

Abele, E., Bäßler, R. *et al.* (1984) *Studie zur Untersuchung der Einsatzmöglichkeiten von flexibel automatisierten Montagesystemen in der industriellen Produktion unter Beachtung der technischen, arbeitsorganisatorischen und sozialen Vorausseztungen einer menschengerechten Gestaltung der Arbeit*, VDI-Verlag (grüne Reihe), Düsseldorf.

Adamczyk, J. (1969) *Der Fließzusammenbau elektrotechnischer Bauteile und Erzeugnisse*, Beuth Vertrieb GmbH.

Dolezalek, C.M. and Warnecke, H.-J. (1981) *Planung von Fabrikanlagen*, Springer, Berlin.

Drexel, P. (1983) 'Flexibles Montagesystem "FMS" von Bosch. Alternativkonzete zur Montage. "Montage in der industriellen Produktion",' *Ergebnis einer Fachkonferenz*, (ed. F.-V. Kuenzer), Projektträger Humanisierung der Arbeitswelt DFVLR Karlsruhe.

Hoßman, J. (1992) *Methodik zur Planung der automatischen Montage von nicht formstabilen Bauteilen*, Zugl. München Univ. Diss. 1991, Springer, Berlin.

Lotter, B. (1982) *Arbeitsbuch der Montagetechnik*, Vereinigte Fachverlage Krausskopf, Ingenieur Digest, Mainz.

Patent DE 39 41 376, *Verfahren und Vorrichtung Zum Befestigen einer Leiste auf einem Befestigungsflansch*, 21 June 1990.

Patent DE 40 34 212, *Vorrichtung zum Befestigen einer U-förmigen Profilleiste auf einem Flansch o. dgl*, 16 May 1991.

Patent EP 0 253 599, *Installing weather stripping in a door or like opening*, 21 January 1988.

Patent EP 0 449 044, *Vorrichtung zum Aufdrücken einer Dichtung*, 14 March 1991.

Patent EP 0 451 023, *Procédé et dispositif pour la mise en place d'un profile à section en U sur un rebord d'un encadrement d'une carrosserie d'automobile*, 9 October 1991.

Petri, H. (1982) *Montageautomaten für große Serien*, Tagungsband No. 3, ICAA, Böblingen.

Schweizer, M. (1988) *Japanische Montagestrategien*, in 8th Deutscher Montagekongreß, 22–23 September, pp. 128–50.

Spingler, J. and Bäßler, R. (1983) 'Montage mit Industrierobotern: Heutiger Stand und Entwicklungstendenzen', *Maschinenmarkt*, Vol, 89, No. 84, pp. 1914–17.

Spingler, J. and Bäßler, R. (1984) 'Industrieroboter in der Montage', *Schweizer Maschinenmarkt*, Vol. 84, No. 24, pp. 24–9.

Warnecke, H.-J. and Haaf, D. (1982) *Auslegung programmierbarer Montagesysteme für den Bereich kleiner Stückzahlen und hoher Variantenvielfalt*, Tagungsband No. 3, ICAA, Böblingen.

Warnecke, H.-J., Löhr, H.-H. and Kiener, W. (1975) *Montagetechnik, Schwerpunkt der Rationalisierung*, Krausskopf-Verlag, Mainz.

5

Assembly system design and planning

5.1 Introduction

This chapter deals with key aspects of designing and planning assembly systems. Until recently, assembly has been considered by many to be a low technology process, and its costs and importance may have been underestimated. However, with the advent of automated assembly – especially programmable, automated assembly – fundamental issues are being rethought and new methods are being devised to assist designers and planners.

World-class operations must be carefully designed and planned to achieve business goals. The production environment is characterized by medium-to-high variety, medium-to-low volume production. This environment places a premium on agile manufacturing methods which allow new assembly systems to be designed and planned within shorter and shorter lead-times. In addition, product lifecycles are becoming shorter as a result of rapid technological advancement and international competition. This accelerating trend requires the redesign and replanning of assembly systems more frequently and within shorter lead-times.

This chapter describes methods that industry might apply in such settings to improve the competitiveness of their assembly operations. The body of this chapter is organized into five sections. Section 5.2 describes assembly line balancing methods for both single- and mixed-model assembly. Line balancing is a means of designing a system and forms a foundation of concepts that are used to formulate and solve the more comprehensive problems involved in programmable automated assembly treated in section 5.3. Sections 5.4 and 5.5 address planning issues: section 5.4 gives an overview of process planning methodologies and section 5.5 relates methods of planning an operation sequence in robotic assembly. Section 5.6 provides a summary and conclusions.

The models described in this chapter are based on the assumption that there are no part shortages. This is a reasonable assumption in dealing with static

Table 5.1 Notation for Chapter 5

i	= index for predecessors of task j; $i \in P_j^*$
j	= index for tasks; $j \in J$
k	= index for successors (followers) of task j; $k \in F_j^*$
s	= index for stations
C	= cycle time
E_j	= earliest station to which task j might be assigned
L_j	= latest station to which task j might be assigned
p_j	= processing time (i.e. duration) of task j
S_j	= station to which task j is assigned
S	= total number of stations in a design
S^*	= optimal number of stations in a design
τ	= amount of cycle time unused by the current assignment of tasks to a station
F_j	= set of tasks that immediately follow task j (j is an immediate predecessor of each)
F_j^*	= set of all tasks that must follow task j
J	= set of tasks
$J(s)$	= set of tasks that can be assigned to station s
J_s	= set of tasks assigned to station s
P_j	= set of tasks that immediately precede task j (j is an immediate successor of each)
P_j^*	= set of all tasks that must precede task j

design and planning issues. However, coordinating part flows is a task that is unique to assembly, and it is crucial to achieving effective material flow, allowing control of in-process inventory and promoting performance to schedule. The following chapters address these dynamic issues. Except for a brief discussion of stochastic line balancing, the prescriptive models described in this chapter are based on the assumption that operations are deterministic. Complementary models which evaluate the performance of stochastic systems are the topic of the next chapter. Some elements of notation used throughout this chapter are summarized in Table 5.1.

5.2 Assembly line balancing

Henry Ford devised the assembly line to simplify the jobs of workers in the high volume automotive business. Ford and Sorensen implemented the first assembly line in 1913. Ford continued to evolve the concept, introducing line balancing to produce the Model T Ford in 1919. The resulting assembly line design reduced the work input required to produce each car from 12 hours 28 minutes to 93 minutes and the cost of a car from $900 to $350; further enhancements allowed a car to be completed every 10 seconds (Boothroyd, 1992). The success of this production configuration was to reshape industry. Its primary advantage over earlier assembly methods was that it allowed each worker to specialize in simplified work assignments, increasing productivity. To compensate for the routinization of jobs and for the work effort required, workers were well paid.

Line balancing is a traditional formulation of the assembly system design problem. However, in manual lines, stations are identical (i.e. the same tooling is used at all stations) so that designs can be easily changed to accommodate the desired throughput rate. Thus, the line can be rebalanced daily to compensate for demand fluctuations, so that line balancing may also be considered to be a scheduling technique for production control. The problem remains important today, even though it has been researched for some time (e.g. see literature surveys by Baybars (1986) and Ghosh and Gagnon (1989)).

This section is organized into six subsections. It begins by describing basic concepts underlying the traditional, single-model line balancing problem. Next, it describes several heuristics which are based on different solution strategies, and then it presents several branch-and-bound algorithms for prescribing optimal solutions. Computational experience with these methods is then summarized. The last two subsections describe extensions to deal with practical features and mixed models.

5.2.1   Basic concepts

Assembly line balancing begins by analyzing the assembly of a product, identifying the elementary tasks (those which cannot be further subdivided) required to complete assembly and the time required to complete each. In addition, technological considerations determine precedence relationships that the assembly sequence must observe. Task precedence relationships may be formulated as a graph $G(N, A)$ in which node $n_j \in N$ represents task $j \in J$, and (directed) arc $a_{ij} \in A$ (i.e. from node n_i to node n_j) indicates that task i is an immediate predecessor of task j. Nodes 1 and $|N|$ may be 'dummy tasks' included to represent the initial and terminal nodes in the graph.

Given the production rate (or throughput rate) R (units per period) that the line must achieve, the required cycle time C is:

$$C = 1/R. \tag{5.1}$$

In a synchronized system, subassemblies at all stations are indexed one station down the line and one completed product leaves the line each C time units, achieving the required throughput R over the time period.

Given a set of tasks, task times and precedence relationships, a feasible assignment of tasks is one that observes precedence relationships. There are actually two versions of the line balancing problem and both are known to be NP-hard (Garey and Johnson, 1979):

- ALB-1 requires a feasible assignment of tasks to stations so that the workload assigned to each station does not exceed the cycle time and the number of stations is minimized; and

- ALB-2 fixes the number of stations and requires a feasible assignment of tasks to stations that minimizes cycle time.

ALB-2 might be encountered, for example, in an automotive assembly line in which a number of robots perform spot welding to assemble the body of a car. If one robot goes down for maintenance, the number of stations is fixed and the cycle time should be minimized by the (re)assignment of tasks to the remaining robots. In general, ALB-2 smooths the workloads assigned to all stations.

Hackman, Magazine and Wee (1989) describe two methods for solving ALB-2 using an algorithm that solves ALB-1. Both methods start by applying the algorithm to solve ALB-1, determining the optimal number of stations S^* for the given cycle time. The first method for solving ALB-2 assumes that a lower bound on the optimal cycle time $C_L < C^*$ is known. Another ALB-1 problem is solved with this cycle time and if the optimal number of stations does not exceed S, C_L is the optimal solution to ALB-2. Otherwise, the algorithm is applied iteratively, increasing C_L one time unit at each iteration until a feasible, and therefore optimal cycle time is determined for S stations. The second method applies the algorithm, using a known upper bound on cycle time $C_U > C^*$. If the number of stations in this ALB-1 problem exceeds S, the optimal cycle time is C_U. Otherwise, the procedure continues iteratively, each time reducing C_U to the maximum workload assigned to a station at the previous iteration minus 1. If $[C_U - C_L]$ is large, a binary search could be used to reduce the number of computations to $O(\log(C_U - C_L))$.

In either case, line balancing applies to paced lines in which the material handling system indexes work from one station to the next at the end of each cycle time C. Thus, all workers must work to maintain this pace. If task times are variable or if the human element introduces variability, an unpaced line may be preferred. On an unpaced line, each worker secures the next workpiece from the handling system, performs assigned tasks and replaces the workpiece on the handling system to be conveyed to the next station. This asynchronous system decouples stations along the line, improving productivity in systems that must deal with variability. Further descriptions of assembly line operations and criteria for selecting a particular structure are given by Buxey, Slack and Wild (1973) and Wild (1975).

Example 5.1 Throughout this section we demonstrate design methodologies using the following example, which deals with the lower case subassembly for a (hypothetical) limited edition notebook computer as sketched in Fig. 5.1.[1] Required parts (i.e. the bill of materials) are listed in Table 5.2, the assembly tasks are described in Table 5.3, and attributes of the tasks are itemized in Table 5.4. The precedence graph, which resulted from analyzing technological requirements, is depicted in Fig. 5.2.

---

[1] We are indebted to Mr Dennis Allen, micro computer specialist in the Department of Industrial Engineering at Texas A&M University, for his assistance in formulating this example.

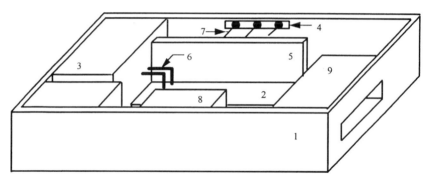

Fig. 5.1 Product sketch (numbers refer to Table 5.2).

Table 5.2 Parts list (bill of materials) (Fig. 5.1*)

Part number	Part description
0	fixture
1	lower case
2	processor board
3	power supply
4	i/o ports
5	i/o board
6	printed wiring connector (i.e. ribbon cable) #1
7	printed wiring connector (i.e. ribbon cable) #2
8	hard disk
9	floppy drive
10	keyboard/top subassembly
11	pointing device (mouse)

*Parts 0, 10 and 11 are not shown in Fig. 5.1.

Table 5.3 Task description

Task j	Description
1	fixture lower case
2	mount processor board
3	mount power supply
4	install i/o ports
5	install i/o board
6	install printed wiring connector (i.e. ribbon cable) #1
7	install printed wiring connector (i.e. ribbon cable) #2
8	mount hard disk
9	mount floppy drive
10	install keyboard/top subassembly
11	install pointing device (e.g. mouse)

Table 5.4 Task attributes

j	p_j	P_j	P_j^*	F_j	F_j^*
1	5	–	–	2, 3	2, 3, 4, 5, 6, 7, 8, 9, 10, 11
2	35	1	1	4–6	4, 5, 6, 7, 8, 9, 10, 11
3	25	1	1	6	6, 7, 10, 11
4	60	2	2, 1	8	8, 9, 10, 11
5	30	2	2, 1	8	8, 9, 10, 11
6	10	3, 2	3, 2, 1	7	7, 10, 11
7	60	6	6, 3, 2, 1	10	10, 11
8	25	5, 4	5, 4, 2, 1	9	9, 10, 11
9	35	8	8, 5, 4, 2, 1	10	10, 11
10	70	9, 7	9, 8, 7, 6, 5, 4, 3, 2, 1	11	11
11	30	10	10, 9, 8, 7, 6, 5, 4, 3, 2, 1	–	–

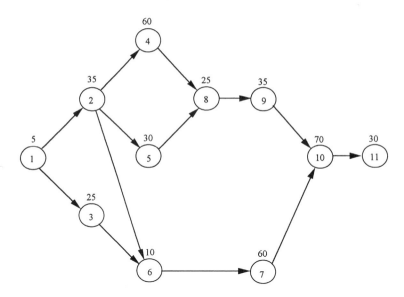

Fig. 5.2 Precedence graph (single model).

Typically, either the set of immediate predecessors P_j or the set of immediate successors F_j is given. If P_j (F_j) is given, F_j (P_j) can easily be determined along with P_j^* and F_j^*, the sets of all predecessors and followers, respectively.

In the example, we assume a required throughput rate of $R = 100\,000$ notebook computers per year, and equation (5.1) can be used to determine the required cycle time C:

$$C = \left[(60\,\text{s/min})(60\,\text{min/h})(40\,\text{h/week})(50\,\text{week/year})\right] / \left[100\,000\right] = 72\,\text{s}$$

Bounds and problem formulation Each task j in the set of J (elementary) tasks has known duration p_j; a set of known predecessor tasks P_j; and must be assigned to one station. A lower bound on the number of stations required in the line design is given by

$$B_1 = \left\lceil \frac{\sum_{j=1}^{J} p_j}{C} \right\rceil. \tag{5.2}$$

The earliest station E_j and latest station L_j to which task j might be assigned are given by (Patterson and Albracht, 1975):

$$E_j = \max\left(1, \left\lceil \frac{p_j + \sum_{i \in P_j^*} p_i}{C} \right\rceil\right)$$

$$L_j = \max\left(1, UB + 1 - \left\lceil \frac{p_j + \sum_{k \in F_j^*} p_k}{C} \right\rceil\right)$$

in which UB is an upper bound on the required number of stations. In Example 5.1:

$$\sum_j p_j = 385 \text{ s, so that}$$
$$B_1 = \lceil 385/72 \rceil = \lceil 5.3472 \rceil = 6,$$
$$E_4 = \max\{1, \lceil (60+40)/72 \rceil\} = \max\{1, \lceil 1.389 \rceil\} = 2,$$

and if $UB = 7$:

$$L_4 = \max\left(1, 7 + 1 - \left\lceil \frac{60 + (25 + 35 + 70 + 30)}{72} \right\rceil\right) = 4$$

Earliest and latest station bounds are listed in Table 5.5, along with other computed values.

The ALB-1 problem may be formulated using decision variables:

$x_{sj} = 1$ if task j is assigned to station s, 0 otherwise $\left(\text{for } E_j \leq s \leq L_j, j \in J\right)$.

The objective, minimizing the number of stations, may be achieved by minimizing the penalty associated with the station number at which the terminal task is assigned. Following Pinnoi and Wilhelm (1994d), we formulate ALB-1 as a 0–1 integer program:

Table 5.5 Computed values for Example 5.1

j	$\sum_{i \in P_j^*} p_i$	$\sum_{k \in F_j^*} p_k$	E_j	PW_j	L_j
1	0	380	1	385	2
2	5	320	1	355	2
3	5	170	1	195	5
4	40	160	2	220	4
5	40	160	1	190	5
6	65	160	2	170	5
7	75	100	2	160	6
8	130	135	3	160	5
9	155	100	3	135	6
10	285	30	5	100	6
11	355	0	6	30	7

$$\text{Minimize}: \quad \sum_{s=B_1+1}^{S} sx_{s,|J|} \tag{5.3}$$

$$\text{Subject to}: \quad \sum_{s=E_j}^{L_j} x_{sj} = 1 \qquad j \in J, \tag{5.4}$$

$$\sum_{s=E_i}^{L_i} sx_{si} - \sum_{s=E_j}^{L_j} sx_{sj} \leq 0 \qquad j \in J, i \in P_j, \tag{5.5}$$

$$\sum_{j \in J(s)} p_j x_{sj} \leq C \qquad s = 1, \ldots, S, \tag{5.6}$$

$$x_{sj} = \{0,\ 1\} \qquad j \in J;\ E_j \leq s \leq L_j. \tag{5.7}$$

The objective function (5.3) assesses a penalty equal to the station number at which the terminal task is assigned. This cost function is increasing in s, so the optimal solution gives the minimum number of stations. Constraint (5.4), known as the **occurrence constraint** (Baybars, 1986) or the **assignment constraint**, requires that each task be assigned to exactly one station. Constraint (5.5), the **precedence constraint** (Baybars, 1986), invokes technological ordering so that if $i \in P_j$, i cannot be assigned to a station downstream from task j. Constraint (5.6), known as the **cycle time constraint** (Baybars, 1986) or the **capacity constraint**, ensures that the total processing time for all tasks assigned to a station does not exceed the cycle time. The summation is over all tasks that can be assigned to station s: $\{j \mid E_j \leq s \leq L_j\}$. The **non-divisibility constraint** (5.7) does not allow a task to be assigned to more than one station.

At least three alternative formulations of the objective function have been

proposed to facilitate solution, and several different ways have been proposed
to invoke the precedence and occurrence requirements with fewer constraints
(Ghosh and Gagnon, 1989).

Metering stations in flow-lines   The relationship of the balance of a deterministic line that assembles a single product to its operating efficiency is easy
to rationalize. Suppose each station operated with a unique cycle time C_s
($s = 1, \ldots, S$). Then the station with the longest cycle time is the bottleneck
and meters the output rate of the entire line. The station with the longest cycle
time upstream of the bottleneck would meter the flow of work to it. In fact, the
line may contain a number of metering stations (Wilhelm and Ahmadi-
Marandi, 1982). Metering stations can be identified knowing that:

$$m(1) = \text{first order metering station} = \text{bottleneck station} = \text{argmax}_s C_s$$

and that the rth order metering station is defined by:

$$m(r) = r\text{th order metering station} = \max\{C_1, C_2, \ldots, C_{m(r-1)}\}$$

so that station 1 is always a metering station. As an example, in a five-station
flow-line with cycle times 5, 3, 6, 8 and 2, station 4 is the bottleneck or first
order metering station, and stations 3 and 1 are the second and third order
metering stations, respectively.

The importance of metering stations is that they are the only stations at
which queuing can occur in deterministic unpaced flow-lines. The effects of
metering stations in both deterministic and stochastic systems are discussed
further in section 8.4.

If workloads are balanced (i.e. the cycle time is the same for all stations so
that they are all metering stations) the throughput rate of the line is maximized
because $R = 1/ \max_s C_s$.

The makespan to assemble a lot of L assemblies is the time required for the
first assembly to reach the bottleneck, plus the time to process all assemblies
at the bottleneck station, plus the time to process the last assembly downstream of the bottleneck:

$$\text{Makespan} = \sum_{s=1}^{m(1)} C_s + LC_{m(1)} + \sum_{s=m(1)+1}^{S} C_s$$

Line balancing also applies to assembly (as well as production) in automatic
transfer lines. Again, the balanced line can be expected to achieve the highest
possible throughput rate. Variability can even occur in automated lines due to
parts jamming or other random events. In stochastic systems, queuing may
occur at stations due to the variability of operation times. Research has shown
(e.g. Muth and Alkaff (1987)) that, in principle, the throughput rate in such

cases may be optimized by assigning expected operation times in a bowl configuration, although the effects of this 'bowl phenomenon' are typically negligible in actual systems that do not exhibit high levels of variability. Scheduling techniques have been devised to exploit imbalances (i.e. bottlenecks) in systems that assemble a mix of products, each with deterministic operation times.

Line efficiency: balance delay Given the cycle time C and S = the number of stations, a measure of the efficiency of the design, the balance delay $BD(S, C)$, is given by:

$$BD(S, C) = 100 \left[\frac{SC - \sum_j p_j}{SC} \right]. \tag{5.8}$$

To demonstrate this function, consider Example 5.1. If six stations are used, a cycle time of 64.163 seconds would result in $BD(6, 64.167) = 0$. In general, $BD(S, C) = 0$ if $S\,C = \sum_j p_j$, so it is easy to determine what cycle time would result in zero balance delay for a selected number of stations, S. For instance, for 6, 7, 8, 9, 10 and 11 stations, cycles times of 64.167, 55, 48.125, 42.78, 38.5 and 35 would result in zero balance delay, respectively. For a given S, one can pick a second cycle time and use equation (5.8) to compute the associated balance delay. The resulting balance delay function is shown in Fig. 5.3. The lower saw-toothed envelope indicated by the solid line shows, for a given S, the balance delay that would be incurred for a given cycle time. Thus, if one

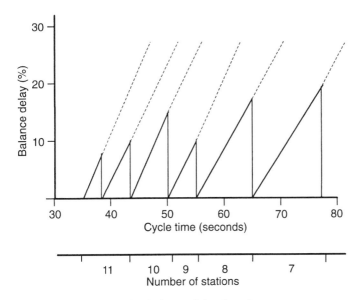

Fig. 5.3 Balance delay function.

picked a cycle time, the minimum number of stations that might be used is indicated below the cycle time axis and the associated balance delay is described by the balance delay function.

For example, a cycle time of 72 seconds requires six stations (if precedence relationships do not render this design infeasible), and the associated balance delay is 10.88%. If precedence relationships force a seven-station design, the balance delay would be 23.61%. Notice that the balance delay function for seven stations goes beyond the solid line, as indicated by the dotted line. If precedence relationships force an eight-station design, the balance delay would increase to 33.16%.

5.2.2   Heuristic solution methods

Since the line balancing problem is NP-hard (Garey and Johnson, 1979), optimizing a large-scale system may require prohibitive runtime. Therefore, heuristics that require little runtime and give 'good' solutions may be used. This section describes a set of heuristics, each of which is based on a different solution strategy.

The Ranked Positional Weight heuristic of Helgeson and Birnie (1961) is task-oriented. It calculates the positional weight of each task, ranks tasks in order according to these weights, then assigns each task to a station. It can be summarized in three steps.

1.   Compute the positional weight of each task $j \in J$, PW_j:

$$PW_j = p_j + \sum_{k \in F_j^*} p_k$$

2.   Rank order tasks with positional weights in non-increasing order.
3.   Assign tasks in the ranked order, each to the lowest numbered station that is feasible relative to precedence relationships and cycle time constraints.

Positional weights for the tasks in Example 5.1 are listed in column 5 of Table 5.5. Step 2 ranks jobs in order – only jobs 3 and 4 are not in their natural order. Table 5.6 summarizes the task assignments made by step 3. Notice that this is a job-based procedure, since jobs are assigned in order according to their ranking.

The combination of precedence relationships and task times forces each of the last four stations to process a single task. Idle times at the stations are 7, 2, 17, 12, 37, 2, 42, respectively, and tend to increase at downstream stations (task 10 must be assigned to a station by itself no matter where it is assigned). The total idle time is $S\,C - \sum_j p_j = 7\,(72) - 385 = 119 = 7 + 2 + 17 + 12 + 37 + 2 + 42$.

If the number of stations prescribed by a heuristic is equal to a lower bound on the optimal number of stations (B_1, for instance), the solution is optimal. However, the number of stations prescribed by the positional weight heuristic

Table 5.6 The ranked positional weight heuristic

Task	Order	Station
1	1	1
2	2	1
4	3	2
3	4	1
5	5	3
6	6	2
7	7	4
8	8	3
9	9	5
10	10	6
11	11	7

Table 5.7 The COMSOAL heuristic

1. Form list A of tasks. Set the trial station $s = 1$ and $\tau = C$.
2. Promote all tasks from list A that have no immediate predecessors to list B, the 'candidate list'.
3. Scan list B, promoting each task with $p_j \leq \tau$ to list C, the 'fit list'.
4. IF list C is empty, increment the trial station number $s \leftarrow s + 1$, set $\tau = C$, and GOTO Step 3.
5. Select task j from list C and assign it to station s. Set $\tau \leftarrow \tau - p_j$.
6. Eliminate task j from all lists.
7. IF all lists are empty, STOP.
8. Update the immediate successors of task j on list A (i.e. reduce the number of immediate predecessors of each by 1 since j has been eliminated) and GOTO Step 2.

is seven and the lower bound $B_1 = 6$, so it is not yet possible to say which is optimal. However, this feasible solution gives an upper bound on the optimal number of stations, and it is now possible to calculate an upper bound on the latest station to which each task can be assigned, L_j for $j \in J$, as shown in Table 5.5. Notice that $UB = 7$ was used to calculate L_4 on page 212.

The COMSOAL (Computer Method of Sequencing Operations for Assembly Lines) heuristic of Arcus (1966) is a station-based, biased random sampling procedure. It tallies the amount of cycle time unused by the current assignment of tasks to a trial station τ, and assigns as much processing time as precedence relationships and the τ limitation permit, then goes on to the next trial station. It is described by the eight steps itemized in Table 5.7.

Alternative criteria could be used to select task j in Step 5. For example, a task could be selected according to minimum operation time, maximum operation time, the number of following tasks or the sum of times for following tasks. Alternatively, Monte Carlo procedures could be used to select task j at random, perhaps biasing the selection according to one of these criteria. Using such a sampling procedure, the problem is resolved a number of times – perhaps 20 times – and the best of these solutions is the one prescribed.

We demonstrate COMSOAL using data from Example 5.1. At Step 5 we selected the task with the largest $|F_j^*|$ (Table 5.8) and broke ties by selecting the one with max p_j. The results are listed in Table 5.8.

At iteration 3, precedence relationships allow only tasks 3, 4 and 5 to be candidates. Since task 4 has $p_4 = 60$, it will not 'fit' on station 1, which has only $\tau = 32$ more seconds available. Tasks 3 and 5 are promoted to the fit list; they have the same $|F_j^*|$, but task 5 has the larger p_j of the two tasks and is assigned to station 1. This assignment reduces τ to 2 and no unassigned task will fit. Thus, a line is drawn under this assignment to indicate that the next trial station is opened. The heuristic proceeds to find a seven-station solution which is different than the one determined by the ranked positional weight method.

The Hoffman heuristic (Hoffman, 1963) is another station-oriented method. It determines all combinations of tasks that might be assigned to the current trial station and selects the one with the least idle time. It is summarized by the following three steps.

1.   Set $s = 1$.
2.   Generate all feasible task assignments for the current trial station and select the one with the least idle time.
3.   Eliminate tasks assigned at the current trial station, increment the trial station number ($s \leftarrow s + 1$), and GOTO Step 2.

For example, with $s = 1$, Step 2 generates assignments (1), (1, 2), (1, 3), (1, 2, 3) and (1, 2, 5) with total processing times of 5, 40, 30, 65 and 70, respectively,

Table 5.8 An example demonstrating the COMSOAL heuristic

Iteration	Available list A	Candidate list B	Fit list C	Assignment station	task	Assignment criterion	τ after assignment
1	1–11	1	1	1	1	–	67
2	2–11	2, 3	2, 3	1	2	$|F_j^*|$	32
3	3–11	3, 4, 5	3, 5	1	5	max p_j	2
4	3, 4, 6–11	3, 4	3, 4	2	4	max p_j	12
5	3, 6–11	3, 8	3, 8	3	3	$|F_j^*|$	47
6	6–11	6, 8	6, 8	3	8	max p_j	22
7	6, 7, 9–11	6, 9	6	3	6	=	12
8	7, 9–11	7, 9	7, 9	4	7	max p_j	12
9	9–11	9	9	5	9	=	37
10	10, 11	10	10	6	10	=	2
11	11	11	11	7	11	=	42

and selects assignment $(1, 2, 5)$ with a time of 70. With $s = 2$, Step 2 generates assignments (3), $(3, 6)$ and (4) with total processing times of 25, 35 and 60, respectively, and selects assignment (4). With $s = 3$, Step 2 generates (3), $(3, 6)$ and $(8, 9)$ with total processing times of 25, 35 and 60, respectively, and selects $(8, 9)$. With $s = 4$, Step 2 generates (3) and $(3, 6)$ and selects $(3, 6)$. Subsequently, assignments (7), (10) and (11) are selected with $s = 5, 6$ and 7, respectively, so that Hoffman's heuristic also prescribes a seven-station line.

Because it considers all possible combinations at each trial station, this heuristic tends to prescribe low idle times at upstream stations. However, longer idle times tend to result at downstream stations because fewer tasks remain for consideration. This condition may result in an inequitable distribution of workloads among workers. To avoid this problem, Gehrlein and Patterson (1975) proposed a modification of Hoffman's heuristic that allows the idle time at a station to be 'acceptably close' to the minimum.

The theoretical minimum total idle time is $B_1 C - \sum_j p_j$, and dividing this amount by B_1 gives an estimate of the amount of idle time at the average station. The modification curtails the search for a better assignment of tasks to the current trial station s, when the set of tasks assigned to station s, J_s, results in an idle time that satisfies the following relationship:

$$0 \leq C - \sum_{j \in J_s} p_j \leq \theta \left[\frac{B_1 C - \sum_j p_j}{B_1} \right]$$

in which parameter θ may be assigned the value 0.5, 1.0 or 2.0. This modification tends to result in solutions that can be found in less runtime and also smooth workloads (Gehrlein and Patterson, 1975, 1978).

New meta heuristics, tabu search and genetic algorithms have recently been applied to the line balancing problem. Meta heuristics provide a search strategy and a structure that may be 'tuned' for application to different types of problems.

Voß and Scholl (1994) test a number of heuristics on both ALB-1 and ALB-2 problems. They use 12 single-pass rules and consider both station- and task-oriented procedures. In addition, they use forward, backward and bidirectional construction procedures, starting at the beginning of the precedence graph, the end, or both, and assigning tasks according to priority. Their goal was to compare these heuristics with an appropriately 'tuned' tabu search (see Glover (1989, 1990) and Glover and Laguna (1993)).

Their tabu search solves ALB-1 by setting $S = B_1$ and attempting to find a feasible solution with cycle time less than or equal to C. If this problem is deemed infeasible by the search, S is increased by 1 and the procedure is restarted. They found that the heuristics which construct the best initial solutions for use in conjunction with the tabu search were bidirectional, assigning priority to the largest $|F_j^*|$ for ALB-2 and to the largest value of $p_j/(L_j - E_j)$

divided by the current number of forward and backward candidate tasks for ALB-1.

Given a current solution, a critical station is one to which the largest total processing time is assigned. Their tabu search employed shift moves, reassigning one task from a critical station to another, and swap moves, exchanging the assignment of one task on a critical station with a task from another station. Moves were implemented using a steepest descent/mildest ascent approach in which the move leading to the lowest cycle time was made. However, if no improving move was possible, the move leading to the smallest increase in cycle time was made. The assignment involving the task moved and the critical station was set tabu to prevent an immediate reversal of the move. Their aspiration criterion allows a tabu move only if it gives the best known solution. They tested different mechanisms for managing the length of the tabu list, setting it to 7, then to 15, then randomly and dynamically changing the length, and, finally, monotonically reducing the length of the list. These four methods gave improving solutions in the order listed above. Their stopping criterion was 50 000 iterations, 500 cpu seconds or upon reaching a solution equal to the lower bound, whichever came first. They also employed moving gap, intensification and diversification strategies. Tests showed that the tabu search outperformed the list-processing heuristics and, in fact, was competitive with optimizing approaches, frequently prescribing optimal solutions.

Anderson and Ferris (1994) present genetic algorithms (e.g. see Michalewicz (1994)) for the line balancing problem, tuning the method by conducting an extensive set of tests. Their findings emphasize the importance of (1) rejecting 'individuals who would decrease fitness of the worst member of a population', (2) correctly choosing a scaling parameter and (3) mutation rate. They describe a parallel implementation and compare parallel and serial implementations. Both give good solutions to ALB-2 with appropriately selected parameters. These genetic algorithms outperformed a neighborhood improvement search that used multiple initial solutions.

This section describes only a few of the heuristics that researchers have proposed. These heuristics represent different solution strategies, and reports indicate that they are effective solution methods. We now describe several optimizing methods.

5.2.3   Optimizing methods

Branch-and-bound methods are now able to resolve problems with up to a few hundred tasks, even certain problems with 1000 tasks. This section describes three recent branch-and-bound methods, but it concentrates on one method because it integrates a number of solution techniques, providing insight into problem structure.

Johnson (1988) presents the Fast Algorithm for Balancing Lines Effectively (FABLE). It incorporates several preprocessing methods and a number of fathoming criteria in a 'laser-type' (i.e. depth-first) search procedure. The following discussion presents topics in that order, since some of the fathoming techniques require certain calculations at the preprocessing stage. A numerical example follows a description of the algorithm itself.

Tasks must first be renumbered to facilitate selecting a 'favorable' task in the search tree. Johnson's procedure for renumbering tasks is summarized by four rules.

1. Renumber a task after its predecessors have been renumbered.
2. Break ties in favor of the task with the largest p_j.
3. Break second level ties in favor of the task with the largest $|F_j|$.
4. Break third level ties arbitrarily.

Task numbering in Example 5.1 conforms with this convention (see Fig. 5.2).

Next, certain task times may be increased to facilitate fathoming. We assume that $p_j \le C$ for $j \in J$, but if $p_j + \min_{k \ne j} p_k > C$, the time for task j should be increased to $p_j = C$ to strengthen the effects of fathoming rules (Johnson, 1988). In Example 5.1, no other task can be assigned along with task 10, since $p_{10} = 70$, $\min_{k \ne j} p_k = 5$, and $p_j + \min_{k \ne j} p_k = 70 + 5 = 75 > C = 72$. Thus, p_{10} should be reset to 72 before implementing the algorithm.

Johnson (1988) incorporates four node-fathoming methods. The following description uses the following terms: **candidate task**, a task whose predecessors have all been assigned to stations so that it is eligible for assignment; **trial station**, the one at which tasks are currently being assigned; and **complete a trial station** whereupon a new trial station is **opened**.

The first node-fathoming criterion is for 'Jackson rule-1 dominance' which requires that a trial station, s, not be completed if some candidate task fits. Jackson (1956) proves that a current assignment at station s cannot lead to a solution better than the one which includes a candidate task that fits. Johnson (1988) invokes the criteria to check if a trial station is completed, fathoming the associated node if some candidate task still fits.

The second node-fathoming criterion is for 'Jackson rule-2 dominance' (Jackson, 1956) which requires that a node representing a completed trial station be fathomed if some assigned task j (i.e. j with $S_j = s$) is dominated by another candidate task i. Task i dominates task j if it has a larger time and its set of **immediate** followers contains the set of **immediate** followers of j. More specifically:

$$i \text{ dominates } j \text{ if } p_i > p_j, \text{ and } F_i \supseteq F_j.$$

Johnson (1988) implements this criterion in two stages. The preprocessing stage checks all pairs of tasks to identify dominance relationships. Example 5.1 admits three relationships:

Task	i	dominates	Task	j	p_i	p_j	F_i	?	F_j
	2	dominates		3	35	25	F_3	$\supseteq$	F_2
	4	dominates		5	60	30	F_5	$=$	F_4
	7	dominates		9	60	35	F_9	$=$	F_7

Then, upon completing trial station s during the search procedure, Johnson checks each task j with $S_j = s$ to see if it is dominated by a candidate task i. If such a dominance relationship exists, the node representing the complete trial station is fathomed.

The third node-fathoming criterion is for 'first station dominance' which requires that a node representing a trial station be fathomed if it has the same set of tasks that were assigned to the first station at some other node. This criterion is based on the principle that if the tasks in the first completed assignment for station 1, or a subset of those tasks, are duplicated in the assignment for another station, the node representing the second assignment can be fathomed, since it is dominated by the first-station assignment (Johnson, 1988). Johnson implements this criterion upon completing each trial station by checking the first few first-station assignments for dominance. This type of dominance is most likely to occur in cases for which several tasks have no predecessors. Since Example 5.1 has a single task with no predecessors (i.e. task 1), it cannot be used to demonstrate this form of dominance.

The fourth node-fathoming criterion is for 'labeling dominance' which requires that a node representing a trial station, s, be fathomed if s is greater than or equal to the station number at which the same set of tasks were assigned at a previous node. The method assigns a label to each task so that the sum of labels for each possible subset of tasks is unique. Then, the minimum number of stations (including a fraction of a station) for each subset of tasks is stored in a one-dimensional array. If a new node has a number of stations that equals or exceeds that of a previous assignment of the same set of tasks, the new node can be fathomed. The preprocessing stage computes task labels using the Schrage and Baker (1978) labeling scheme, which assumes that task numbers are consistent with predecessor relationships. Using notation:

$\mathbf{L}(j)$ = label for task j

$\mathbf{p}(j)$ = sum of labels for lower numbered tasks that are predecessors of task j

$\ell(j)$ = sum of labels for tasks that are numbered lower than tasks j,

the scheme involves the following calculations:

> Initialization: Set $\mathbf{p}(j) = 0$ and $\ell(j) = 0$ for $j \in J$
>
> For $j = 1, \ldots, |J|$ Do
>
> $\qquad \mathbf{L}(j) = \ell(j) - \mathbf{p}(j) + 1$
>
> $\qquad \ell(j+1) = \ell(j) + \mathbf{L}(j)$
>
> $\qquad$ Add $\mathbf{L}(j)$ to $\mathbf{p}(k)$ for each successor, k, of task j
>
> $\qquad \left(\text{i.e. } \mathbf{p}(k) \leftarrow \mathbf{p}(k) + \mathbf{L}(j) \text{ for each } k \in F_j^* \right)$
>
> Next j

Table 5.9 Computation of task labels

j	$\ell(j)$	p(j)	L(j)	j + 1	$\ell(j+1)$	p(1)	p(2)	p(3)	p(4)	p(5)	p(6)	p(7)	p(8)	p(9)	p(10)	p(11)
0	0	0				0	0	0	0	0	0	0	0	0	0	0
1	1	1	1	2	1		1	1	1	1	1	1	1	1	1	1
2	2	1	1	3	2			1	2	2	2	2	2	2	2	2
3	4	2	2	4	4				2	2	4	4	2	2	4	4
4	7	2	3	5	7					2	4	4	5	5	7	7
5	13	4	6	6	13						4	4	11	11	13	13
6	23	14	10	7	23							14	11	11	23	23
7	33	11	10	8	33								11	11	33	33
8	56	34	23	9	56									34	56	56
9	79	79	23	10	79										79	79
10	80	80	1	11	80											80
11			1													

Labels of the tasks in Example 5.1 are shown in Table 5.9. For example, for task j = 6:

$$\mathbf{L}(6) = \ell(6) - \mathbf{p}(6) + 1 = 13 - 4 + 1 = 10$$

$$\ell(j+1) = \ell(7) = \ell(6) + \mathbf{L}(6) = 13 + 10 = 23$$

Add $\mathbf{L}(6) = 10$ to $\mathbf{p}(k)$ for k = 7, 10 and 11 (i.e. $k \in F_6^*$)

$$\mathbf{p}(k) \leftarrow \mathbf{p}(k) + \mathbf{L}(j) = \mathbf{p}(k) + 10, \text{ so } \mathbf{p}(7) = 14, \text{ and } \mathbf{p}(10) = \mathbf{p}(11) = 23$$

FABLE checks for label dominance when assigning a new task to a trial station and upon completing a trial station. This scheme has an inherent limitation, since the sum of labels may exceed the largest integer that a computer can address. Therefore, this check is made only for sets of tasks that have sums of labels that are less than that largest integer value; Johnson (1988) used 32 000. Thus, this check can be made only in initial portions of the search, but this helps to fathom nodes high in the tree, improving overall performance.

FABLE also computes four types of bounds during the preprocessing stage and uses them during the search to fathom nodes. The first three bounds deal with the total number of stations required.

Bound 1 is the traditional bound calculated by equation (5.2).

Bound 2 is a similar bound based on task durations relative to C/2 (Johnson, 1981). Bound 2 proves helpful if cycle time is small so that only three or fewer tasks can be assigned to a station. Strong bounds are very important in such cases. The underlying logic is that two tasks each with durations > C/2 cannot be assigned to the same station, but two tasks with durations = C/2 can be. First, let $K = \lceil |J|/B_1 \rceil$, then the bound can be calculated for $k = 2, \ldots, K$:

$$B_k = \left\lceil \sum_{j=1}^{k-1} \frac{\text{Number of tasks with } p_j > jC/k}{k-1} \right.$$
$$+ \sum_{j=1}^{k} j \frac{\text{Number of tasks with } p_j = jC/k}{k} \qquad (5.9)$$
$$\left. - \sum_{j=1}^{k-1} j \frac{\text{Number of tasks with } p_j = (j+1) C/k}{k-1} \right\rceil.$$

In Example 5.1, $K = \lceil |J|/B_1 \rceil = \lceil 11/6 \rceil = 2$ and:

$$B_2 = \left\lceil \sum_{j=1}^{1} \frac{\text{Number of tasks with } p_j > 36j}{1} \right.$$
$$+ \sum_{j=1}^{2} j \frac{\text{Number of tasks with } p_j = 36j}{2}$$
$$\left. - \sum_{j=1}^{1} j \frac{\text{Number of tasks with } p_j = 36(j+1)}{1} \right\rceil$$

so $B_2 = \lceil (3) + (2)(1) / (2) - (1)(1) / 2 \rceil = 3$ and, in this case, is not as tight as B_1.

Bound 3 is a similar bound based on task durations relative to $C/3$ (Johnson, 1981). The underlying logic is that no more than two tasks with durations $> C/3$ can fit on one station and none can be assigned to the same station as a task with duration $>2C/3$. This bound results when $k = 3$ in equation (5.9).

Bound 4 deals with the number of stations that must follow task j. Let

R_j = bound on the number of stations that must follow task j.

The preprocessing stage can calculate R_j for each task j in order from $|J|$ to 1 using:

$$R_j = \sum_{j \in P_j^*} (p_j/C)$$

in which task times are in terms of number of stations. In Example 5.1 we obtain $R_{11} = 0$, $R_{10} = 30/72$, $R_9 = 2$, $R_8 = 2\ 35/72$, $R_7 = 2$, $R_6 = 2\ 60/72$, $R_5 = 2\ 60/72$, $R_4 = 2\ 60/72$, $R_3 = 3$, $R_2 = 5\ 4/72$, $R_1 = 5\ 64/72$. We see that $R_9 = R_7 = 2$ because task 10 forces the assignment of task 11 to a station of its own. Johnson (1988) invokes this bound just before assigning a task to a station. Let:

B_4 = bound on the total number of stations required, then
B_4 = current number of stations + $p_j/C + R_j$

and the node can be fathomed if $B_4 \geq$ incumbent solution value.

Any solution which becomes the new current incumbent during the search can be deemed optimal if the number of stations it uses is equal to B_1, B_2 or B_3.

The FABLE branch-and-bound algorithm (Johnson, 1988) consists of seven steps which are described in Table 5.10 using the additional notation:

β = backtrack task

η = number of tasks currently assigned

κ = newly selected task

$T(\kappa)$ = array giving the κth task assigned.

In summary, the FABLE algorithm (Johnson, 1988) consists of three main stages.

1. Read data.
2. Preprocessing stage:
    (a) increment task durations;
    (b) renumber tasks;
    (c) check all pairs of tasks for Jackson rule-2 dominance;
    (d) determine R_j, the bound on the number of stations that must follow task j;
    (e) compute Schrage and Baker labels for all tasks.
3. Invoke the tree enumeration procedure.

Table 5.10   The FABLE branch-and-bound algorithm

1. Initialization
    Input C, J, p_j and P_j for j ∈ J
    Set   current station: s = 1
          unused cycle time at station s: τ = C
          newly selected task: κ = 1
          number of tasks currently assigned: η = 0
          S_j = 0 for j ∈ J
          T(j) = 0 for j ∈ J

2. Assign newly selected task κ to the current station s
    Record the assignment of task κ to station s: $S_κ$ = s
    Reduce the amount of cycle time remaining at station s: τ ← τ − $p_κ$
    Increment the number of tasks currently assigned: η ← η + 1
    Record this, then ηth task assignment: T(η) = κ
    Set backtrack task: β = 0
    If η = |J|, GOTO Step 6.

3. Select j as the next task to assign to station s if
    j is the lowest numbered task satisfying
          j has not already been assigned (i.e. S_j = 0)
          j > κ
          j > β
          j can fit within remaining cycle time on station s: p_j ≤ τ
          all predecessors of j [i ∈ P_j] have been assigned
          j is not excluded by label dominance or Bound 4
    IF task j satisfies all of these requirements: set κ = j and GOTO Step 2
    ELSE no task satisfies all of these requirements
          IF κ > 0,
              Invoke all fathoming rules
              IF the node can be fathomed, GOTO Step 7; ELSE, GOTO Step 4
          If β > 0, GOTO Step 7

4. Open a new station where additional tasks may be assigned
    Increment the current station: s ← s + 1
    Initialize the unused cycle time for this current station: τ = C

5. Select j as the next task to assign to station s if
    j is the lowest numbered task satisfying
          j has not already been assigned (i.e. S_j = 0)
          all predecessors [i ∈ P_j] of j have been assigned
          j > β
          j is not excluded by label dominance or Bound 4
    IF task j satisfies all of these requirements: set κ = j and GOTO Step 2
    ELSE no task satisfies these requirements
          IF η > 0, GOTO Step 7
          IF η = 0, enumeration is complete: **STOP**

Table 5.10 *Continued*

6. A new solution has been attained
    Save this solution as the (new) current incumbent solution
            IF it is the first solution attained or
            IF it is better than the previous incumbent solution
        IF the lower bound on the optimal number of stations = incumbent solution,
            STOP (the incumbent solution is an optimal solution).

7. Backtrack
    IF $\tau = C$ (i.e. station s has no assigned tasks),

    $$\text{decrement } s \leftarrow s-1, \text{ and set } \tau = C - \sum_{j \in J_s} p_j$$

    Eliminate the last task assignment
        Identify the last task assigned: $\beta = T(\eta)$
        Remove task β from station s: $S_\beta = 0$
        Update unused cycle time at station s: $\tau \leftarrow \tau + p_\beta$
        Update the last task assignment: $T(\eta) = 0$
        Decrement number of tasks currently assigned: $\eta \leftarrow \eta - 1$
        Set newly selected task: $\kappa = 0$
    If $\tau < C$, GOTO Step 3
    If $\tau = C$, GOTO Step 5.

Source: Johnson, R.V., *Management Science*, 1988.

According to the laser search strategy, FABLE constructs the search tree one branch at a time with no parallel branches. We now demonstrate FABLE using the data of Example 5.1. In Fig. 5.4 nodes are numbered in the order in which they are searched.

Processing at each step proceeds as shown in Table 5.11. If the solution noted at Step 6 agreed with the lower bound $B_1 = 6$, the algorithm could stop, knowing that this is an optimal solution. However, further enumeration is required in this case, either to prove that a seven-station line is optimal or to find a six-station design. Backtracking steps of the algorithm are demonstrated in the following steps. After backtracking, the algorithm can continue its search, considering lower levels in the search tree. It is left to the reader to determine the optimal solution for this problem.

The complete search tree is depicted in Fig. 5.4. The symbol $S_j = s$ indicates that task j is assigned to station s at that node. There are a total of 12 alternative optimal solutions with seven stations. Each of the four nodes that can be fathomed according to the Jackson rule-2 dominance conditions is noted with the symbol '----i/j', indicating that task i dominates task j. Nodes are numbered from 1 to 44 in the order in which they would be searched by the algorithm. The fathoming rules facilitate the search by pruning much of the tree: only 44 of the 85 nodes must be searched. Note that the laser search always progresses by penetrating as deeply as possible and never creates active branches in parallel.

Table 5.11 An example demonstrating the FABLE algorithm

---

1. Input C, J, p_j and P_j for $j \in j$. $s = 1$, $\tau = C$, $\kappa = 1$, $\eta = 0$, $S_j = T(j) = 0$ for $j \in J$.
2. $S_1 = 1$, $\tau = 67$, $\eta = 1$, $T(1) = 1$, $\beta = 0$.

3. $\kappa = j = 2$.   Candidates are tasks 2 and 3.
2. $S_2 = 1$, $\tau = 32$, $\eta = 2$, $T(2) = 2$, $\beta = 0$.

3. $\kappa = j = 3$.   Candidates are tasks 3, 4 and 5.
2. $S_3 = 1$, $\tau = 7$, $\eta = 3$, $T(3) = 3$, $\beta = 0$.

3. GOTO Step 4.   Candidates are tasks 4, 5 and 6; none fit.
4. $s = 2$, $\tau = 72$.
5. $\kappa = j = 4$.
2. $S_4 = 2$, $\tau = 12$, $\eta = 4$, $T(4) = 4$, $\beta = 0$.

3. $\kappa = j = 6$.   Candidates are tasks 5 and 6.
2. $S_6 = 2$, $\tau = 2$, $\eta = 5$, $T(5) = 6$, $\beta = 0$.

3. GOTO Step 4.   Candidates are tasks 5 and 7; neither fits.
4. $s = 3$, $\tau = 72$.
5. $\kappa = j = 5$.   Candidates are tasks 5 and 7.
2. $S_5 = 3$, $\tau = 42$, $\eta = 6$, $T(6) = 5$, $\beta = 0$.

3. $\kappa = j = 8$.   Candidates are tasks 7 and 8.
2. $S_8 = 3$, $\tau = 17$, $\eta = 7$, $T(7) = 8$, $\beta = 0$.

3. GOTO Step 4.   Candidates are tasks 7 and 9; neither fits.
4. $s = 4$, $\tau = 72$.
5. $\kappa = j = 7$.   Candidates are tasks 7 and 9.
2. $S_7 = 4$, $\tau = 12$, $\eta = 8$, $T(8) = 7$, $\beta = 0$.

3. GOTO Step 4.   Candidate is task 9; it does not fit.
4. $s = 5$, $\tau = 72$.
5. $\kappa = j = 9$.   Candidate is task 9.
2. $S_9 = 5$, $\tau = 37$, $\eta = 9$, $T(9) = 9$, $\beta = 0$.

3. GOTO Step 4.   Candidate is task 10; it does not fit.
4. $s = 6$, $\tau = 72$.
5. $\kappa = j = 10$.   Candidate is task 10.
2. $S_{10} = 6$, $\tau = 0$, $\eta = 10$, $T(10) = 10$, $\beta = 0$.

3. GOTO Step 4.   Candidate is task 11; it does not fit.
4. $s = 7$, $\tau = 72$.
5. $\kappa = j = 11$.   Candidate is task 11.
2. $S_{11} = 7$, $\tau = 42$, $\eta = 11$, $T(11) = 11$, $\beta = 0$, $\eta = |J| = 11$; GOTO Step 6.

6. **Save solution:** $S_1 = S_2 = S_3 = 1$, $S_4 = S_6 = 2$, $S_5 = S_8 = 3$, $S_7 = 4$, $S_9 = 5$, $S_{10} = 6$, $S_{11} = 7$.

7. $\beta = T(11) = 11$, $S_{11} = 0$, $\tau = 72$, $T(11) = 0$, $\eta = 10$, $\kappa = 0$.

5. GOTO Step 7.   Candidate is task 11.
7. $s = 6$, $\beta = T(10) = 10$, $S_{10} = 0$, $\tau = 72$, $T(10) = 0$, $\eta = 9$, $\kappa = 0$.

5. GOTO Step 7.   Candidate is task 10.

Table 5.11 *Continued*

7. $s = 5$, $\beta = T(9) = 9$, $S_9 = 0$, $\tau = 72$, $T(9) = 0$, $\eta = 8$, $\kappa = 0$.

5. GOTO Step 7.   Candidate is task 9.

7. $s = 4$, $\beta = T(8) = 7$, $S_7 = 0$, $\tau = 72$, $T(8) = 0$, $\eta = 7$, $\kappa = 0$.

5. GOTO Step 7.   Candidates are tasks 9 and 7.

7. $s = 3$, $\beta = T(7) = 8$, $S_8 = 0$, $\tau = 42$, $T(7) = 0$, $\eta = 6$, $\kappa = 0$.

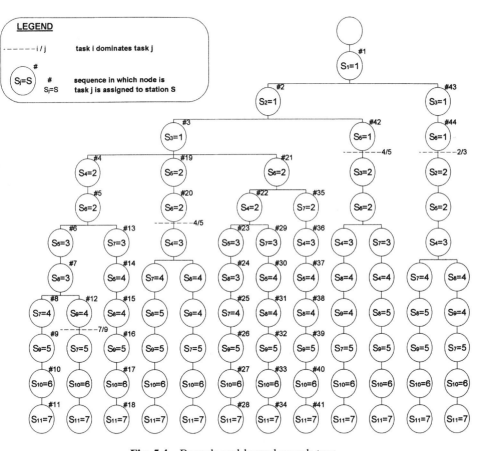

Fig. 5.4 Branch-and-bound search tree.

It should be noted that this example can be resolved more readily by starting with task 11 and proceeding in the backward direction. Because task 10 'fills' a station, it must be assigned to a station by itself. Task 11 is 'trapped' and must also be assigned to a station by itself. Task 7 can only be assigned along with task 6. Tasks 8 and 9 must be assigned together at a station. Task 4 must then be assigned to a station by itself (it could be combined with task 6, but this would require that task 7 be assigned by itself and would gain nothing). Tasks 1, 2, 3 and 5 remain with a total processing time of 95, requiring two additional stations. This analysis shows that no design with less than seven stations is feasible. Thus, the seven-station designs proposed by the heuristics and this algorithm all describe optimal solutions.

Hackman, Magazine and Wee (1989) propose another effective branch-and-bound method that relies upon their Immediate Update First-Fit heuristic (IUFF). Another generic method, the IUFF heuristic involves three steps.

1.   Assign a numerical score to each task j.
2.   Update the list of candidate tasks. If the list is empty, **STOP**.
3.   Assign the candidate task with the largest numerical score on the first station on which both cycle time and precedence constraints are satisfied; GOTO Step 2.

A number of alternative score functions can be used, as indicated in Table 5.12.

In addition, Talbot, Patterson and Gehrlein (1986) give other single pass criteria that might be used. The IUFF heuristic prescribes solutions with two important properties:

1.   the work content of each consecutive pair of stations exceeds C;
2.   the work content of stations $1, \ldots, S - 1$ exceeds $C - \max_j p_j$.

The branch-and-bound algorithm employs the IUFF heuristic at each node in the search tree to determine upper and lower bounds on the optimal number of stations. Descendants of a node in the search tree are determined by enumerating all maximal assignments of tasks to the next station. Given node n in the search tree and heuristic (i.e. score function) h, let:

$$\Gamma(n) = \text{tasks assigned at node } n$$

$$d(n) = \text{number of stations used for tasks in set } \Gamma(n)$$

$$D(n) = \text{total idle time at node } n = d(n)C - \left\lceil \frac{\sum_{j \in \Gamma(n)} p_j}{C} \right\rceil$$

$$S(n) = \text{number of stations prescribed by heuristic } h \text{ for tasks not assigned}$$
$$\text{at node } n,$$

then:

$$U_h(n) = d(n) + S_h(n) = \text{upper bound on the optimal number of stations}$$

$$L_h(n) = d(n) + \left\lceil \sum_{j \in \Gamma(n)} P_j / C \right\rceil = \text{lower bound on the optimal number of stations}$$

Table 5.12 Numerical score functions

Positional weight Helgeson and Birnie (1961)	Sum of times for task j and $k \in F_j^*$
Reverse positional weight	Sum of times for task j and $i \in P_j^*$
Number of followers Talbot and Patterson (1984)	Number of tasks that follow task j, $|F_j^*|$
Number of immediate followers Tonge (1961)	Number of immediate followers of task j, $|F_j|$
Number of predecessors Hackman, Magazine and Wee (1989)	Number of predecessors of task j, $|P_j^*|$
Task time Hackman, Magazine and Wee (1989)	p_j Effectiveness demonstrated by computations
Backward recursive positional weight Hackman, Magazine and Wee (1989)	Sum of times for task j and all tasks in paths with j as their root
Backward recursive edges Hackman, Magazine and Wee (1989)	Number of tasks in all paths with j as their root

The branch-and-bound algorithm then uses the stopping rule:

Stop at a complete solution with $U_h(n) = B_1$

and two dominance tests to fathom nodes in the search tree:

Test 1: fathom node n if there is another node n' such that

$$U_h(n') = L_h(n') \leq L_h(n).$$

Test 2: fathom node n if there is another node n' such that

$$\Gamma(n) \subset \Gamma(n') \text{ and } d(n) \geq d(n').$$

For Test 1, it is sufficient to check if $D(n) \geq D(n')$ (Talbot, Patterson and Gehrlein, 1986, page 919); and for Test 2, if (a) $d(n) = d(n')$ and (b) n and n' have different predecessors (ibid., page 919).

Heuristics are also used to fathom nodes with the purpose of reducing runtime. Since each node may have a large number of descendants, several heuristics are used to select nodes that will be considered further (others are fathomed).

1.   Select a predetermined number of maximal assignments.
2.   Upon branching, select <u>m</u> nodes with the least accumulated idle time.
3.   At any tree level, select <u>m</u> nodes with least accumulated idle time.
4.   If <u>m</u> nodes are branched at a level, fathom all unbranched nodes higher in the tree.

Table 5.13   The EUREKA algorithm

---

1. Determine the theoretical minimum number of stations, B_1, and set $S = B_1$
    Compute the theoretical minimum total slack $= B_1C - \sum_j p_j$.

2. Set s = 1.

3. Generate a feasible task assignment that has not been considered previously for the current trial station s.

4. IF there are no more feasible assignments for trial station s and s = S
       Increase target slack by permitting one more station: $S \leftarrow S + 1$
       Increase the theoretical minimal total slack by C
       Begin anew to determine if S stations are feasible: GOTO Step 2.

5. IF there are no more feasible assignments for trial station s and s < S
       Backtrack by decrementing the trial station number: $s \leftarrow s - 1$, then GOTO Step 3.

6. IF Step 3 generated a feasible assignment for station s, compute the cumulative slack time on all completed stations.

7. IF this cumulative slack exceeds the theoretical minimum total slack time fathom the node, and GOTO Step 3.

8. IF this is trial station S a complete feasible solution with slack less than or equal to the theoretical minimum total slack has been found; **STOP**.

9. ELSE, increment the trial station number $s \leftarrow s + 1$, and GOTO Step 3.

In addition, if runtime exceeds a predetermined bound, the smallest known upper bound is used as the solution.

We now describe a third optimizing method, an implicit enumeration algorithm called EUREKA (Hoffman, 1992). This method employs a depth-first laser search called ICE, and is summarized by the nine steps itemized in Table 5.13.

Clearly, this method is an extension of the enumeration procedure (Step 3) used in the Hoffman (1963) heuristic so it is able to exploit the numerical characteristics of an instance. ICE is embedded in the following overall approach.

1.   Apply ICE, solving the problem in the forward direction.
2.   IF, after $\underline{M}$ minutes, an optimal solution has not been found, apply ICE, solving the problem in the backward direction.
3.   IF, after $\underline{M}$ minutes, an optimal solution has not been found, apply the Hoffman heuristic.

The parameter $\underline{M}$ is selected with the goal of finding an optimal solution quickly, either in the forward or backward direction. Step 2 starts with the best lower bound on the number of stations determined by the forward search and is based on computational experience which has demonstrated that some instances are more easily solved in the backward direction than in the forward direction. If the instance cannot be solved quickly in either direction, Step 3 resorts to the Hoffman (1963) heuristic, which can use any bounds determined by ICE to identify an optimal solution.

These three optimizing methods are based on quite different principles. Two of them return heuristic solutions if a predetermined runtime is exceeded. The computational efficacy of both heuristic and optimizing methods is discussed in the next section.

5.2.4   Computational evaluation

While the line balancing problem is known to be NP-hard, computational results have led to further insights, suggesting that actual computational results may be sensitive to the total number of tasks, the average number of tasks assigned to a station, and the order strength, the total number of predecessors for all tasks (including redundant precedence relationships) divided by the maximum total number of predecessor relationships, $|J|$ ($|J| - 1)/2$. The optimal solution to many test problems is either B_1 or $B_1 + 1$, so procedures frequently stop by finding a solution equal to the bound without exploring the vast number of combinations that are possible. Since there may be a large number of alternative optimal solutions, it is most helpful if a procedure can stop when it finds that any one of them is optimal (i.e. by comparison with a lower bound). Talbot and Patterson (1984) point out that problems which require more that B_1 stations are more challenging to solve. Because of the form of the problem itself, numerous iterations may be required to find a solution with a different number of stations. Hoffman (1990) indicates that problems are more challenging when they have theoretical minimum total idleness close to zero or, equivalently, when $\sum_j p_j/C$ is approximately an integer. In addition, the value of task times relative to the cycle time C may affect the degree of difficulty of an instance. In a wide range of test problems, Hoffman (1990) found that neither cycle time, precedence relationships, or task times by themselves could determine how challenging a problem might be. Rather, the combination of these factors and the solution method determines runtime.

A library of 'standard' line balancing test problems has been accumulated over the years. However, many originated when computers and algorithms were less capable than they are today and no longer pose computational challenges. Hoffman (1990) recently created a number of more challenging problems that can be used in more discerning tests.

Comparative studies of line balancing heuristics have been undertaken periodically (e.g. Mastor (1970), Johnson (1981), Hoffman (1990)). For example, Talbot, Patterson and Gehrlein (1986) tested 26 heuristics, including 13 single-pass rules such as those in Table 5.12, concluding that the structure of the precedence network does not appear to affect the performance of heuristics and that the modified Hoffman heuristic was, perhaps, the best of the heuristics they tested. Through an extensive set of tests, Voß and Scholl (1994) showed that their tabu search procedure outperformed the alternative heuristics they tested.

Computational tests (Johnson, 1988) demonstrate that FABLE is able to solve, in reasonable time, problems with up to 1000 tasks of low order strength and with an average of six or more tasks per station. In instances with an average of two tasks per station, FABLE performed better on those problems with a high order strength (i.e. for which fewer sequences are feasible). In instances with an average of four or more tasks per station, FABLE performed better when order strength was low. FABLE does best on instances for which the optimum number of stations is close to B_1. This often occurs in cases for which a number of stations have zero idle time, a situation which occurs more often with low order strength, since there are more combinations of task assignments that will achieve low idle time. These conditions most often allow FABLE to identify an optimal solution by a bound argument rather than taking the time to enumerate the tree.

FABLE is least effective in instances that incur large amounts of idle time. In these cases, the optimal solution tends to have more stations than B_1 and enumeration must be relied upon to identify an optimal solution.

In testing the EUREKA algorithm, Hoffman (1992) concluded that problems are more difficult if the optimum number of stations exceeds B_1. This is likely to occur when $\sum_j p_j/C$ is approximately an integer so that the theoretical minimum total slack will be close to zero. In addition, problems with $\max_j p_j = C$ tend to be harder to solve.

Runtimes reported for both heuristic and optimizing methods are often inconsequential, perhaps because the library contains mostly small test problems that can be solved easily or because solution methods are well matched to exploit the instances on which they are tested. Optimizing methods may require prohibitive runtimes on challenging problems, especially 'worst cases'.

Line balancing procedures tend to result in low idle times at upstream stations and higher idle times at downstream stations, so they do not necessarily smooth workloads effectively. These and other practical considerations are addressed in the next subsection.

5.2.5 Practical features

While it may be possible to solve many instances of the traditional ALB-1 and ALB-2 problems effectively, actual applications may involve extended features representing practical needs. For example, if a long task is involved (i.e. with $p_j > C$), parallel machines must be provided at a station (e.g. Pinto, Dannenbring and Khumawala (1981), Sarker and Shanthikumar (1983) and Bard (1989)) and if processing alternatives exist, machines must be selected appropriately (e.g. Pinto, Dannenbring and Khumawala (1983)). Dealing with an extensive set of features, Johnson (1983) permits planned imbalances, allows tasks to be assigned to specified types of stations, treats stochastic processing times, deals with required task groupings according to assembly skills (positive zoning), limits task assignment to the right (or left) side of the line, imposes task separations (negative zoning) and addresses mixed model considerations.

This subsection describes two types of feature that make important extensions to the basic ALB problem. The first involves smoothing workloads on stations, and the second addresses cases in which operation times are stochastic rather than deterministic.

Workload smoothing That ALB-1 be called the assembly line 'balancing' problem is somewhat of a misnomer, since it admits solutions that assign unequal workloads to stations, resulting in unequal work assignments to operators. A measure of line efficiency is given by the smoothness index (Elsayed and Boucher, 1994):

$$\text{Smoothness index} = \left[\sum_{s=1}^{S} \left(\min_s \sum_{j \in J_s} p_j - \sum_{j \in J_s} p_j \right)^2 \right]^{1/2}$$

in which J_s is the set of tasks assigned to station s. A large value of the smoothness index indicates a high degree of variability in the workloads assigned to various stations and should be avoided. A low value of the smoothness index indicates a uniform assignment of workloads to stations, the goal of workload smoothing.

A number of methods can be used to smooth workloads, including solving ALB-2 with the optimal number of stations or transferring workers among stations. Deckro (1989) proposes a mixed 0/1 integer model with assignment, precedence, cycle time and (compatible and incompatible) zoning constraints along with others to assure that the minimum number of stations is used. He proposes an objective which is a weighted combination of line length and cycle time, simultaneously designing and smoothing the line.

Rachamadugu and Talbot (1991) propose a heuristic that attempts to smooth the workloads assigned to all stations, compensating for ALB-1 solutions that assign heavier workloads to stations upstream and larger idle times to stations downstream. Let:

W_s = workload assigned to station s = $\sum_{j \in J_s} p_j$

$\underline{W}$ = average workload per station = $(1/S) \sum_{j \in J} p_j$

MAD = mean absolute deviation = $(1/S) \sum_s |W_s - \underline{W}|$

E_j^f = earliest station to which one (or more) of $k \in F_j$ has (have) been assigned

L_j^f = latest station to which one (or more) of $k \in F_j$ has (have) been assigned.

Table 5.14 The Rachamadugu and Talbot heuristic

---

1. Set s = S.

2. IF s = 0, **STOP**. ELSE, IF $W_s > \underline{W}$, GOTO Step 5.

3. Find the task j which has the largest operation time among tasks that satisfy
   $W_{S_j} \geq \underline{W}$ and $W_{S_j} - t_j \geq W_s$
   and either $E_j^f \geq s$ if $s > S_j$ or $L_j^f \leq s$ if $s < S_j$.
   IF such a job j exists
   Transfer it from station S_j to station s; update S, E, L, and W; and GOTO Step 5.

4. ELSE find the task j, which has the largest operation time among tasks that satisfy
   $W_{S_j} - t_j < W_s$ and $W_{S_j} \geq \underline{W}$.
   IF such a task exists
   Interchange tasks between station S_j and station s invoking principle (5), subject to precedence relationships. Update S, E, L, and W.

5. Decrement $s \leftarrow s - 1$, then GOTO Step 2.

They establish five principles by which workload smoothing can be improved.

1. Transfer task j from station S_j to station s $(s > S_j)$ if $E_j^f \geq s$ and $W_s + p_j < C$.
2. Transfer task j from station S_j to station s if $L_j^f \leq s$ and $W_s + p_j < C$.
3. If the workloads assigned to two stations are both above (below) $\underline{W}$, reassigning tasks among these stations will not improve *MAD*.
4. *MAD* can be improved by switching task j from station S_j to station s if:

$$W_{S_j} - p_j > W_s, \qquad W_{S_j} > \underline{W} \qquad \text{and } W_s < \underline{W}.$$

5. When task j is transferred from station S_j to station s, there may be a set of tasks, J_s, on station s that can be transferred to station S_j. Such an interchange improves *MAD* if:

$$W_{S_j} > \underline{W} > W_s \qquad \text{and } W_s - W_{S_j} + p_j < \sum_{i \in J_s} p_i < p_j.$$

Their heuristic employs the five principles in the procedure itemized in Table 5.14.

   Their tests demonstrate that this heuristic improves on the modified Hoffman heuristic of Gehrlein and Patterson (1975, 1978). While the parameter θ may tend to balance workloads, it may also require more than an optimal number of stations, forcing excessive idleness.

   Pinnoi and Wilhelm (1994c) recently devised an optimizing approach based on strong cutting plane methods. Given the number of stations, S, they address the Workload Smoothing Problem (WSP), requiring tasks to be assigned to stations in order to minimize the maximum idle time incurred by any station. Using continuous decision variable:

$$z_{max} = \text{the maximum idle time at a station,}$$

the WSP may be formulated as the following mixed 0/1 integer program:

Problem WSP: Minimize z_{max}
              Subject to: (5.3) – (5.6), and

$$\sum_j p_j x_{sj} + z_{max} \geq C \quad s = 1, \ldots, S \qquad (5.10)$$

$$z_{max} \geq \left\lceil \frac{CS - \sum_{j \in J} p_j}{C} \right\rceil \qquad (5.11)$$

Constraint (5.10) guarantees that z_{max} will take on the value of the maximum idle time at any station. Since C and p_j are integers, z_{max} automatically attains an integer value in each feasible solution. Inequality (5.11) gives a lower bound for z_{max} and facilitates solution.

Pinnoi and Wilhelm (1994c) present a branch-and-cut approach for resolving the WSP, including specialized preprocessing procedures, families of valid inequalities and facets (see Pinnoi and Wilhelm (1994a, 1994e) for underlying proofs), polynomial time separation heuristics and computational evaluation. They base test problems on the set of Sawyer 'standard' 30-task problems and initiate each with the minimum number of stations as given by Hoffman (1990). To evaluate their approach, they compare runtimes with those required by a standard implementation of the Optimization Subroutine Library (OSL), which incorporates supernode processing. Two runs were made for each problem: in the first, cuts were added at the root node before invoking branch-and-cut; and, in the second, cuts were not added at the root node. The better of the two results was compared with the OSL solution.

On average, the branch-and-cut approach achieved an order of magnitude improvement in runtime, primarily since it enumerated significantly fewer nodes. However, runtime was very sensitive to cycle time; a slight change in C can make an easy problem very challenging. Most studies employing a strong cutting plane approach use gap reduction as a measure of efficacy, assuming that a large gap reduction indicates the strength of the cuts. The gap is defined as the difference between the optimal values of the LP relaxation and the integer solution. However, the WSP typically poses a small gap and improving values of (most) variables toward integrality often did not contribute directly to improving the objective value or to reducing the gap. Thus, gap reduction is not a good indicator of efficacy in this case.

Stochastic line balancing   Most prescriptive models for line balancing assume that the assembly system operates deterministically. However, the actual system may be stochastic. For example, human operators may introduce variability in repeating the same set of tasks on different subassemblies, automated assembly machines may be subject to random jams and robots may require a random number of attempts to place parts.

While a number of studies have addressed stochastic line balancing, we focus discussion on a fundamental approach that employs chance constraints, requiring:

Pr [workload assigned to station $s \leq C] \geq \alpha$,

which may be expressed as:

$$Pr\left[\sum_{j \in J_s} p_j \leq C\right] \geq \alpha. \tag{5.12}$$

For cases in which task times are independent normally distributed random variables (e.g. see Wilhelm (1987b) and pages 374–6), this criterion may be expressed as:

$$\mu + \sigma Z_\alpha \leq C \tag{5.13}$$

in which:

$E[p_j]$ = expected value of task time p_j

μ = $E[\text{workload assigned to station } s] = \sum_{j \in J_s} E[p_j]$

$V[p_j]$ = variance of task time p_j

σ = standard deviation of workload assigned to station $s = \left(\sum_{j \in J_s} V[p_j]\right)^{1/2}$

Z_α = value of the standard normal evaluated at probability α

The requirement of equation (5.12) – in particular of equation (5.13) if task times are independent and normally distributed – can be invoked by checking appropriately as tasks are assigned to stations.

5.2.6 *Mixed-model line balancing*

Many assembly systems produce a variety of products, so multi- and mixed-model line balancing procedures are important to industry. Lots of different products are assembled sequentially on a multi-model line and products are assembled in any order on a mixed-model line. With a lot size of one unit, these two types of line are identical; in general, lot sizing and model sequencing is a primary factor in determining efficiency, especially for mixed-model lines.

Additional notation is needed to formulate the mixed-model line balancing problem:

m = index for models $m \in M$

p_{jm} = time for task j of model m

T = duration of the period (e.g. a shift)
U_m = number of units of model m to be produced during the period
W_m = workload associated with model m during the period
M = set of models
M_j = set of models which require task j

and A_m, F_{jm}, F^*_{jm}, J_m, P_{jm}, P^*_{jm}, v_{jm}, V_m are defined as before but relative to model $m \in M$. The precedence graph $G(N, A)$ for the mixed-model problem may be defined (as described by Thomopoulus (1970) and formalized by Macaskill (1972)) as the union of the precedence graphs of the individual models:

$$N = \bigcup_{m \in M} N_m \text{ and } A = \bigcup_{m \in M} A_m$$

Any arc $a_{ij} \in A$ is redundant and can be eliminated if there is some additional chain of arcs connecting node n_i and node n_j.

The mixed-model problem introduces some new issues. For example, should the same number of stations be used for each model? In the assignment of task j for models $m \in M_j$, each of these related tasks should, in most applications, be assigned to the same station or, at least, to nearby stations (Thomopoulus, 1967, 1970). To assure that tasks j for all $m \in M_j$ are assigned to the same station, define an aggregated task with processing time:

$$\overline{P}_j = \sum_{m \in M} p_{jm} U_m$$

Other sets of tasks might be aggregated in the same manner, taking care not to violate precedence restrictions. Balance efficiency must now be defined using the total workload associated with model m. Defining the workload associated with model m as:

$$W_m = U_m \sum_{j \in J_m} p_{jm}$$

the total workload over planning period T, W_T, may be expressed as:

$$W_T = \sum_m W_m$$

so that total operator idle time is $T - W_T$, and the balance efficiency E is:

$$E = M_T / T.$$

The optimal mixed-model solution assigns tasks observing precedence relationships and assuring that a given task is assigned to one station for all $m \in M_j$ in such a way that all stations can complete their assigned workloads during time T. Line balancing methods (e.g. Labach (1990)) can be applied, using the aggregated precedence graph and aggregated task processing times, replacing the cycle time constraint (5.6) with its equivalent:

$$\sum_{j \in J} \overline{P}_j x_{sj} \leq T \qquad \text{for } s = 1, \ldots, S.$$

Even though this solution balances workloads over all models, workloads for particular models may be far from balanced. Therefore, model sequencing, a topic addressed in section 7.4, is important to assure a smooth flow. If the model mix changes, the problem must be resolved, since the mix dramatically influences the relative workload of each model.

Various practical features have been considered, including minimizing line length by balancing and sequencing models (Dar-El and Cother, 1975; Dar-El and Curry, 1977), permitting parallel work stations (Kumar, Gurnani and Kekre, 1990), and smoothing workloads associated with each model (Thomopoulos, 1970). Somewhat broader views of assembly systems have led to models which include lot sizing decisions to minimize the costs of setup, inventory and station idleness (Chakravarty and Stubb, 1985) and to the selection of a robot for each station (Chakravarty and Stubb, 1986). Prior work has dealt with selecting from assembly alternatives (Roberts and Villa, 1970) and productivity-improving, after-design decisions such as lot sizing (Afentakis, Gavish and Karmarker, 1984) and allocating models to lines (Lehman, 1969). A few studies have addressed stochastic, mixed-model line balancing issues (Arcus, 1966; Vrat and Varani, 1976; Johnson, 1983).

5.3   Assembly system design

While the assembly line was one of the most productive configurations known to history, by the 1970s difficulties arose due to psychological, social and economic issues. Workers became dissatisfied because of the monotony of the work and because they were not intellectually challenged. As a consequence, job enrichment, work structuring and other approaches were developed to humanize the workplace.

Since 1980, the capabilities of robots and other automated equipment and their controls have evolved to the point at which many intricate assembly operations can be automated. These systems give rise to new types of quantitative design problems in which equipment and tooling must be selected for each station. The objective in such problems is to minimize total cost, consisting of investment and operating costs. Pinnoi and Wilhelm (1994d) have recently devised a family of hierarchical models to represent various types of assembly system design problems. This section describes solution approaches for certain types of design problems, including enumeration methods and branch-and-cut.

5.3.1   Enumeration methods

Initial studies focused on equipment selection for single-product assembly systems. Pinto, Dannenbring and Khumawala (1983) allowed limited selection of equipment at stations. Graves and Whitney (1979) devised what is perhaps the first optimizing approach; however, it does not prescribe tool changeovers, an important aspect of robotic assembly. Subsequently, Graves and Lamar

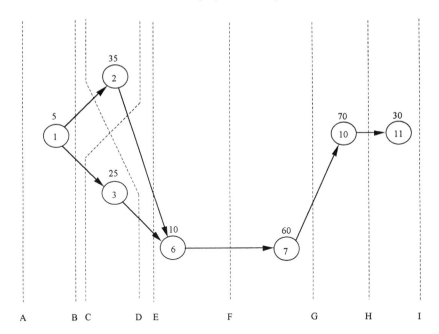

Fig. 5.5 Candidate workstation generation by cut sets in the precedence graph.

(1983) addressed tool changeover using a heuristic that was demonstrated to be effective in their tests. Gustavson (1986) also developed an effective heuristic for both single- and mixed-model cases.

Graves and Redfield (1988) devised a heuristic that addresses the mixed-model case and explicitly prescribes tooling and tool changeover. They use the combined precedence graph $G(N, A)$ for all models and define the target cycle time for each model. Their method enumerates all candidate workstations, defined as an assignment of tasks that observes precedence relationships; a selection of equipment and tooling; and a cycle time composed of load, unload, assembly and tool changeover times. Each candidate workstation must be able to achieve target cycle times for all assigned models.

Candidate workstations are generated by considering cut sets in the precedence graph. Graves and Redfield (1988) define a cut set as 'a group of tasks such that, for any task in the set, all the predecessors of that task are also in the set'. For example, cut sets associated with the precedence graph of Fig. 5.2 are shown in Fig. 5.5 – here, tasks 4, 5, 8 and 9 have been omitted to depict the fundamental concepts without excessive detail. Nine cut sets, labeled *A–I*, are associated with these seven tasks; had all 11 tasks of Fig. 5.2 been used, there would have been 31 cut sets to deal with . Candidate stations are generated by considering 'all pairs of cut sets in which one cut set is a subset of the other'. For instance, $D = (1, 3)$; $F = (1, 2, 3, 6)$; $D \subset F$; and $F–D = (2, 6)$, an assignment which has a total task time of 45 and is feasible with respect to cycle time ($C = 72$). In an actual design context, each subset of tasks which is feasible relative to precedence relationships must be tested to determine the best combination of equipment and tooling. The minimum cost alternative that meets the target cycle times is selected as a candidate workstation.

After enumerating all possible candidate workstations, a shortest path problem is solved to determine the best combination of stations in the system. A network is constructed with nodes representing the cut sets and arcs connecting pairs of cut sets that define candidate stations as shown in Fig. 5.6. Each arc is labelled with the cost of the minimum cost design alternative for the candidate station. The shortest path in this network traverses the set of candidate workstations that comprise the minimum cost system design. For example, consider path A–E–G–H–I in Fig. 5.6, which defines a four-station design:

$$E–A = (1, 2, 3) \qquad \text{with total task time} = 65$$
$$G–E = (6, 7) \qquad \text{with total task time} = 70$$
$$H–G = (10) \qquad \text{with total task time} = 70$$
$$I–H = (11) \qquad \text{with total task time} = 30$$

Implementing this procedure on a PC, Graves and Redfield (1988) were able to resolve problems consisting of 28 tasks within a few minutes of computing time. However, since this is basically an enumerative procedure, its application is limited to problems of modest size.

Another recently developed, rough-cut approach to devising alternative designs for flexible assembly systems combines branch-and-bound to prescribe line balance and queuing analysis to analyze stochastic performance (Lee and Johnson, 1991). The next subsection describes a branch-and-cut approach for assembly system design.

5.3.2   Branch-and-cut approach

Pinnoi and Wilhelm (1994b) devised a branch-and-cut approach to an Assembly System Design Problem (ASDP) that involves determining the number of stations, selecting the type of machine at each station and assigning tasks so that total cost is minimized. Their model is based on the following seven assumptions:

A1. C, $\{p_j\}$ and precedence relationships are known deterministically;
A2. no processing time is larger than the cycle time;
A3. task times assigned to a machine are additive and independent of the sequence;
A4. a line consists of a series of stations;
A5. setup time for a task is negligible (or included in task time);
A6. there are no zoning restrictions or other special-case constraints; and
A7. task time p_{mj} depends upon the machine m to which task j is assigned.

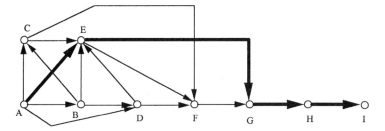

Fig. 5.6 Shortest path problem (over cut sets).

Employing decision variables:

$$x_{msj} = \begin{cases} 1 & \text{if task } j \text{ is assigned to machine } m \text{ at station } s \\ 0 & \text{otherwise} \end{cases}$$

$$y_{ms} = \begin{cases} 1 & \text{if machine } m \text{ is located at station } s \\ 0 & \text{otherwise} \end{cases}$$

their model is a 0/1 integer program:

Problem ASDP: Minimize $\displaystyle\sum_{m}\sum_{s} f_m y_{ms} + \sum_{m}\sum_{s}\sum_{j} g_{mj} x_{msj}$ $\qquad(5.14)$

Subject to $\displaystyle\sum_{m}\sum_{s} x_{msj} = 1 \qquad j \in J$ $\qquad(5.15)$

$\displaystyle\sum_{m}\sum_{s} sx_{msi} \leq \sum_{m}\sum_{s} sx_{msj} \quad j \in J;\ i \in P^*(j)$ $\quad(5.16)$

$\displaystyle\sum_{j} p_{mj} x_{msj} \leq C y_{ms} \qquad \forall m,\ s$ $\qquad(5.17)$

$\displaystyle\sum_{m} y_{ms} \leq 1 \qquad \forall s$ $\qquad(5.18)$

$x_{msj},\ y_{ms} = \{0,\ 1\} \qquad \forall m,\ s,\ j.$ $\qquad(5.19)$

The objective function (5.14) defines total cost as the sum of fixed and variable components. The fixed cost for a machine of type m, f_m, may include annualized capital cost and costs for installation, maintenance and training. If machines are already owned by the company, the fixed cost may include, for example, costs for relocation and re-installation rather than initial purchase cost. Variable operating cost g_{mt} may include assembly cost, labor cost and other costs that depend on task assignment.

Equality (5.15) requires that each task be assigned to one machine type at only one station. Inequality (5.16) invokes task precedence relationships, and (5.17) assures that the workload assigned to the machine located at each station does not exceed than the cycle time, C. Inequality (5.18) guarantees that at most one machine type will be assigned to each station; and, finally, constraint (5.19) requires that all decision variables be binary.

Pinnoi and Wilhelm (1994b) present a branch-and-cut approach for resolving the ASDP, including specialized preprocessing procedures, a heuristic to generate a 'good' initial upper bound, families of valid inequalities and facets (see Pinnoi and Wilhelm (1994a, 1994e) for underlying proofs), polynomial time separation heuristics and computational evaluation. Computational tests were conducted on a set of 20 problems which were based on five 'standard' precedence graphs from the literature and an actual one taken from industry. An additional 16 randomly generated problems were based on two levels of each of four factors: number of alternative machines, order strength, distribution from which processing times were generated and distribution from which cycle time was generated. The heuristic performed well, averaging within 13% of the optimal solution. The branch-and-cut approach was compared with OSL's branch-and-bound routine, which incorporates supernode processing. The branch-and-cut approach achieved an order of magnitude improvement in runtime on the more challenging problems; cuts made significant reductions in the number of nodes enumerated. Runtimes are typically larger in cases that require more stations (since the number of stations is one factor that determines problem size), in those which have larger idle times at stations (affecting the assignment of tasks), and in those with a larger number of alternative machines that might be located at each station. Problems with high order strengths were found to be easier to solve.

Having addressed methods for designing assembly systems, this chapter now focuses on important aspects of process planning for assembly.

5.4   Process planning in assembly

The automation of assembly has emphasized the need for more structured and rigorous methods for planning the sequence in which parts are assembled. The assembly sequence influences a number of vital design factors, including the part fixturing that is necessary, the degrees of freedom required to effect assembly and the feasibility of conducting in-process testing. It also affects the productivity of assembly operations, since it determines the number of delays necessary to change tooling, for example the end-of-arm effector used by a robot. Thus, operation sequence is a primary factor in determining the cost of assembly. In Chapter 3, section 3.6 explained assembly process planning in the context of computer-aided methods for design for assembly. Section 3.7 approached assembly planning in the context of artificial intelligence for DFA. This section reviews recent results related to sequence planning and its modeling for both mechanical assembly and for circuit card assembly.

5.4.1   Sequence planning for mechanical assembly

A number of alternative means for representing feasible assembly sequences have been proposed, including contact graphs (Jentsch and Kaden, 1984), liaison graphs (Bourjault, 1984), partial assembly trees (Wolter, 1992), subassembly trees (So and Scott, 1994), digraphs (Park and Chung, 1993), state space sequence graphs (De Fazio and Whitney, 1987, 1988), AND/OR graphs (Homem de Mello and Sanderson, 1986a, 1986b, 1988, 1991a, 1991b) and specialized languages (Wolter, Chakrabarty and Tsao, 1994). These representations provide convenient structures for analysis and reflect modern assembly requirements more fully than do the simple precedence relationships used in line balancing. However, the complexity of these methods is generally exponential in the number of parts (Wolter, 1992) because parts can usually be combined in a vast number of ways.

Based on these representations, a variety of methods have been proposed to generate assembly sequences. The output of such a process is a task-level program that specifies the step-by-step actions of, say, a robot. In fact, methods have been devised to prescribe 'optimal' assembly sequences (Jentsch and Kaden, 1984; Homem de Mello and Sanderson, 1986b; De Fazio and Whitney, 1987; Baldwin *et al.*, 1991; Wolter, 1991; Milner, Graves and Whitney, 1994; So and Scott, 1994). One study (Whitney *et al.*, 1986, 1987) developed a method for optimizing the design of a robotic assembly system, minimizing cost by prescribing the types and number of workstations, assigning tasks and incorporating tool-change times (see pages 241–2).

In related work, De Fazio and Whitney (1987) present a method for identifying feasible assembly sequences. One must begin with a product drawing or an actual assembly and analyze the parts, identifying 'liaisons' or relationships between pairs of parts (Bourjault, 1984). While the planner has flexibility to define liaisons as necessary, each is intended to represent 'a close bond or connection' (Bourjault, 1984) and typically involves contact relationships such as 'force fits, threaded fits, adhesion, compressional contact, . . . a part resting on another part' (De Fazio and Whitney, 1987). The results of this analysis are depicted in a liaison graph in which nodes represent parts, and arcs liaisons. If there are n parts, there must be at least $n - 1$ liaisons so that all parts are related in the assembly.

The second step establishes relationships between pairs of liaisons and is completed by posing and answering two questions about each liaison (De Fazio and Whitney, 1987):

1. What liaisons must be done prior to doing liaison i?
2. What liaisons must be left to be done after doing liaison i?

These pragmatic questions must be answered correctly in order to achieve the goal of identifying all feasible assembly sequences. Because of the vast number of combinations in which parts may be assembled, planners cannot identify all feasible sequences without some systematic procedure such as this. Formulating these specific questions was a primary contribution of De Fazio and Whitney (1987), because it requires only $2L$ questions to be answered (where L is the total number of liaisons).

The third step is to construct a state transition graph in which each node represents a state, a set of completed liaisons (equivalently, a subassembly), and each arc represents a transition or the operation that effects a liaison (e.g. assembling two parts or subassemblies). Each node is labelled to indicate the state of assembly, i.e. the liaisons that are complete in that state. The initial state has no completed liaisons, and the final state has all liaisons completed. Some transitions may establish more than one liaison; for example, one part may be in contact with several others. Each state is represented by a single node, and there may be a number of transitions into a state from other states. One may consider each transition as adding another part so nodes may be arranged in horizontal rows according to the number of parts assembled. With this construction, the number of rows of nodes will be equal to the number of parts to be assembled. Any path from the initial node to the final node passes through n states and represents a feasible sequence.

As the graph is constructed, the planner must consider assembly alternatives carefully, imposing any necessary constraints. For example, certain states may be undesirable because they would require assembly of more subassemblies than could be handled effectively. Other states may be eliminated because they represent unstable subassemblies. Yet others may be eliminated in order to require a particularly good subsequence or to avoid especially bad ones. Imposing constraints such as these reduces the size of the state transition graph and facilitates planning because it will lead to dramatic reductions in the number of sequences that the procedure identifies for detailed consideration.

Finally, the planner must analyze the feasible sequences by tracing paths through the state transition graph and evaluating each. During this process, the planner may identify other constraints that should be imposed to reflect practical limitations. This important step can also be facilitated by quantitative means (Milner, Graves and Whitney, 1994).

We can now demonstrate this procedure using the data from Example 5.1. Table 5.2 itemizes the set of parts in the bill of materials, and Fig. 5.1 gives a product sketch, showing the relationships between the parts. This information allows the liaison graph to be constructed as shown in Fig. 5.7. Most parts mount on the lower case, establishing the primary liaisons. The i/o board mounts on the processor board, but is held in place by screws that thread into the lower case. Printed wiring connectors, parts 5 and 6, each connect two parts, and the pointing device mounts on the keyboard/top subassembly. In Fig. 5.7 parts are numbered according to the parts list (Table 5.2) and liaisons are lettered A through N.

In Step 2, the two questions are answered for each liaison by referring to the precedence graph in Fig. 5.2. In actual applications, other considerations may influence answers to these questions. The state transition graph, as constructed in Step 3, is shown in Fig. 5.8. The typical state transition graph is diamond-shaped, first spreading out to represent an expanding set of combinations, then narrowing again as the number of alternatives is reduced near the end of assembly. Each state (i.e. node in the graph) is labeled, indicating the liaisons that are established at that state. Each arc in Fig. 5.8 represents a state

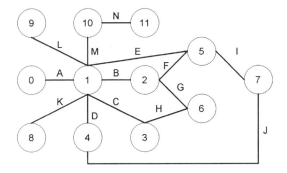

Fig. 5.7 Liaison graph.

transition and is labeled with the part number assembled in making the transition. Each path from the initial node to the terminal node identifies a unique sequence in which liaisons are established (and, equivalently, parts are assembled).

It seems that in this example there are no practical considerations that limit the assembly sequences depicted by Fig. 5.8. However, the Jackson rule-2 dominance relationships described in section 5.2.3 give a basis for closely examining states AC; ABC EFG, ABC EFGHIJ; and ABCDEFGH KL, ABCDEF KL to assure that the dominant task in each case is not assigned to a downstream station.

Circuit cards are mechanical assemblies, but they require specialized planning methods because they typically require placement of a large number of different components. The next subsection addresses these planning needs.

5.4.2 *Process planning for circuit card assembly*

Circuit cards are a primary means of packaging electronics and are used in an ever-increasing variety of products. The technology for assembling circuit cards is unique, but well known (Kear, 1987; Hinch, 1988; Riley, 1988; Mangin and Mclelland, 1989; Noble, 1989). Research in electronics assembly has increased exponentially in the last decade, covering the full gamut of issues from concurrent engineering to setup planning. Trends are toward increasing levels of miniaturization and component density, shorter product lifecycles, and higher variety, lower volume production. Wilhelm and Fowler (1992) give a succinct summary of continuing research needs in the area of electronics assembly (see Chapter 10).

Sequence planning methods for circuit card assembly (McGinnis *et al.*, 1992) involve four types of interrelated decisions at three hierarchical levels. The first level identifies families of circuit cards and assigns them to appropriate groups of machines. Second-level decisions allocate components to insertion (onsertion) machines. The third level assigns each component to a feeder location on the machine and prescribes the insertion (onsertion) sequence of

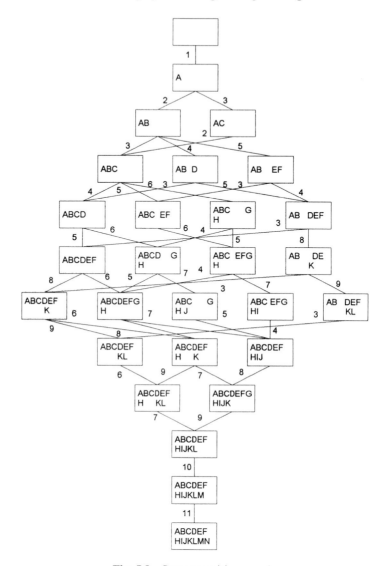

Fig. 5.8 State transition graph.

components on each card. The goal of process planning is to minimize assembly time, the total time to complete all cycles necessary to populate a circuit card, but the structures of these problems depend upon the types of machines used in the assembly process so there are no generic models that can be applied in all settings.

Circuit card assembly machines perform five basic operations (McGinnis *et al.*, 1992): position to retrieve a component from a feeder, retrieve a component from a feeder, transport the component from the feeder to the circuit card (and perhaps prepare the component for assembly), position to insert (onsert) the component on the card, and insert (onsert) the component on the card. On each machine cycle a component is retrieved, and one is inserted (onserted). Machines may operate in a sequential mode, performing all five operations in series, or in a concurrent mode in which several operations are performed simultaneously.

Typically, a series of different types of machines are used to populate a circuit card with different types and sizes of components. While there is a variety of component types, they may be categorized as requiring either pin-through-hole insertion or surface-mount onsertion, the more modern design that can accommodate greater component densities. In facilities that produce higher volumes, each card is typically mounted on a fixture and transported from one machine to the next by an automated conveyor. In facilities that produce in lower volumes, cards may be transported between machines in some type of container, decoupling the machines. Even though the machines may be different, it may be possible to assemble some components using several different types of machines, so the problem of allocating components to insertion (onsertion) machines may be non-trivial.

A machine may have several heads where components are temporarily stored after retrieval. A turret with multiple positions is often used. A component is retrieved as a turret position indexes under a feeder, and the component is transported as the turret rotates, one position during each cycle, until the component is in position for insertion (onsertion). The card may be mounted on an x–y table which positions the insertion (onsertion) location for assembly. The sequence in which components are inserted (onserted) thus determines the time required between insertions (onsertions), affecting machine cycle time.

There are several types of component feeders, including stick (tube), reel (tape-mounted components) and vibrating devices that separate and orient components correctly. Feeders must be selected to accommodate the types and sizes of components. Since each feeder is at a different location, the assignment of components to feeders affects the cycle time; feeders for the most frequently used components should be in the locations from which retrieval can be accomplished in the least amount of time. The strategy for managing setups that assign components to feeders thus affects the cycle time.

If feeders are set up with components for one type of circuit card, it is relatively easy to optimize the feeder location for each type of component. If several types of cards are to be assembled, it may be more efficient to set up all components once instead of initiating a new setup for each type of card. However, such a single setup may result in inefficiencies for some types of cards; for example, one type of component may qualify for a favored feeder location by virtue of its high usage on one card, but another card may not use the component at all and incur extra cycle time to retrieve components from more distant locations. Because the number of feeder locations on a machine is finite, it may not be possible to assemble many types of cards with a single setup. Thus, the setup strategy may group cards so that each family can be accommodated by a single setup or so that the incremental setups required by successive families is minimized. Each strategy entails different implications for in-process inventories of components and staged circuit cards.

The level 1 problem is similar to the group technology problem by which part families are identified and machines are grouped into appropriate cells. Apparently, little research has been directed specifically to the problem in the context of circuit card assembly.

The goal of allocating components at level 2 is to balance the resulting workloads on the machines. Bin packing, set covering or set packing models can usually be used to represent the essential features of the level 2 problem.

Models applicable to the level 3 problem commonly invoke simplifying assumptions to facilitate prescribing solutions. For instance, models usually allow a given component type to be assigned to just one feeder at just one machine. In addition, any variation in the sizes of feeders needed by different types of components is usually neglected.

The goal of level 3 is to minimize assembly time. The problem of assigning components to feeder locations can typically be modeled as a Quadratic Assignment Problem and the related problem of sequencing component placement as a Traveling Salesman Problem. However, the two problems interact because the cost incurred in one problem depends upon the solution to the other. Thus, heuristics commonly iterate between these two problems, converging to a final solution.

Yet another aspect of process planning is determining the quality control plan to achieve desired levels of product quality. Wilhelm (1992) presents a structured approach for analyzing the effects of test and rework operations on material flow in systems that assemble circuit cards. He identifies stages that affect quality (e.g. vendor preparation of components, pooling, assembly processes, test and rework) and devises models of each. Each product is treated as a group of components, each of which is vulnerable to a variety of defects. Testing is treated as fallible and rework as imperfect. Several procedures by which rework can be accomplished are considered.

Examples demonstrate that it is possible to quantify the effects of component quality, process capability, test fallibility, replacement parts and imperfect rework on material flow at test/rework operations. For example, in an environment with high-quality parts (e.g. with one defect per million parts), test fallibility may lead to more rejects than will bad components. Furthermore, poor workmanship at rework may degrade the quality of a circuit card, requiring more and more rework at each cycle. Thus, in some cases, a good card could be falsely rejected initially, and its quality could be degraded by successive rework operations. It is important to balance the quality capabilities of the various elements in the quality control plan. For instance, only incremental gains may be achieved by improving component quality without also improving test fallibility and rework workmanship. Each of these elements has an important influence on material flow, upon the creation of bottlenecks at test/rework operations and upon the outgoing quality of the product.

The next section presents approaches to the level 3 problem (Drezner and Nof, 1984; Nof and Drezner, 1993; Francis *et al.*, 1994). While these methods can be applied to circuit card assembly, they can also be used in more general settings.

5.5 Sequence planning for robotic assembly

Robots allow flexible, programmable assembly to be accomplished economically, a crucial characteristic in the current manufacturing setting which emphasizes the flexible assembly of a variety of products each in low volume. In addition, robots may be well suited for certain high-volume assembly operations, facilitating model changeovers for instance, and for some low-volume assembly operations, assuring accuracy and repeatability. Several examples of robotic, flexible assembly are described in Chapter 4.

This section describes the results of two studies that address sequence planning for robotic assembly. The first deals with a rectangular workpiece such as a circuit card, and the second provides a structure to analyze the complexity of sequencing problems in robotic cells. Sequence planning in a more generic setting of an assembly station is described in Chapter 1, section 1.9.

5.5.1 *Assembly sequence and bin locations for rectangular workpieces*

Francis *et al.* (1994) provide an interesting solution for the problem in which n different parts are to be assembled on a rectangular workpiece a units long and b units wide, for example a circuit card. Parts are to be assigned to n bins located at points in set B on the perimeter of the workpiece, and a Cartesian robot is to place parts one at a time. The problem is to minimize assembly time by prescribing the optimal bin locations and placement sequence for the parts.

As a robot moves from one point to another, it accelerates, then moves with constant velocity, then decelerates. Modeling these equations of motion leads to a non-linear problem which can be approximated by minimizing the distance between two points. Since a Cartesian robot moves simultaneously and independently from one point to another, the Tchebyshev distance between any two points $X = (x_1, x_2)$ and $Y = (y_1, y_2)$ is given by

$$D(X,\ Y) = \max\{|x_1 - y_1| + |x_2 - y_2|\},$$

a piecewise linear function. The robot travels in a straight line path from X to Y, but the Euclidean distance should not be used, since the actual travel time is functionally related to the Tchebyshev distance.

The robot is fitted with a single gripper, so after placing one part, it must retrieve the next part from its bin. Thus, if the robot places part i at point Q_i then part j at point Q_j, it should conform to the 'on-the-way principle' (Francis et al., 1994), selecting a bin location for part j that minimizes the travel time from Q_i to Q_j, t_{ij}:

$$t_{ij} = \min\left\{D(Q_i,\ X) + D(X,\ Q_j)\ \middle|\ X \in B\right\}. \tag{5.20}$$

With these assumptions, the problem of optimizing assembly sequence and bin locations can be addressed by solving a Traveling Salesman Problem (TSP) defined on a graph G with n nodes representing the parts and with travel time on the arc from node i to node j ($i \neq j$) given by t_{ij}. Since Tchebyshev distance is involved, matrix $\mathbf{T} = \{t_{ij}\}$ obeys the triangle inequality, so that:

$$t_{ij} \le t_{ik} + t_{kj} \qquad \text{for } 1 \le i, j, k \le n;$$

and, fortunately, this structured TSP can be resolved more readily than the general case. The shortest tour in G gives the assembly sequence as well as bin locations.

To calculate t_{ij}, the points in B that minimize (5.20) must be identified. This can be done by constructing a Tchebyshev ellipse with foci at Q_i and Q_j, identifying a contour line, the locus of all points that give the same value of t_{ij}, say K. The smallest ellipse that intersects point(s) in B, define (alternate) optimal bin location(s) for part j. Ellipses can be drawn as shown in Fig. 5.9, first drawing ± 45 lines through Q_i and Q_j to partition the plane into nine segments. The inner rectangle with Q_i and Q_j at its corners is a degenerate ellipse that gives the smallest possible value for t_{ij}. Larger ellipses are composed of linear segments with slope 0, ∞ or ± 1 in different segments as shown in the diagram. The ellipse that defines t_{ij} may include a number of alternative optimal locations for bin j.

Finally, a diagonal element of T, t_{ij}, is defined as the minimum Tchebyshev distance incurred by the robot arm traveling from Q_i to the perimeter of the workplace and back to Q_i. A lower bound on the solution to the TSP, LB, is given by these elements:

$$LB = \sum_i t_{ii}.$$

To see that this is a valid lower bound, notice that the robot must travel from some point in B to Q_i, then to another point in B. The lower bound assumes that this travel is always to the point in B that is closest to Q_i. An easily solved special case occurs if there is a point in B that is simultaneously closest to Q_i and Q_j for each i and j in sequence on the tour.

Francis et al. (1994) present the 'clock heuristic', an efficient $O(n \ln n)$ method for prescribing a solution. First, each point Q_i is projected either horizontally or vertically to the closest point in B. The sequence is then generated by the order in which a clock hand points to the projected points while rotating clockwise about the center of the workpiece. For example, Fig.

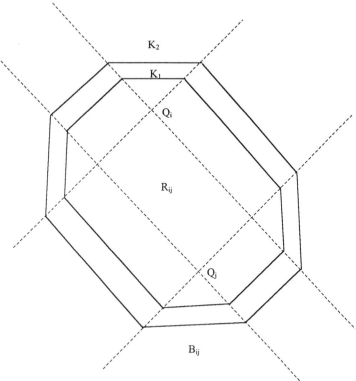

Fig. 5.9 Tchebyshev ellipses (adapted from Francis *et al.* (1994) reproduced with permission from IIE, Georgia, USA).

5.10 shows eight part positions on a rectangular workpiece. Each point is projected to the nearest point on the perimeter of the workpiece (dashed lines) and the clock hand generates the sequence (1, 2, 3, 4, 5, 6, 7, 8).

Francis *et al.* (1994) show that the clock heuristic is asymptotically optimal. Their computational tests show that the heuristic gives solutions within about 5% of optimum for $n \geq 20$. For smaller problems, an optimizing algorithm can easily be used to solve the TSP.

So far, bins have been assumed to be points. If points Q_i are spread out, optimal locations for bins will also be dispersed, and this assumption will be innocuous. In other cases, it is desirable to have a means of assigning bins to locations that are separated far enough to accommodate the actual size of each bin. An assignment problem can be formulated to accomplish this. Knowing the sequence in which parts are placed, the cost of assigning a bin to each feasible location $B = \{b_1, \ldots, b_{|B|}\}$ can be calculated. Given the sequence i–j, let:

$$c_{jk} = \left\{ D\left(Q_i, b_k\right) + D\left(b_k, Q_j\right) \right\} \qquad \text{for } j = 1, \ldots, n; \text{ and } k = 1, \ldots |B|$$

and:

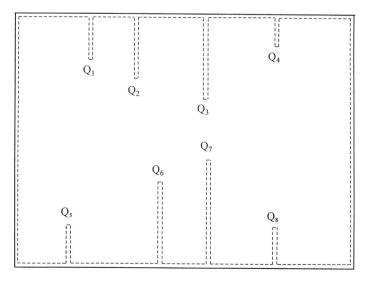

Fig. 5.10 Example demonstrating the clock heuristic (adapted from Francis *et al.* (1994) reproduced with permission from IIE, Georgia, USA).

$$c_{jk} = 0 \qquad \text{for } j = n + 1, \dots, |B|; \text{ and } k = 1, \dots, |B|.$$

The $|B| \times |B|$ assignment problem with parameters c_{jk} prescribes an optimal set of bin locations.

The results of the analysis have interesting practical implications. Computational tests indicate that the lower bound, LB, is 'usually attained' if points Q_i are arranged in a grid such as the rows that are commonly used to position components on digital (as opposed to analog) circuit cards. Francis *et al.* (1994) perform a probabilistic analysis to establish the asymptotic optimality of the clock heuristic. This analysis assumes that points Q_i are located randomly and uniformly on the surface of the workpiece and calculates the expected value of LB. Using this analysis, they were able to rank strategies for locating bins on long (*a*) or short (*b*) sides of the workpiece, showing that *b* is the worst strategy, *a* is somewhat better, *ab* is not much better, *aa* is better than *ab* and twice as good as *a*, and *aabb* is better than *aab* but not twice as good as *aa*. They conclude that a 'best buy' is to use *aa*, since practical requirements may preclude *aab* and *aabb*. They note that if the area of the workpiece is fixed, the square is the worst shape for the workpiece. They also provide a detailed analysis, showing that most results derived for the linear case also go through for the non-linear case in which acceleration and deceleration are considered. They conclude that the optimal solutions to the linear and non-linear problems are the same. Finally, they note that, since the Cartesian robot travels according to the Tchebyshev distance metric, if many moves parallel the workpiece

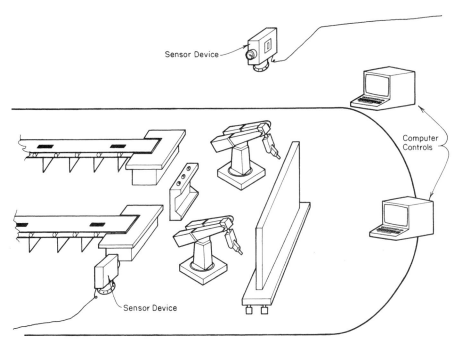

Sensor Device

Computer
Controls

Sensor Device

Fig. 5.11 A robotic assembly cell (Cash and Wilhelm (1988) reproduced with permission from the Society of Manufacturing Engineers).

sides, a 29% reduction in travel time can be achieved by rotating the workpiece 45 degrees.

5.5.2 *Operation sequencing in robotic cells*

The technological capabilities of robots and their controls now allow robotic cells to undertake sophisticated assembly operations. For example, consider a hypothetical cell such as the one depicted in Fig. 5.11. The workpiece is a sheet metal panel to which a number of components must be riveted, for example an airframe subassembly (Huber, 1984). A robot retrieves a part, displays it so that the sensing device can identify it, then positions that part on the workpiece where it is held in place temporarily by, say, an adhesive. After positioning a number of parts, the robot must change tools to drill holes through the positioned parts and the panel. Upon completing drilling operations, the robot must change tools again to ream each hole. The robot must then change tools to rivet the parts into place. A second robot is required to work in cooperation with the first, for example to provide resistance from the back side to facilitate each operation. The robots exchange roles to mount parts on the back side of the workpiece.

This hypothetical cell represents a variety of cases in which robots must change tools and work in cooperation with one another. Sequence planning in such applications must deal with a number of considerations that determine the efficiency of cell operations. For example, how may parts can be positioned temporarily before they are secured? It would be desirable to position a number of parts before losing time to change tools; however, the technological means by which parts are held temporarily may limit the time until parts must be attached permanently. The order in which parts are sequenced is also important, since it determines the time required for the robot to travel from one part location to another on the workpiece. In addition, parts that require the same tool 'size' should be 'batched' in sequence to avoid unnecessary tool changes.

Since each part requires several operations, each involving different tools, the process planning methods described in section 5.4.1 must be adapted. In addition, process plans may be designed and evaluated by applying the models described below.

Robotic operations are computer controlled, so one can argue that operation times may be considered to be deterministic. Wilhelm (1987a) devised a structure to analyze the computational complexity of deterministic sequencing problems in assembly cells attended by one or two robots. The structure can be used to identify certain sequencing problems that are 'easy' to optimize and cell features that affect complexity, leading to 'hard' problems that should be resolved either by a heuristic, which can prescribe a 'good' solution within reasonable runtime, or by a redesign to avoid complicating features. Minimizing cycle time, the makespan to complete a lot of assemblies, is a useful objective, since it leads to maximizing the throughput rate of the cell.

Sequencing problems can be categorized according to the number of robots, the allowable assignment of tasks to robots, the task precedence relationships and attributes associated with tooling, sensors and controls, and processing times. An operation may be assigned to robot 1 (R1), to robot 2 (R2), to robot 1 **or** 2 (R1oR2) or to robots 1 **and** 2 (R1aR2) to complete an operation cooperatively. Certain results apply to general task precedence relationships (G), but others may be restricted to special types of graphs (e.g. chain, parallel chain, assembly tree). Figure 5.12 depicts a series/parallel precedence graph in which each node represents an operation. The greek letters indicate tooling requirements and are used to represent a feasible tool sequence: α, β, γ, δ (e.g. positioning, drilling, reaming, riveting). Subscripts indicate part number and superscripts specify which robot(s) must be used for the operation.

Sequencing in cells attended by a single robot tends to be easy. The primary requirements are to keep the robot busy continuously and to observe precedence relationships. If all tool changes require the same amount of time, even problems like the one in Fig. 5.12 can be solved easily by a greedy approach, performing all operations that have completed predecessors and require a particular tool before changing tools. However, if tool-change time is sequence dependent, the optimal sequence must be prescribed by a TSP, an NP-hard problem.

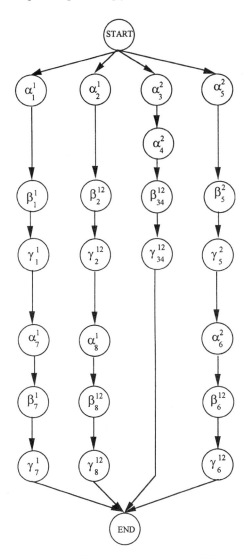

Fig. 5.12 Series/parallel precedence graph (Wilhelm, 1987a).

Sequencing problems in cells with two robots tend to be hard, but easy special cases can be identified. For example, if each operation is pre-assigned to a specific robot and the robots do not interact (e.g. by sharing tools), the problem decomposes into two problems, each with one robot in a cell. If both robots must perform all operations in concert, the problem again involves essentially one 'robot'. If each operation can be performed by either robot (R1oR2) and the robots do not interact, the problem reduces to a parallel machine sequencing problem. In general, minimizing makespan on two parallel machines is NP-hard, but if all operations take the same time, the problem is easy. Even if some operations require the robots to cooperate (i.e. R1aR2) sequencing is easy if the cooperative operations can be done without delaying one of the robots (e.g. at the beginning of the sequence).

These special cases may not allow the full capability of the robots to be utilized; in particular, it seems that the flexibility offered by robots may be most advantageous if the process plan allows a mix of R1, R2, R1aR2 and R1oR2 operations. However, these problems are NP-hard for general precedence graphs, whether or not tools must be changed.

Even though robotic operations are computer controlled, one may also argue that cell operations are subject to random variations, introduced, for example, by breakdowns, by the sensor system requiring a number of views to identify a part or by the robot requiring a number of adjustments to position a part accurately. Adopting this view, Cash and Wilhelm (1988) propose a structure for simulating the operations of such a cell, including appropriate variance reduction techniques. In addition, Wilhelm, Kalkunte and Cash (1988) devise recursion models (of the type described in Chapter 8) that can be used to evaluate such process plans. Either of these models could be used to evaluate alternative designs in order to identify productive combinations of equipment (e.g. robots, tools and sensors) and configuration (e.g. cell layout and operating procedures).

5.6   Summary

Methods to balance assembly lines have been researched for some time and represent a maturing area in the realm of assembly planning and design. A number of effective heuristics are available. In addition, optimizing methods have demonstrated the capability of solving problems with up to 1000 tasks. Researchers have extended the traditional line balancing problem to address a variety of the practical features found in many plants. The development of solution methods has paralleled the development of integer programming methods in general, progressing from special-purpose heuristics to meta heuristics and from branch-and-bound optimizing algorithms to branch-and-cut approaches. Even though line balancing has been the focus of much research and numerous solution methods have been proposed, the literature holds few accounts of the application of these methods in industry.

As the technology of robots and their controls has evolved, robots have been able to undertake assembly tasks that often require high accuracy and intricate maneuvers, perhaps calling for the coordinated activities of several robots. A new generation of system-design problems has been spawned by the application of robots to assembly and represents the type of design problem of most interest to industry today. Relatively few solution methods have been developed for these more sophisticated design problems.

Process planning for fabrication has been the focus of much interest for some time, but analogous planning for assembly has received relatively less attention. On the one hand, the human worker can make 'low tech' assembly appear to be an easy process, but, on the other hand, many aspects of assembly are difficult to quantify in order to provide a foundation on which to base automated process planning methods.

Tools now appear to be available to undertake product design for assembly as well as process design for assembly, starting at the earliest stage of product design. In particular, the modern environment, which is characterized by higher variety, lower-volume production and short product lifecycles resulting from rapid technological advances and vigorous international competition, requires a higher level of agile manufacturing in which the lead-time required to plan and design new assembly systems is dramatically reduced and replanning and redesigning are required much more frequently.

Except for a brief discussion of stochastic line balancing, this chapter assumed that the assembly system would operate deterministically. The next chapter relaxes that assumption, presenting models that describe the long-term effects of variability in processing time, station reliability and yield.

5.7 Review questions

1. Describe important differences between paced and unpaced lines.
2. Discuss the advantages and limitations of assembly line balancing.
3. What is the significance of metering stations?
4. Why should the designer seek to smooth workloads assigned to stations?
5. Why would one use a heuristic to balance an assembly line and what difficulties might be encountered in applying an optimizing method?
6. Describe problem features that have been shown to influence solution runtime.
7. Line balancing procedures tend to result in low idle times at upstream stations and higher idle times at downstream stations and may not smooth workloads effectively: true or false?
8. Describe the importance of sequence planning in automated assembly.
9. Describe sequence planning issues in circuit card assembly and outline a hierarchical approach to process planning.
10. Describe how to draw a Tchebyshev ellipse and how it might be used in sequence planning, say for circuit card assembly.
11. Discuss the complexity of sequencing problems in robotic cells.
12. Select a product of your choice, select a production rate, sketch the task precedence graph and demonstrate the following numerical methods:
    (a) Compute the cycle time and the balance delay function.
    (b) Compute B_1, E_j and L_j for all tasks j.
    (c) Apply four heuristics to balance the assembly line for this product: positional weight, COMSOAL, Hoffman, and Hoffman with Gehrlein and Patterson's extension.

(d) Apply two methods to optimize the line balance: FABLE, EUREKA.
(e) Develop a liaison graph and the state transition network for sequence planning.
(f) Use the method involving cut set analysis and shortest path solution to determine an optimal assembly sequence.

References

Afentakis, P., Gavish, B. and Karmaker, U. (1984) 'Computationally efficient optimal solutions to the lot-sizing problem in multistage assembly systems', *Management Science*, **30** (2), 222–38.
Anderson, E.J. and Ferris, M.C. (1994) 'Genetic algorithms for combinatorial optimization: the assembly line balancing problem', *ORSA Journal on Computing*, **6** (2), 161–73.
Arcus, A.L. (1966) 'A computer method of sequencing operations for assembly lines', *International Journal of Production Research*, **4** (4).
Baldwin, D.F., Abell, T.E., De Fazio, T.L. and Whitney, D.E. (1991) 'An integrated computer aid for generating and evaluating assembly sequences for mechanical products', *IEEE Transactions on Robotics and Automation*, **7** (1), 78–94.
Bard, J. (1989) 'Assembly line balancing with parallel workstations and dead time', *International Journal of Production Research*, **27** (6), 1005–18.
Baybars, I. (1986) 'A survey of exact algorithms for the simple assembly line balancing problem', *Management Science*, **32** (8), 909–32.
Boothroyd, G. (1992) *Assembly Automation and Product Design*, Marcel Dekker, New York.
Bourjault, A. (1984) 'Contribution a une approache methodologue de l'assemblage automatise: Elaboration automatique des sequences operatoires', Thesis: Grade de Docteur es Sciences Physiques at L'Universit de Franche-Comte.
Buxey, G.M., Slack, N.D. and Wild, R. (1973) 'Production flow line system design – a review', *AIIE Transactions*, **5**, 37–48.
Cash, C.R. and Wilhelm, W.E. (1988) 'A simulation model for use in designing robotic assembly cells', *Journal of Manufacturing Systems*, **7** (4), 279–91.
Chakravarty, A.K. and Stubb, A. (1985) 'Balancing mixed-model lines with in-process inventories', *Management Science*, **31** (9).
Chakravarty, A.K. and Stubb, A. (1986) 'Integration of assembly robots in a flexible assembly system', *Flexible Manufacturing Systems: Methods and Studies*, (ed. A. Kusiak), Elsevier Publishers, North Holland.
Dar-El, E.M. and Cother, R.F. (1975) 'Assembly line sequencing for model mix', *International Journal of Production Research*, **13**, 463–77.
Dar-El, E.M. and Curry, S. (1977) 'Optimal mixed-model sequencing for balanced assembly lines', *Omega*, **5**, 333–41.
Deckro, R.F. (1989) 'Balancing cycle time and workstations', *IIE Transactions*, **21**, 106–11.
De Fazio, T.L. and Whitney, D.E. (1987) 'Simplified generation of all mechanical assembly sequences', *IEEE Journal of Robotics and Automation*, **3** (6), 640–58.
De Fazio, T.L. and Whitney, D.E. (1988) 'Correction to: Simplified generation of all

mechanical assembly sequences', *IEEE Journal of Robotics and Automation*, **4** (6), 705–8.

Drezner, Z. and Nof, S.Y. (1984) 'On optimizing bin picking and insertion plans for assembly robots', *IIE Transactions*, **16** (3), 262–70.

Elsayed, E.A. and Boucher, T.O. (1994) *Analysis and Control of Production Systems*, Prentice-Hall, Englewood Cliffs, NJ.

Francis, R.L., Hamacher, H.W., Lee, C.Y. and Yeralan, S. (1994) 'Finding placement sequences and bin locations for Cartesian robots', *IIE Transactions*, **26** (1), 47–59.

Garey, M.R. and Johnson, D.S. (1979) *Computers and Intractability: A Guide to the Theory of NP-Completeness*, Freeman, New York.

Gehrlein, W.V. and Patterson, J.H. (1975) 'Sequencing for assembly lines with integer task times', *Management Science*, **21** (9), 1064–70.

Gehrlein, W.V. and Patterson, J.H. (1978) 'Balancing single model assembly lines: comments on a paper by E.M. Dar-El', *AIIE Transactions*, **10** (1), 109–12.

Ghosh, S. and Gagnon, R.J. (1989) 'A comprehensive literature review and analysis of the design, balancing and scheduling of assembly systems', *International Journal of Production Research*, **27** (4), 637–70.

Glover, F. (1989) 'Tabu search: Part I', *ORSA Journal on Computing*, **1**, 190–206.

Glover, F. (1990) 'Tabu search: Part II', *ORSA Journal on Computing*, **2**, 190–206.

Glover, F. and Laguna, M. (1993) 'Tabu search', *Modern Heuristic Techniques for Combinatorial Problems* (ed. C.R. Reeves), Blackwell, Oxford, 70–150.

Graves, S.C. and Lamar, B.W. (1983) 'An integer programming procedure for assembly system design problems', *Operations Research*, **31** (3), 522–45.

Graves, S.C. and Redfield, C.H. (1988) 'Equipment selection and task assignment for multiproduct assembly system design', *International Journal of Flexible Manufacturing Systems*, **1**, 31–50.

Graves, S.C. and Whitney, D.E. (1979) 'A mathematical programming procedure for equipment selection and system evaluation in programmable assembly', *Proceedings of the 18th IEEE Conference on Decision and Control*, 531–6.

Gustavson, R. (1986) 'Design of cost-effective assembly systems', *Proceedings of the Design for Assembly Conference*, April, 2661.

Hackman, S.T., Magazine, M.J. and Wee, T.S. (1989) 'Fast, effective algorithms for simple assembly line balancing problems', *Operations Research*, **37** (6), 916–24.

Helgeson, W.B. and Birnie, D.P. (1961) 'Assembly line balancing using the ranked positional weight technique', *Journal of Industrial Engineering*, **12**, 394–8.

Hinch, S.E.W. (1988) *Handbook of Surface Mount Technology*, Longman Scientific and Technical, New York.

Hoffman, T.R. (1963) 'Assembly line balancing with a precedence matrix', *Management Science*, **9**, 551–62.

Hoffman, T.R. (1990) 'Assembly line balancing: a set of challenging problems', *International Journal of Production Research*, **28** (10), 1807–15.

Hoffman, T.R. (1992) 'EUREKA: a hybrid system for assembly line balancing', *Management Science*, **38** (1), 39–47.

Homem de Mello, L.S. and Sanderson, A.C. (1986a) 'AND/OR graph representation of assembly plans', *AAAI-86 Proceedings of the Fifth National Conference on Artificial Intelligence, American Assoc. of Artificial Intelligence*, Morgan Kaufmann, 1113–19.

Homem de Mello, L.S. and Sanderson, A.C. (1986b) 'Automatic generation of mechanical assembly sequences', The Robotics Institute, Carnegie Mellon Univ., Pittsburgh, PA.

Homem de Mello, L.S. and Sanderson, A.C. (1988) 'Task sequence planning for assembly', *Proceedings IMACS World Congress '88 on Scientific Comp.*, Paris, France, 390–2.

Homem de Mello, L.S. and Sanderson, A.C. (1991a) 'A correct and complete algorithm for the generation of mechanical assembly sequences', *IEEE Transactions on Robotics and Automation*, **7** (2), 228–40.

Homem de Mello, L.S. and Sanderson, A.C. (1991b) 'Representations of mechanical assembly sequences', *IEEE Transactions on Robotics and Automation*, **7** (2), 211–27.

Huber, J. (1984) 'Technology requirements to support aerospace plant modernization', *Proceedings, Annual International Industrial Engineering Conference*, 262–70.

Jackson, J.R. (1956) 'A computing procedure for a line balancing problem', *Management Science*, **2**, 261–71.

Jentsch, W. and Kaden, F. (1984) 'Automatic generation of assembly sequence', *Artificial Intelligence and Information-Control Systems of Robots* (ed. I. Plander), Elsevier Science Publishers, North Holland, 197–200.

Johnson, R.V. (1981) 'Assembly line balancing algorithms: computational comparison', *International Journal of Production Research*, **19** (3), 277–87.

Johnson, R.V. (1983) 'A branch and bound algorithm for assembly line balancing problems with formulation irregularities', *Management Science*, **29** (11), 1309–24.

Johnson, R.V. (1988) 'Optimally balancing large assembly lines with "FABLE"', *Management Science*, **34**, 240–53.

Kear, F.W. (1987) *Printed Circuit Assembly Manufacturing*, Marcel Dekker, New York.

Kumar, A., Gurnani, H. and Kekre, S. (April 1990) *Design Choices for Flexible Parallel Assembly: Strategic and Tactical Considerations*, Working Paper, Carnegie Mellon Univ., Pittsburgh, PA.

Labach, E.J. (1990) *A Comparative Evaluation of Heuristic Sequencing and Line Balancing Techniques for Mixed Model Assembly Lines*, Working Paper, School of Management, Boston Univ., Boston, MA.

Lee, H.F. and Johnson, R.V. (1991) 'A line balancing strategy for designing flexible assembly systems', *International Journal of Flexible Manufacturing Systems*, **3**, 91–120.

Lehman, M. (1969) 'On criteria for assigning models to assembly lines', *International Journal of Production Research*, **7** (4).

Macaskill, J.L.C. (1972) 'Production-line balances for mixed-model lines', *Management Science*, **19** (4), 423–34.

Mangin, C.H. and Mclelland, S. (1989) *Surface Mount Technology: the Future of Surface Mount Technology*, IFS Publications, London.

Mastor, A.A. (1970) 'An experiment investigation and comparative evaluation of production line balancing techniques', *Management Science*, **16**, 728–46.

McGinnis, L.F. and Ammons, J.C. *et al.* (1992) 'Automated process planning for printed circuit card assembly', *IIE Transactions*, **24** (4), 18–30.

Michalewicz, Z. (1994) *Genetic Algorithms and Data Structures*, Springer-Verlag, New York.

Milner, J.M., Graves, S.C. and Whitney, D.E. (1994) 'Using simulated annealing to select least-cost assembly sequences', *Proceedings IEEE International Conference on Robotics and Automation*, May 1994, San Diego, CA, 2058–63.

Muth, E.J. and Alkaff, A. (1987) 'The bowl phenomenon revisited', *International Journal of Production Research*, **25** (2), 161–73.

Noble, P.J.W. (1989) *Printed Circuit Board Assembly*, John Wiley & Sons, New York.

Nof, S.Y. and Drezner, Z. (1993) 'The multiple-robot assembly plan problem', *Journal of Intelligent and Robotic Systems*, **5**, 57–71.

Park, J.H. and Chung, M.J. (1993) 'Automatic generation of assembly sequences for multi-robot workcell', *Robotics and Computer-Integrated Manufacturing*, **10**, 355–63.

Patterson, J.H. and Albracht, J.J. (1975) 'Assembly line balancing: 0–1 programming with Fibonacci search', *Operations Research*, **23**, 166–74.

Pinnoi, A. and Wilhelm, W.E. (1994a) *Assembly System Design: Part I, Valid Inequalities for the Underlying Line Balancing Problem*, Working Paper, Department of Industrial Engineering, Texas A&M University, TX.

Pinnoi, A. and Wilhelm, W.E. (1994b) *Assembly System Design: Part II, a Branch-and-Cut Approach*, Working Paper, Department of Industrial Engineering, Texas A&M University, TX.

Pinnoi, A. and Wilhelm, W.E. (1994c) *A Branch-and-Cut Approach for Workload Smoothing on Assembly Lines*, Working Paper, Department of Industrial Engineering, Texas A&M University, TX.

Pinnoi, A. and Wilhelm, W.E. (1994d) *A Family of Hierarchical Models for Assembly System Design*, Working Paper, Department of Industrial Engineering, Texas A&M University, TX.

Pinnoi, A. and Wilhelm, W.E. (1994e) *Valid Inequalities for the Assembly Line Balancing Problem*, International Journal of Production Research, (in press).

Pinto, P.A., Dannenbring, D.G. and Khumawala, B.M. (1981) 'Branch and bound and heuristic procedures for assembly line balancing with paralleling of stations', *International Journal of Production Research*, **19**, 565–76.

Pinto, P.A., Dannenbring, D.G. and Khumawala, B.M. (1983) 'Assembly line balancing with processing alternatives: an application', *Management Science*, **29** (7), 817–30.

Rachamadugu, R.M.V. and Talbot, B. (1991) 'Improving the equality of workload assignments in assembly lines', *International Journal of Production Research*, **29** (3), 619–33.

Riley, F. (1988) *The Electronics Assembly Handbook*, IFS Publications, London.

Roberts, S. and Villa, C. (1970) 'A multiproduct assembly line balancing problem', *AIIE Transactions*, **2** (4), 361–4.

Sarker, B.R. and Shanthikumar, J.G. (1983) 'A generalized approach for serial or parallel line balancing', *International Journal of Production Research*, **21**, 109–33.

Schrage, L. and Baker, K.R. (1978) 'Dynamic programming solution of sequencing problems with precedence constraints', *Operations Research*, **26** (3), 444–9.

So, K.C. and Scott, C.H. (1994) 'Optimal production sequence for a product with matching components', *Operations Research*, **42** (4), 694–708.

Talbot, F.B. and Patterson, J.H. (1984) 'An integer programming algorithm with network cuts for solving the assembly line balancing problem', *Management Science*, **30**, 85–99.

Talbot, F.B., Patterson, J.H. and Gehrlein, W.V. (1986) 'A comparative evaluation of heuristic line balancing techniques', *Management Science*, **32** (4), 430–54.

Thomopoulos, N.T. (1967) 'Line balancing-sequencing for mixed-model assembly', *Management Science*, **14**, B.59–B.75.

Thomopoulos, N.T. (1970) 'Mixed-model line balancing with smoothed station assignments', *Management Science*, **16** (9), 593–603.

Tonge, F.M. (1961) *A Heuristic Program for Assembly Line Balancing*, Prentice-Hall.

Voß, S. and Scholl, A. (1994) *Simple Assembly Line Balancing – Heuristic Approaches*, Working Paper 2/94, Technische Hochschule Darmstadt, Institut für Betriebswirtschaftslehre, Darmstadt, Germany.

Vrat, P. and Varani, A. (1976) 'A cost model for optimal mix of balanced stochastic assembly line and the modular assembly system for a customer-oriented production system', *International Journal of Production Research*, **14** (4).

Whitney, D.E., De Fazio, T.L., Gustavson, R.E., Graves, S.C., Holmes, C.A. and Klein, C.J. (1986) *Computer-Aided Design of Flexible Assembly Systems*, CSDL-R-1947, The Charles Stark Draper Laboratory, Cambridge, MA.

Whitney, D.E. *et al.* (1987) 'Computer aided design of flexible assembly systems', Progress Reports at the 14th NSF Conference on Production Research and Technology, Univ. of Florida, 1986, 205–212; and 15th NSF Conference on Production Research and Technology, Univ. of Michigan, 389–97.

Wild, R. (1975) 'On the selection of mass production systems', *International Journal of Production Research*, **13** (5), 443–61.

Wilhelm, W.E. (1987a) 'Complexity of sequencing tasks in assembly cells attended by one or two robots', *Naval Logistics Research Quarterly*, **34**, 721–38.

Wilhelm, W.E. (1987b) 'On the normality of operation times in small-lot assembly systems', *International Journal of Production Research*, **25** (1), 145–9.

Wilhelm, W.E. (1992) 'The effects of test and rework operations on the flow of materials in circuit card assembly', *Journal of Manufacturing Systems*, **11** (3), 167–78.

Wilhelm, W.E. and Ahmadi-Marandi, S. (1982) 'A methodology to describe operating characteristics of assembly systems', *IIE Transactions*, September, 204–13.

Wilhelm, W.E. and Fowler, J. (1992) 'Research directions in electronics manufacturing', *IIE Transactions*, **24** (4), 6–17.

Wilhelm, W.E., Kalkunte, M.V. and Cash, C. (1988) 'A modeling approach to aid in designing robotized manufacturing cells', *Journal of Robotic Systems*, **4** (1), 25–48.

Wolter, J.D. (1991) 'On the automatic generation of assembly plans', *Computer Aided Mechanical Assembly Planning* (eds H. Homem de Mello and S. Lee), Kluwer Academic Press, Boston, MA, 263–88.

Wolter, J.D. (1992) 'A combinatorial analysis of enumerative data structures for assembly planning', *Journal of Design and Manufacturing*, **2** (2), 93–104.

Wolter, J.D., Chakrabarty, S. and Tsao, J. (1994) *Mating Constraint Languages for Assembly Sequence Planning*, Working Paper, Department of Computer Science, Texas A&M University, TX.

6

Performance evaluation of stochastic assembly systems

6.1 Introduction

The engineer designing an assembly system is typically concerned with comparing alternative designs and must be able to evaluate the performance that each might achieve in the long run in order to select the most appropriate design. Decisions such as these are, perhaps, best made using a model of steady-state performance.

Stochastic models, especially those based on queuing theory, are typically used for this purpose. Models can describe system performance and, thus, be used to develop insights into operations that can be used to make trade-offs in selecting an alternative, leading to the most effective design consisting of station configuration, machine selection, tooling provision, buffer capacities, material handling devices and operating procedures. It is most important that stochastic models identify the effects of randomness in system operations, including variabilities in operation times as well as part arrivals (i.e. representing uncertain delivery from vendors or from upstream production processes), since all systems must somehow coordinate part inputs to the assembly process.

The traditional assembly line balancing problem results in a flow-line (or transfer line), a configuration that has been thoroughly researched (Dallery and Gershwin, 1992) because it is also important in many production settings. A primary concern has been determining the effectiveness of buffers in preventing blocking and starving and prescribing optimal buffer capacities. Models have been developed for cases with reliable or unreliable machines and with synchronous or asynchronous operations. The effects of variability on design, of line length on performance and of scrapping or rejecting bad parts on throughput have been studied (Kala and Hitchings, 1973; Hillier and Co, 1991). Fundamental characteristics of flow-lines – the reversibility property and the bowl phenomenon – have also been established. Furthermore, queuing network models have been developed for more complex manufacturing systems, especially flexible manufacturing systems. These results are well known (Bitran and Dasu, 1992; Buzacott and Shanthikumar, 1992; Suri, Sanders and Kamath, 1993) and will not be detailed here, since they do not relate exclusively to assembly systems which uniquely involve the joining of part flows to form assemblies.

One primary use of analytical models is to determine conditions under which an assembly system would attain a steady state. Mathematical analysis shows that, surprisingly, assembly systems may not attain a steady state unless certain restrictions are placed on the system design. This information is crucial to assuring effective designs.

Simulation models could be used to support the design engineer, but they may not be the most appropriate in the early stages of design, since they require considerable runtimes and may thus discourage the designer from evaluating an adequate number of potential design alternatives. In such a situation, an analytical model that can quickly provide good estimates of system performance may be preferred. The overall approach would be to use the analytical model to screen a large number of alternatives quickly, then use a simulation model to study the few most promising designs in greater detail and with greater accuracy.

This chapter provides an overview of models that can be used to evaluate the performance of assembly system design alternatives. Since assembly systems present formidable challenges to analytical modeling techniques, many models provide only bounds for, or approximations of, system performance. Section 6.2 describes some analytical models of push systems, including an exact analysis of the kitting process whereby parts of different types are accumulated to initiate assembly. Kitting is unique to assembly and must be well understood since it has a significant influence on system performance. Models of pull systems are presented in section 6.3. Flexible assembly systems are addressed in section 6.4, including closed-loop, flexible assembly systems and programmable, robotic systems. Section 6.5 provides some insights into the capabilities that simulation models provide for analyzing the performance of assembly systems, including an analysis of buffer capacities and the design and operation of a cellular configuration. Section 6.6 gives a summary and conclusions. Because researchers have only recently turned their attention to modeling assembly systems and because such systems are difficult to analyze, relatively few models exist today, even though there is virtually an unlimited number of configurations in which assembly systems can be organized.

6.2   Push systems

Based on forecasts of demand, materials are delivered to a push system and are subsequently 'pushed' through the required operations in order to produce end products to satisfy customer demand. This traditional philosophy of material management has been in wide use for some time and is basic to many production systems.

Push systems are often implemented within a Materials Requirements Planning (MRP) environment (Orlicky, 1975). MRP is a means of relating demand for a product to the parts and subassemblies that compose it. Knowing the due date on which a customer requires a product, the due dates for part deliveries can be calculated by 'backing off' the 'planned lead-times' for constituent subassemblies and parts. Parts suppliers then deliver according to these due dates, and materials are subsequently 'pushed' through the production system.

The assembly line in which no station ever waits for parts operates like any other flow-line, and a substantial body of literature describes models that can be used to evaluate the performance of such a line (Buxey, Slack and Wild, 1973; Magazine and Silver, 1978; Sarker, 1984). Hunt (1956) initiated this work, which has led to descriptive models of general flow-line configurations (Muth, 1973) as well as special types of flow-line such as synchronized automatic transfer lines (Buzacott, 1967; Buzacott and Hanifin, 1978).

A number of fundamental properties of flow-lines have been established, including the bowl phenomenon and the reversibility property (Muth, 1979). The bowl phenomenon is that the production rate of a flow-line is maximized by placing stations with longer expected processing times at the center of the line (Rao, 1976). However, this phenomenon is most pronounced when processing times are highly variable and, thus, may not be important in actual systems (Muth and Alkalf, 1987). The reversibility property of flow-lines is that 'any serial line has a dual which is identical except that the direction of material flow is reversed . . . the production capacities of a line and its dual are identical', and an important corollary is that two serial lines that are mirror images of each other have identically the same production rate (Yazamaki and Sakasegawa, 1975). This property allows the flow-line to be analyzed by assuming material flow occurs in either of the two directions.

The role of buffers in preventing blocking of upstream stations and starving of downstream stations has been researched extensively. A recent simulation study provides an excellent, pragmatic analysis of the role of in-process inventories as well as guidelines for designing buffer capacities (Conway *et al.*, 1988).

Even for the flow-line configuration, most stochastic models are limited to two- or (certain) three-station lines, because the state space required to describe system operation grows exponentially with the number of stations. However, recent work has developed approaches for approximating the performance of longer flow-lines (Gershwin, 1987; Dallery, David and Xie, 1988; de Koster, 1988).

Work on models specifically related to assembly systems was apparently initiated quite recently by Harrison (1973). This class of models explicitly treats the arrival streams of all parts required to complete an assembly and lends considerable insights into the performance of assembly systems, which are characterized uniquely by this process. These assembly-like queues are the topic of section 6.2.1. Section 6.2.2 describes models that approximate the performance of networks involving both disassembly and assembly, a very general configuration. Of course, it is not necessary that a system perform disassembly, so that the model also relates to complex assembly networks. Section 6.2.3 presents an exact analysis of the kitting process and describes a means of analyzing downstream assembly and end product inventory operations, completing analysis of an entire assembly system. Finally, section 6.2.4 discusses a model of stochastic MRP control. Again, few exact models of system operations are available, since assembly operations present formidable challenges to analysis. However, a number of useful models that give bounds for, or approximations of, system performance are related in this section.

6.2.1   Assembly queues

Apparently, Harrison (1973) formulated the earliest queuing model of an assembly system. Specifically, he studied a system in which P parts are assembled in one operation to produce a product, invoking the following assumptions:

H1. the times between successive arrivals of parts p ($p = 1, \ldots, P$) are independent, identically distributed, non-negative random variables, so that the input processes for parts are mutually independent renewal processes;
H2. one part of each type is required to assemble each product (although this assumption could easily be relaxed for the case in which several parts of each type are required to assemble a product);
H3. parts of type p (for $p = 1, \ldots, P$) are queued in a separate buffer and used on a first-come first-served basis;
H4. after completing a product, the server starts another assembly immediately if a part of each type is available. If a set of 'matched' parts is not on hand, the server becomes idle until a set of required parts becomes available;
H5. assembly times are independent and identically distributed and independent of the arrival processes.

These assumptions are described in detail because they lead to an important but surprising result which holds for a broad variety of assembly systems.

Harrison (1973) was able to show that a system operating according to these apparently innocuous assumptions is unstable if no limitations are placed on the buffers where parts are held in a queue awaiting assembly. The distribution of time for which parts of each type must wait for assembly does not converge to a non-defective limit. Thus, this rather general queuing model will attain a steady state only if a finite limit is placed on the capacity of each part buffer.

In general, queuing models of assembly operations are unwieldy, if not intractable. Even for the case with restricted assumptions

R1. each part arrives according to a Poisson process (with mean rate λ); and
R5. assembly times are exponentially distributed (with mean rate μ);

it is difficult to derive models in closed form. Thus, several studies have sought bounds on important measures of performance.

Lipper and Sengupta (1986) study an actual, in-plant assembly operation. They model a system according to the restrictive assumptions (R1), (H2)–(H4) and (R5) along with the additional assumptions that

A6. the buffer for part p has finite capacity K (for $p = 1, \ldots, P$); and
A7. a part of type p, which arrives when buffer p is full, is 'lost'.

Because some parts are 'lost', the rate of output from the system is less than the rate of input. Thus, the system output rate is relatively small when K is small. As K increases, the output rate increases, but at the cost of carrying more in-process inventory in the buffers.

As $\lambda \to \infty$, buffers each contain K parts and the assembly station is busy with probability 1. In this case, the sojourn (i.e. the queuing plus processing) time of each part is distributed as an Erlang with parameters $(K + 1)$ and μ.

In another simple case, $K = 0$, so that no queue can develop. When the assembly station completes one product, it must accumulate parts before it can begin the next assembly operation. The time required to accumulate the matched set of P parts is the maximum of P independent exponential (according to assumption (R5)) random variables. The distribution function of this random variable is:

$$F(t) = \left(1 - e^{-\lambda t}\right)^P$$

and its expected value is:

$$E(t) = \sum_{p=1}^{P} \left(\lambda p\right)^{-1}$$

so that the throughput rate for the system is:

$$\left[\sum_{p=1}^{P} \left(\lambda p\right)^{-1} + \left(\mu\right)^{-1}\right]^{-1}.$$

The waiting time for a randomly chosen part has distribution:

$$\frac{1 - \left(1 - e^{-\lambda t}\right)^P}{P e^{-\lambda t}}$$

and expected value:

$$\sum_{p=1}^{P-1} \frac{(P - p)}{\lambda P p}.$$

The sojourn time for a randomly chosen part is the convolution of this waiting time and the exponentially distributed assembly time, so it is easily described.

Other cases are more complicated, so Lipper and Sengupta (1986) devised an iterative numerical procedure to approximate throughput rate and mean sojourn time. By comparison with a simulation model, they found that their method gives good approximations for small to moderate K, moderate to large λ and moderate to large P. Since these values are typical in many industrial applications, the approximations may be useful in practice.

Hopp and Simon (1989) propose bounds for throughput and inventories in a system that assembles $P = 2$ parts to form a product as shown in Fig. 6.1. An inexhaustible source supplies part p to machine M_p, which processes parts individually until a bin of type p is full. Each bin holds enough parts of type p to assemble one product. When full, the bin moves instantaneously to buffer B_p ahead of the assembly machine M_A. When emptied at machine M_A, a bin of type p returns instantly to machine M_p, which requires an empty bin to be available before it can process the next part. When assembly is complete, machine M_A can initiate another operation as soon as one bin of type 1 and one bin of type 2 are available. Interestingly, Hopp and Simon (1989) show that this assembly system is equivalent to a certain production flow-line with blocking. This equivalence is important, since it may ultimately allow the copious research on serial lines to be exploited to better understand assembly systems.

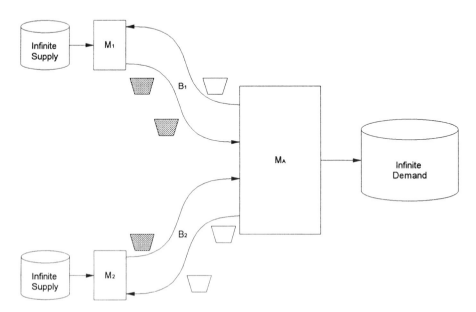

Fig. 6.1 Three-station assembly with part bins (adapted from Hopp and Simon, (1989)).

Hopp and Simon (1989) assume that the service times at all three machines are independent identically distributed (i.i.d.) exponential random variables with mean rates λ_1, λ_2 and μ, respectively. They limit the capacity of buffer B_p to the finite value of K_p. Thus, parameters λ_1, λ_2, μ, K_1 and K_2 completely describe their system.

They show that certain bounds for the assembly system can be obtained from the $M/M/1/K$ queuing system, which provides a single server and a finite queue capacity of K to process customers that arrive according to a Poisson process (with rate λ) and receive service according to an exponential distribution with mean rate μ. Performance measures that are used to develop the bounds for the assembly system include the expected throughput rate:

$$T[\lambda/\mu/1/K] = [1 - (1-\rho)/(1-\rho^{K+1})]\mu \qquad \text{if } \rho \neq 1$$
$$= [K/(K+1)]\mu \qquad \text{if } \rho = 1,$$

the expected queue length:

$$I[\lambda/\mu/1/K] = \rho[1 - (K+1)\rho^K + K\rho^{K+1}]/[(1-\rho)(1-\rho^{K+1})] \quad \text{if } \rho \neq 1$$
$$= K/2 \qquad \text{if } \rho = 1,$$

and the probability that the system will be empty in the steady state:

$$p_0[\lambda/\mu/1/K] = (1-\rho)/(1-\rho^{K+1}) \qquad \text{if } \rho \neq 1$$
$$= 1/(K+1) \qquad \text{if } \rho = 1.$$

A valid upper bound on the throughput rate for the assembly system is given by:

$$T[\lambda_1/\lambda_2/\mu/K_1/K_2] \leq \min\{T[\lambda_1/\mu/1/K_1], \ T[\lambda_2/\mu/1/K_2],$$
$$T[\lambda_1/\lambda_2/1/K_1 + K_2]\}.$$

This bound becomes tight if $\lambda_1 \to \infty$, $\lambda_2 \to \infty$ or $\mu \to \infty$, so it can be expected to give a good approximation if one parameter is large compared to the other two.

They also give a valid lower bound on the throughput rate:

$$T[\lambda_1/\lambda_2/\mu/K_1/K_2] \geq \max\{\mu[1 - p_0[\lambda_1/\mu/1/K_1] - p_0[\lambda_2/\mu/1/K_2]\},$$
$$q/E[\max\{\text{Erlang } \lambda_1, q),$$
$$\text{Erlang}(\lambda_2, \text{Erlang}(\mu, q)\}]\}. \qquad (6.1)$$

The first term on the right-hand side of (6.1) is tight if $\lambda_1 \to \infty$ or if $\lambda_2 \to \infty$;

$$q = \left\lfloor \frac{\min[K_1, K_2]}{2} \right\rfloor$$

in the second term and the authors give a numerical means of computing the required expected value of the maximum of three Erlangs in their appendix.

A means of approximating the system throughput rate, which often gives a lower bound, is given by:

$$T\big[\lambda_1/\lambda_2/\mu/K_1/K_2\big] \approx \max\Big\{ T\big[\lambda_1/\mu[1 - p_0(\lambda_2/\mu/1/K_2)]/1/K_1\big],$$
$$T\big[\lambda_2/\mu[1 - p_0(\lambda_1/\mu/1/K_1)]/1/K_2\big]\Big\}.$$

A lower (upper) bound on the throughput rate, T_{LB} (T_{UB}), provides an upper (lower) bound on the inventory of bins held in buffer p:

$$I_p\big[\lambda_1/\lambda_2/\mu/K_1/K_2\big] \le K_p - T_{LB}/\lambda_p.$$

Similarly, an upper bound on the throughput rate, T_{UB}, provides a lower bound on the inventory of bins held in buffer p, as do the $M/M/1/K$ system and an assembly system with $\mu = \infty$:

$$I_p\big[\lambda_1/\lambda_2/\mu/K_1/K_2\big] \ge \max\Big\{ [1 - T_{UB}/\lambda_p]K_p, I[\lambda_p/\mu/1/K_p],$$
$$I_p\big[\lambda_1/\lambda_2/\infty/K_1/K_2\big]\Big\},$$

where:

$$I_1\big[\lambda_1/\lambda_2/\infty/K_1/K_2\big] = \frac{\rho(1 - \rho^{K_1}) - K_1\rho^{K_1+1}(1 - \rho)}{(1 - \rho)^2} \qquad \text{if } \rho \ne 1$$
$$= K_1(K_1 + 1)/2 \qquad \text{if } \rho = 1$$

and I_2 $[\lambda_1/\lambda_2/\infty/K_1/K_2]$ results from substituting K_2 for K_1 and ρ^{-1} for ρ.

An approximation for expected inventory levels is given by:

$$AI_1\big[\lambda_1/\lambda_2/\mu/K_1/K_2\big] \approx I_1\big[\lambda_1/\mu[1 - p_0(\lambda_1/\mu/1/K_2/1/K_1)]\big]$$

for $p = 1$ and by exchanging subscripts 1 and 2 for the corresponding approximation for $p = 2$.

These bounds and approximations are easy to compute and may be used, for example, to determine optimal buffer capacities for a two-part assembly system. A computational evaluation provided by Hopp and Simon (1989) shows that their bounds and approximations perform well relative to exact values and to the bounds of Lipper and Sengupta (1986), at least in some instances. The next section considers approximations for large-scale assembly/ disassembly networks.

6.2.2 *Assembly/disassembly networks*

The assembly/disassembly network is, perhaps, the most general system configuration which has been studied. In these networks of queues, some stations may perform disassembly operations, and others assembly operations as shown in Fig. 6.2. An assembly/disassembly network might model automotive production and recycling, while a disassembly/assembly network might model aircraft engine repair operations.

Gershwin (1991) develops a method to approximate expected throughput rate and average in-process inventory in each buffer in acyclic (or tree-structured) assembly/disassembly networks in which machines are subject to failure, buffers have finite capacity and each machine operates whenever none of the upstream buffers that feed it are empty and none of the downstream buffers that it feeds are full. If a machine is down for repair, the buffers that feed it will tend to accumulate workpieces, and those it feeds will tend to be depleted. If a machine is in repair for a lengthy time, it will tend to block upstream stations and starve downstream stations.

By assumption, each station is composed of a single machine, and all machines have fixed and equal processing times. Buffers are assumed to be finite. Variability is introduced by assuming that operational (or up) time and repair times are geometrically distributed. Buzacott (1967) first invoked these discrete-time assumptions in analyzing the transfer line. Other systems that operate according to these discrete-time assumptions have been analyzed, including the three-machine transfer line for which Gershwin and Schick (1983) derived an exact solution, and longer tandem lines for which Gershwin (1987) developed an approximation method. This method is extended to the assembly/disassembly network (Gershwin, 1991) by modeling a set of related,

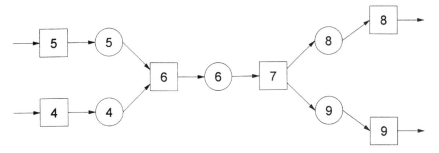

Fig. 6.2 An assembly/disassembly network (adapted from Gershwin (1991)).

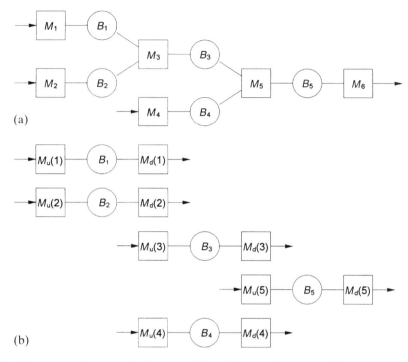

(a)

(b)

Fig. 6.3   Assembly network decomposition: (a)assembly system; (b) decomposition (adapted from Gershwin (1991)).

two-machine tandem lines. For example, each buffer in an original network that is an assembly tree establishes one of these tandem lines by associating it with the machine that feeds it and the one it feeds, as shown in Fig. 6.3.

For each of these two-machine tandem lines it is possible to write a set of equations that represent conservation of flow; flow rate and idle time relationships; the resumption of flow after repair, relating neighboring two-station lines; conditional probabilities that upstream and downstream machines are both in repair at the same time; and boundary conditions. These equations are solved iteratively, proceeding in a special manner from upstream to downstream and then reversing the direction of evaluation. By comparing with simulation estimates, Gershwin (1991) shows that the resulting approximations of expected throughput and average in-process inventory at each buffer are highly accurate. Confirming earlier results, he also shows that the average in-process inventory increases without limit as buffer capacity increases. The approximation is less accurate if buffer capacities are large.

An alternative approach for modeling acyclic assembly networks is based on similar assumptions that processing times are constant, buffers provide finite capacity and machines fail. However, this approach assumes that times to failure as well as repair times are exponentially distributed. In addition, it assumes that work flows in a continuous process rather than in discrete steps. Such a continuous flow model would most closely represent fluid blending, for example in a chemical process, but it may also be used as an approximation to the discrete case. The interested reader may refer to Simon and Hopp (1991) for an analysis of inventory in a simple continuous flow case; to Dallery, David and Xie (1989) and David, Xie and Dallery (1990) for an analysis of the transfer line; and to Di Mascolo, David and Dallery (1991) for a model of more complex assembly networks.

Following this discussion of approximations to assembly queues and networks, the next subsection considers another configuration, an assembly system which produces end products for inventory.

6.2.3   Analysis of an assembly/inventory system

An important and unique aspect of assembly operations is kitting (or accumulating) required components and subsequently releasing the kit to initiate assembly. Due to the stochastic nature of component availability, random events frequently disrupt kitting and, consequently, assembly schedules. This section describes an explicit model of stochastic kitting operations. The model leads to a new understanding of the kitting process and its influence on system performance.

In addition, this section shows that certain systems may be decomposed for detailed study. This result allows that, under certain conditions, the system can be decomposed into two components to facilitate analysis: the upstream kitting process and the downstream assembly and end-product inventory. The following two subsections deal with these respective components.

Exact analysis of the kitting process   Som, Wilhelm and Disney (1994) model the kitting process of an assembly system as a Markov renewal process, assuming that component arrival streams follow independent Poisson distributions. The system is assumed to be an assembly-like queue as shown in Fig. 6.4.

M_1 and M_2 are machines that process components (to prepare them for assembly) and M_3 is the assembly machine. I_1 and I_2 are the buffers for components, I_0 is the buffer for kits and I_3 is the buffer for the end product. M_1 and M_2 work independently, withdraw raw materials from their respective pools of unlimited supply, and deliver processed components to buffers I_1 and I_2 respectively. A component arriving at buffer I_1 (I_2) is immediately kitted with a part from buffer I_2 (I_1) if one is available, and a 'kit' is said to be composed. If a kit cannot be composed, the processed part is held in buffer I_1 (I_2) to await the arrival of a 'matching' part at buffer I_2 (I_1). Once composed, a kit of matching components from I_1 and I_2 is sent immediately to I_0 and the kit is considered to be one arrival at I_0. If the arriving kit finds I_0 empty and M_3 idle, it is immediately placed in the assembly machine M_3. Otherwise, the kit is held in buffer I_0.

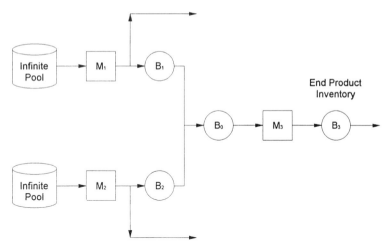

Fig. 6.4 A stochastic assembly/inventory system (Som, Wilhelm and Disney, 1994).

It is assumed that component buffers have limited capacity and that each component is processed according to an exponential distribution (before kitting) to prepare it for assembly. When M_3 completes an assembly, it withdraws a kit (i.e. two matched components) from I_0, whenever available, then assembles another end product and delivers it to buffer I_3. If a kit is not available in I_0 when M_3 completes an assembly, it remains idle until a completed kit arrives. Demands for end products arrive at I_3; each demand is assumed to be for a lot of unit size and is satisfied immediately if stock is available. Unsatisfied demands are backordered, causing the inventory position (defined as the number of parts on hand plus on order minus the number on back order) at I_3 to take on negative values.

Som, Wilhelm and Disney (1994) show that if components arrive according to Poisson processes, the output stream departing the kitting operation is a Markov renewal process. They also derive the distribution of time between kit completions and show that, when component arrival streams are identical Poisson processes and when the component buffers are large enough, the output stream of kits approximates a Poisson process with parameter equal to that of the input stream. This result allows assembly to be (approximately) decoupled from kitting, allowing the assembly operation to be analyzed separately. If component buffer capacities are finite, the long run probability of observing inventory position j at I_1 (I_2) depends on the inventory position j.

Since two components are required to compose a kit, the queues of components form a double-ended queue (Dobbie, 1961; Kashyap, 1965). A double-ended queue is demonstrated by the well known taxi-cab problem in which taxis and passengers form two different queues. A customer waits in queue and leaves it as soon as a taxi is available; taxis wait in queue for customers and leave when a customer is available. The two queues are interdependent and their combination is known as a double-ended queue for which the related queuing process is known to be a random walk on $\{\ldots, -2, -1, 0, 1, 2, \ldots\}$ and is transient or null unless the queues are bounded. The kitting process in Fig. 6.4 can be considered as a double-ended queue of the type examined by Kashyap and Chaudhury (1988). They show that each queue length distribution is independent of occupancy when arrival rates to the double-ended queue are equal. They also derive the distribution of waiting times in double-ended queues but make no attempt to analyze its output process.

Bhat (1986) incorporated limited buffer capacities in assembly-like queues and derived expressions for the stationary probability vector of the queue length. Latouche (1981) considered assembly systems with Poisson procurement processes and exponential processing times and derived conditions required for stability. Assembly networks that represent one-time production (for example the space shuttle, aircraft prototypes, etc.) are discussed in Chapter 8.

The output processes from queues operating according to various disciplines are reviewed by Disney and Konig (1985) in detail, including $GI/D/s$, $M/M/s$, $M/GI/1/L$, $M/E_K/1/L$, $M/GI/\infty$, $GI/GI/1/L$ and $GI/M/1/L$ systems. Apparently, Som, Wilhelm and Disney (1994) were first to analyze the output process of a double-ended queue.

The model of the kitting operation considers upstream operations (Fig. 6.4) and results in arrivals of completed kits at buffer I_0. These operations are discussed below, since they reflect operations in more general systems and lend insight into their performance.

A little thought indicates that it is not possible for both buffers I_1 and I_2 to have positive stock levels at the same time. An arrival which increases the stock level of one of the buffers to a positive value creates a 'virtual backorder' at the other buffer. At any time t $(t > 0)$, the inventory position '#' in one buffer is associated with inventory position '-#' in the other, and equality holds only when the inventory position is zero (0) for both buffers I_1 and I_2. The inventory positions at I_1 and I_2 may thus be viewed as 'mirror images' of one another, a special structure that can be exploited to analyze the kitting process.

The model described in this section relies upon three fundamental assumptions.

1. Processing times at part processing machines M_1 and M_2 are i.i.d., non-negative exponential random variables with rates μ_1 and μ_2, respectively.
2. The capacities of buffers I_1 and I_2 are bounded from above by K_1 and K_2, respectively, representing practical limitations on buffer space, and allowing the system to reach a steady state. No capacity restriction is imposed on I_0.
3. M_1 (M_2) prepares parts exclusively for I_1 (I_2). However, when I_1 (I_2) is filled to capacity K_1 (K_2), additional arrivals are not processed in the

system under analysis (e.g. they may be processed and assembled by a subcontractor).

The inventory positions at I_1 and I_2 change with the arrival and departure of components to and from the respective buffers. Som, Wilhelm and Disney (1994) analyze the **mirror image process** as a **marked point process**, characterizing the inventory positions or states at the arrival and departure epochs. Because of the mirror image relationship between inventory positions in I_1 and I_2, it is necessary to consider only the inventory position at I_1 (or equivalently, at I_2) to completely describe operations. Kits are completed at epochs for which the positive inventory position decreases or the negative inventory position increases. The series of kit completions constitutes the output process from kitting or, equivalently, the arrival process to buffer I_0.

While the details of the stochastic analysis are beyond the scope of this book, the results are described below. In particular, the **output process** is characterized by:

1. the stationary probability vector Π of the underlying Markov chain; and
2. the distribution of time between kit completions.

The vector Π indicates the time-stationary probability distribution of the inventory position at I_1, observed at a randomly selected kit completion epoch.

The output process is an irreducible, non-null, recurrent and persistent Markov renewal process for $K_1, K_2 < \infty$; under these conditions, it possesses a stationary distribution defined as Π (Disney and Kiessler, 1987). Note that the output process will be recurrent null, if K_1 and K_2 are infinite. The stationary probability vector Π of the underlying Markov chain Z is obtained from the set of equations expressed in the matrix form:

$$\Pi = \Pi P$$

along with the normalizing expression:

$$\sum_j \Pi(j) = 1.$$

The solution to these equations gives $\Pi(j)$, the stationary probability of positive (negative) inventory position j in buffer I_1, observed at a kit completion time:

$$\Pi(0) = \frac{(\rho - 1)(\mu_1 + \mu_2)}{\mu_2 \left(\rho^{K_1} - \rho^{-K_2} \right)}$$

$$\Pi(j) = v\rho^j \Pi(0) \qquad j = 1, 2, ..., K_1 - 1$$

$$\Pi(j) = (1 - v)\rho^j \Pi(0) \qquad j = -1, -2, ..., -K_2 + 1.$$

It may be observed that the probability $\Pi(j)$ depends on the inventory position j.

The distribution of time between two consecutive kit completions $(\tau_{n+1} - \tau_n)$ is:

$$L\left[Dp(\tau_{n+1} - \tau_n \leq t)\right] = \left(\frac{\mu_1}{\mu_1 + s}\right)\left[\Pi(0) + \sum_{j=-1}^{-K_2+1}(1-v)\rho^{j}\Pi(0)\right]$$

$$+ \left(\frac{\mu_2}{\mu_2 + s}\right)\left[\Pi(0) + \sum_{j=1}^{K_1-2}v\rho^{j}\Pi(0)\right] - \left(\frac{\mu_1 + \mu_2}{\mu_1 + \mu_2 + s}\right)\left[\Pi(0)\right]$$

in Laplace transform form. If this were inverted (i.e. to the time domain), it is apparent that the distribution of the time between kit completions would be the weighted sum of three exponential distributions with rates μ_1, μ_2 and $\mu_1 + \mu_2$.

Now, consider the special case in which component processing times at machines M_1 and M_2 are independent exponential random variables with the same rates (i.e. $\mu_1 = \mu_2 = \mu$). In practice, this situation may occur when components are obtained from independent suppliers with identical (and independent) lead-time distributions. Also, the same situation may occur during 'in-house' production where the machines employed, M_1 and M_2, are identical (and independent).

In this case, the stationary probabilities of inventory position are given by

$$\Pi(0) = \frac{2}{K_1 + K_2}$$

$$\Pi(j) = \frac{1}{K_1 + K_2} \qquad \forall j \neq 0.$$

These results have striking similarities with those obtained by Bhat (1970) for the limiting distribution of the population in the finite buffer of a double-ended queue.

The distribution of time between kit completions, in Laplace transform form, is

$$L\left[Dp(\tau_n - \tau_{n-1} \leq t)\right] = \left(\frac{\mu}{\mu + s}\right)\left[1 - \left(\frac{2}{K_1 + K_2}\right)\left(\frac{s}{2\mu + s}\right)\right].$$

Clearly, for large values of $K_1 + K_2$, the distribution of time between kit completions is approximately exponential with rate μ. The value of $K_1 + K_2$ necessary to allow this approximation can be determined as a function of the degree of approximation desired. The Laplace transform of the ε-approximate distribution of time between kit completions is $\dfrac{\mu}{\mu + s}$, which is the Laplace transform of an exponential distribution with rate μ. For the output process to be exponential, it is also necessary that consecutive interdeparture intervals be statistically independent. By showing that the joint distribution of m consecutive interdeparture intervals equals the product of the respective m marginal distributions, Som, Wilhelm and Disney (1994) proved that, for sufficiently large $K_1 + K_2$, m consecutive interdeparture intervals become independent to within an error of ε. Statistical independence holds for $m \to \infty$, but this limiting case is not easily evaluated.

These results lead to a theorem which allows the system to be decomposed for analysis:

> **Theorem 1:** The arrival process of kits at buffer I_0 can be approximated by a Poisson process with rate μ, the degree of approximation depending on the value of $K_1 + K_2$.

To illustrate the relationship between the degree of approximation of the arrival rate at I_0 and the buffer capacities K_1 (K_2), consider the following example with equal buffer capacities $K_1 = K_2 = K$ and equal Poisson arrival rates $\mu_1 = \mu_2 = \mu$ at buffers I_1 and I_2, respectively. The density function of the time between kit completions, in Laplace transform form, is:

$$f(s) = \left(\frac{\mu}{\mu+s}\right) - \frac{1}{K}\left(\frac{\mu}{\mu+s}\right)\left(\frac{s}{2\mu+s}\right).$$

Inverting to the time domain, the density function is:

$$f(t) = \mu e^{-\mu t} + \frac{\mu}{K}e^{-\mu t} - \frac{2\mu}{K}e^{-2\mu t}, \quad t \geq 0.$$

Now, define the error $e(t)$ as the absolute difference between the exponential density and the actual density of the interdeparture interval:

$$e(t) = \frac{\mu}{K}\left|2e^{-2\mu t} - e^{-\mu t}\right|.$$

A graph of $e(t)$ against K, plotted for $\mu = 1$ and $t = 0.2$ (Fig. 6.5) indicates that the error term $e(t)$ approaches zero (0) rapidly as K increases.

Using the definition of $e(t)$, it is easy to determine the buffer capacity K that achieves a desired level of accuracy ($\varepsilon > 0$) over all t ($t > 0$) for the component arrival rate μ using:

$$\frac{\mu}{K} \leq \varepsilon.$$

This section gives conditions for which the interarrival times of kits arriving for assembly are approximately i.i.d. exponential random variables. If compo-

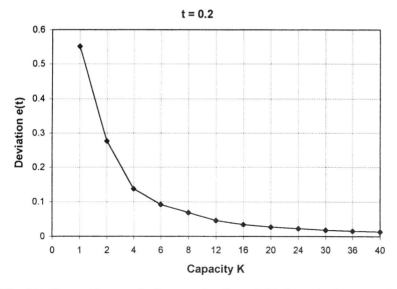

Fig. 6.5 Error $e(t)$ versus buffer capacity (Som, Wilhelm and Disney, 1994).

nents arrive at I_1 and I_2 according to i.i.d. Poisson arrival streams and if $K_1 + K_2$ is sufficiently large, the output stream from kitting approximates a Poisson process. The practical importance of this result is that the assembly process downstream of the kitting operation can be decoupled from kitting for further analysis. The required conditions (for decoupling) are not restrictive and may, in fact, hold in actual applications.

It is also interesting to note that the expressions for the long-term probability distribution of inventory position j at I_1 (I_2), observed at kit completion epochs, depends on the inventory position j. If arrival rates to I_1 and I_2 are equal (i.e. $\mu_1 = \mu_2 = \mu$), all the inventory positions (except zero) become equally likely with a probability that is inversely proportional to the total inventory capacity ($K_1 + K_2$). The incidence of observing both buffers empty is twice as likely as observing a particular inventory position at either of the buffers.

Harrison (1973) showed that a sufficient condition for an assembly-like queue to reach steady state is that buffer capacities must be bounded from above. New results described in this section show that the total buffer capacity $K_1 + K_2$ must be 'sufficiently large' to obtain a Poisson approximation of the output stream of kits. However, from Fig. 6.5, we see that $K_1 + K_2$ need not be impractically large to achieve an approximate Poisson output stream; the value of $K_1 + K_2$ depends upon the degree of approximation desired. Since the arrival process at assembly machine M_3 may be approximated by a Poisson distribution, the downstream assembly operation can be approximated by the much studied $M/G/1$ queue as described next.

Numerical solution for assembly and end-product inventory   In an assembly system, variability occurs due to the probabilistic nature of both production and supplier lead-times. The variability in procurement lead-time depends on the performance of individual suppliers, and that of production lead-time can be attributed to both setup and processing times. The previous subsection describes conditions under which an assembly system such as the one in Fig. 6.4 can be decomposed for analysis and models the effects of procurement lead-time variability on kitting operations. This subsection completes the analysis by dealing with the downstream assembly operation.

A component is processed at station M_1 (or M_2) then moves to buffer I_1 (or I_2, respectively) to be matched with a component from buffer I_2 (I_1) to compose a kit, which is moved to buffer I_0. If the arriving kit finds I_0 empty and M_3 idle, it is immediately placed on the assembly machine M_3. Otherwise, the kit is held in buffer I_0. When M_3 completes an assembly, it withdraws a kit from I_0, whenever available, then assembles another end product and delivers it to buffer I_3. If a kit is not available in I_0 when M_3 completes an assembly, it remains idle until a completed kit arrives. Demands for end products arrive at I_3; each demand is assumed to be for a lot of unit size and is satisfied immediately if stock is available. Unsatisfied demands are backordered, causing the inventory position at I_3 to take on negative values.

The arrival stream of kits at I_0 is assumed to follow a Poisson distribution with parameter μ. The validity of this assumption is justified in the previous section. The processing times of assembly machine M_3 are assumed to be independent and identically distributed non-negative random variables, which follow a general distribution $\Phi(\cdot)$ that is independent of arrivals.

Latouche (1981) considers a system with independent Poisson inputs of two types of components and exponentially distributed assembly times. He shows that the system reaches steady state when the arrival rates depend on the difference between the queue lengths of the two components. He uses the matrix-geometric approach of Neuts (1981) to find the steady-state probability vector. Bonomy (1987) proposes a procedure to approximate average queue length and assembly throughput rate for systems with multiple Poisson inputs.

These cases consider an assembly system with Poisson inputs of components and exponential assembly times. Even with the simplifying assumption that assembly times are exponential, analytical solutions cannot be derived due to the size of the state-space required to model end-product inventory. Approximation methods have, thus, been commonly used to provide bounds on performance measures instead of actual distributional measures. Wilhelm and Som (1993) extend these earlier results by modeling assembly times as G.I. random variables, generalizing earlier assumptions based on the exponential distribution, and resulting in distributional measures, thus providing new capability to analyze assembly systems. They show that the assembly process which produces end products, described as the **work completion process**, is a Markov renewal process.

To assure stability of the assembly system, finite capacity limitations must be placed on I_0 and I_3 (Harrison, 1973). However, when capacity limitations are imposed on both I_0 and I_3, it does not appear possible to derive a closed form solution for the distribution of the end-product inventory position. Hence, the matrix-geometric approach (Neuts, 1981) may be used to obtain a numerical solution for the inventory position distribution. The inventory capacity of I_0 is restricted to zero from below. It is assumed that backorders up to K_3 will be accepted for end products, but there exists no upper limit on the capacity of buffer I_3. The demand for the end product is assumed to follow a Poisson distribution with rate λ. The distribution of time between kit arrivals at I_0 is exponential and is given by $\Psi(\cdot)$.

The inventory position at I_0 may be defined to include the number of parts on hand as well as any work in process at M_3 (i.e. a kit in the server). The inventory position at I_0 thus changes with the arrival of kits at I_0 and their departure from M_3. A departure from the inventory position at I_0 is marked by the completion of a kit on machine M_3 and the subsequent transfer of the end product to I_3. The inventory position at I_3 notes the number of products on hand plus on order minus the number on backorder and changes with the arrival and departure of finished goods to and from I_3. A departure from I_3 occurs when demand for an end product arrives at I_3. The inventory position at I_0 can be distinguished as the **kit inventory position** and the inventory position at I_3 as the **end-product inventory position**.

Observing the state change epochs at I_0 and I_3, it is possible to identify a sequence of times at which state changes occur, and the corresponding inventory positions at I_0 and I_3 are used to describe the work completion process. Whenever M_3 completes an operation and sends an end product to I_3, there is a decrease in the kit inventory position and a simultaneous increase in the end-product inventory position. These epochs, along with the corresponding inventory positions at I_0 and I_3, describe the **work completion process**. If we observe the inventory position at I_0, it is apparent that a particular subset of the epochs, marked by a decrease in the positive inventory position at I_0, indicate the completion of an end product by M_3 as well as a state change in the work completion process. The corresponding set of inventory positions at I_0 and I_3 define the **work completion state process** and the joint random variables which incorporate the epochs define the work completion process.

Wilhelm and Som (1993) show that the work completion process is a Markov renewal process and that the work completion state process forms a Markov chain with one-step probability matrix **P**. They analyze the work completion process producing the following:

1. the one-step transition probability matrix **P** of the underlying Markov chain;
2. the Palm probability vector **v** of the end-product inventory position observed at end-product completion epochs;

3.   the approximate joint distribution of the inventory positions at I_0 and I_3 using the matrix-geometric approach.

Once the state transition matrix **P** and its elements are known, all the required distributional measures on the kit and end-product inventory positions can be obtained. Also, the average kit and end-product inventory positions in the long run can be obtained from:

$$\overline{Y} = \text{average end-product inventory position in the long run;}$$
$$\overline{X} = \text{average kit inventory position in the long run.}$$

To demonstrate the solution approach, consider a numerical example that provides Palm probability distributions of kit and the end-product inventory position, and average inventory positions of kits and end products. From this numerical example, the effect of kitting on end-product inventory can be observed. The following inputs are used in the example:

Distribution function of the time between kit arrivals:
$\Psi(t) = 1 - e^{-\mu t}$ and the corresponding density function, $d\Psi(t) = \mu e^{-\mu t}$

Distribution function of the time between demands of end products:
$\Theta(t) = 1 - e^{-\lambda t}$ and the corresponding density function, $d\Theta(t) = \lambda e^{-\lambda t}$

Distribution function of the processing time on the assembly machine M_3:
$\Phi(t) = 1 - e^{-\nu t}$ and the corresponding density function $d\Phi(t) = \nu e^{-\nu t}$

As in prior studies (Hopp and Simon, 1989; Lipper and Sengupta, 1986), Wilhelm and Som (1993) assume that the processing time on machine M_3 is exponentially distributed. This assumption facilitates evaluating convolution operations. However, this approach is not restricted to a particular distribution for the processing time on assembly machine M_3. Let:

$K_0 = $ finite capacity of buffer I_0; and

$K_3 = $ maximum number of backorders allowed at buffer I_3.

Using the distribution functions $\Psi(t)$, $\Theta(t)$ and $\Phi(t)$, it is possible to derive necessary expressions that we evaluate numerically to model the case in which $\lambda = 0.7$, $\nu = 0.5$, $\mu = 0.3$, $K_3 = 0$ and K_0 has one of three values: 3, 4 and 5.

Tables 6.1–6.3 show the long-term marginal Palm probabilities of kit and end-product inventory positions, respectively. From Table 6.1, it is observed that the probability that kits will stock-out decreases as the capacity of buffer I_0 increases. Also, the probability of observing buffer I_0 filled to capacity decreases sharply as the capacity of I_0 increases. The marginal Palm probability distribution of the end-product inventory position, corresponding to $K_0 = 3$, 4 and 5, is given in Table 6.2. We can observe that the probability of end-

Table 6.1 Long-term marginal probability of kit inventory position

$K_0 = 5$	$K_0 = 4$	$K_0 = 3$
$z_0 = 0.43373$	$z_0 = 0.4596$	$z_0 = 0.5102$
$z_1 = 0.26024$	$z_1 = 0.2757$	$z_1 = 0.3061$
$z_2 = 0.15614$	$z_2 = 0.1654$	$z_2 = 0.1837$
$z_3 = 0.09368$	$z_3 = 0.0993$	
$z_4 = 0.05621$		

z_j = long-term marginal probability of observing j kits.

Table 6.2 Long-term marginal Palm probability of end-product inventory position

	$K_0 = 5$	$K_0 = 4$	$K_0 = 3$
y_0	0.5322	0.5423	0.5689
y_1	0.2684	0.2697	0.2709
y_2	0.1213	0.1174	0.1069
y_3	0.0490	0.0455	0.0354
y_4	0.0186	0.0162	0.0119
y_5	0.0068	0.0057	0.0040
y_6	0.0025	0.0020	0.0013
y_7	0.0009	0.0007	0.0005
y_8	0.0003	0.0003	0.0002

y_n = long-term marginal Palm probability of observing n end products.

Table 6.3 Long-term average inventory positions

		$K_0 = 5$	$K_0 = 4$	$K_0 = 3$
Kits	$\overline{Z}$	1.0784	0.9044	0.6735
End Products	$\overline{Y}$	0.8535	0.7586	0.6714

product stock-out decreases as K_0 increases. This relationship indicates that a large buffer capacity for kits at I_0 reduces the chance of end product stock-out. Also, the likelihood of observing more end products in buffer I_3 increases with an increase in the capacity of kit buffer I_0. Table 6.3 shows the long-term average inventory positions of kits and end products, and it is apparent that average inventory positions for both kits and end products fall as the capacity of buffer I_0 decreases.

It is easy to observe that the buffer capacity for kits plays an important role in designing the assembly system. A larger buffer capacity for kits reduces the risks of kit as well as end-product stock-out. Also, blocking at I_0 can be avoided by increasing the kit buffer capacity.

This subsection describes a model that treats the assembly operation as an $M/G/1$ queue operating in conjunction with the end-product inventory. While it appears intractable to formulate an exact model, system performance can be studied effectively using numerical techniques. Combining the exact model of the kitting process on page 279 with the analysis of downstream operations presented in this subsection, section 6.2.3 describes a new approach for modeling certain types of assembly systems. Next, a model for stochastic MRP control is discussed.

6.2.4   Stochastic MRP control

Materials Requirements Planning (MRP) (Orlicky, 1975) has two primary shortcomings: it is based on the assumptions that adequate capacity is available and that the system operates deterministically. This section deals with the latter assumption, summarizing a model of MRP in stochastic assembly systems.

Actually, relatively few stochastic models of MRP-controlled assembly/inventory systems have been derived because such systems present difficult challenges to analysis. A single-stage production facility serving the demand for a single product was first studied by Gavish and Graves (1980), who assumed a constant production rate. In a companion paper, Gavish and Graves (1981) considered an arbitrary distribution for production time and Poisson demand for the end product. They developed a continuous review policy to minimize expected assembly/inventory cost. However, their approach does not consider MRP control for production; rather, it follows an (R, r) policy.

A two-stage production/inventory system, operating according to a base stock policy, was studied by Zipkin (1989). By assuming that production times at both stages are exponentially distributed and that demand follows a Poisson process, he was able to analyze the system as a simple Markov process. He obtained several performance measures, including the average inventory level and the average number of end product backorders.

In recent years, researchers have devoted much attention to comparing the operating philosophies of MRP with kanban-controlled, just-in-time (JIT) production systems. Buzacott (1989) modeled a kanban-controlled production system with stochastic demand and stochastic processing times as a linked queuing network. He derived various performance measures related to system capacity: delivery performance, inventory levels and number of kanbans. He also showed that a conventional MRP system can be approximated by a kanban-controlled production system if kanbans are triggered by forecast instead of actual demand.

A single-stage, MRP-controlled production/inventory system with sto-chastic demand, stochastic processing times and limited production capacity was studied by Buzacott and Shanthikumar (1994). Their model of the MRP-controlled production system was based on observations at discrete as well as at continuous times. The authors considered the relative importance of the key control variables – lead-time and safety stock – and showed that lead-time is less appropriate than safety stock to manage uncertainties.

A production facility with a single processor producing multiple products was studied by Altiok and Shiue (1994). Assuming Poisson demand, arbitrarily distributed processing and setup times, and backorders, they were able to determine various performance measures, including average inventory level, average backorder level and the probability distribution of backorders. They identified that a single-product system reduces to an $M/G/1$ queue.

Wilhelm and Som (1994) derive a model of the MRP control of a stochastic, single-stage assembly system, explicitly treating processing times and demand as random variables. The model provides various performance measures; in particular, it can be used to obtain insights into the effects of kitting in this stochastic environment – effects which are overlooked by the traditional de-terministic approach.

Wilhelm and Som (1994) identify a special stochastic structure in the process that describes the end-product inventory position. They show that this inven-tory position process is a Markov renewal process and exploit this structure to determine system performance measures such as average inventory level, average backorder level, the probability distribution of the end product inven-tory position and the effect of kitting on the availability of end products. Their novel approach allows them to model systems in which processing times are generally distributed random variables that depend on order quantities.

To best interpret their results, we must review the specific system and ordering strategy they modeled. They considered a single-stage production/ inventory system, consisting of production center P and end-product inven-tory I as shown in Fig. 6.6. In an assembly system, the production center may consist of machines $M_1, M_2, \ldots, M_n$, each of which withdraws raw materials from an unlimited pool and processes a unique part type. When a kit of parts required to assemble one lot of end product is composed, it is moved instantaneously to machine M_A, which assembles all n parts into the end product. After assembly, the lot of end products is moved immediately into inventory I.

Demands arrive at end-product inventory I according to a Poisson process and are satisfied immediately if stock is available. Unsatisfied demands are backordered. A specialized production-ordering strategy is followed: an order for producing one lot of end products is placed on P at the instant a **replenish-ment order** arrives at I. The rationale for this ordering strategy is that, since a single-stage production facility is considered, it can start processing an order only after it finishes the previous one. No orders are placed while the machine is producing, since any such order would needlessly wait in queue and would,

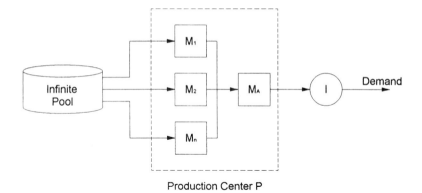

Production Center P

Fig. 6.6   Single-stage production/inventory system (Wilhelm and Som, 1994).

therefore, provide no advantage to the system. This scheme allows the ordering rules to use the latest possible information about the end-product inventory position, and the ordering quantity is determined by looking ahead into the future, which is a key characteristic of MRP control. Production lead-times vary with order quantities and are considered independent and generally distributed.

The inventory position (defined as the number of items on hand plus on order minus the number on backorder) at an instant when a replenishment order arrives at I determines the quantity of the next replenishment order to be placed on P: either the quantity Q or nothing at all will be ordered. This ordering strategy has been used for deterministic assembly systems (Vollman, Berry and Whybark, 1984) in which replenishment order lead-time is assumed to be constant. In deterministic MRP, when an order is released for quantity Q, it arrives after a constant lead-time L. Wilhelm and Som (1994) treat this lead-time as a random variable. Again, in deterministic MRP, when no order is placed in some period (i.e. the order release quantity is zero), the projected available balance is checked at the end of a time bucket of deterministic duration, when a new order release quantity is determined. Equivalently, a stochastic MRP system can check again after a random time for which the replenishment order quantity is zero.

Various MRP lot-sizing techniques are described in the literature (Nahmias, 1989); for example, lot-for-lot, economic ordering quantity (EOQ), the part period balancing (PPB) rule, and the least unit cost (LUC) rule. Wilhelm and Som (1994) assume that order quantity Q is the economic ordering quantity. However, they make no attempt to obtain the optimum batch size, since they do not address the stochastic optimization issue.

To describe the ordering philosophy, we need the following notation:

s = safety stock;
Q = economic ordering quantity;
τ_m = instant of the mth replenishment order arrival;

X_m = inventory position at τ_m (prior to a replenishment order arrival);
R_m = replenishment order quantity arriving at time τ_m (either 0 or Q);
λ = mean rate of the Poisson demand process.

After an order for Q items is placed, the production center P takes random time L_Q to produce the lot, and the replenishment order arrives at time τ_{m+1}, a time duration of L_Q after τ_m. If no order is placed at time τ_m (i.e. replenishment quantity $R_{m+1} = 0$), the inventory position is checked again at time τ_{m+1}, after random time duration L_R following τ_m, to determine if an order should be placed at that time. The random variables L_Q and L_R are G.I. distributed with distribution functions $\Phi(\cdot)$ and $\Psi(\cdot)$, respectively. At time τ_m an order is placed, either for Q or 0 end products, ensuring that the expected safety stock that will exist when the replenishment order (of lot size R_{m+1}) arrives at time τ_{m+1} will be at least s.

Representing practical relationships, assume that:

$$E\big[L_Q\big] \geq E\big[L_R\big].$$

If the order quantity is Q, the following inequality must be satisfied to ensure an expected safety stock of s end products at time τ_{m+1}:

$$X_m + R_m - \lambda E\big[L_Q\big] \geq s$$

in which:

$$E\big[L_Q\big] = \text{the expected value of the random lead time } L_Q.$$

Rearranging, the criterion becomes:

$$X_m + R_m \geq K_1$$

in which:

$$K_1 = s + \lambda E\big[L_Q\big].$$

If the order quantity is zero, the following inequality must be satisfied to ensure an expected safety stock of s end products at time τ_{m+1}:

$$X_m + R_m - \lambda E\big[L_R\big] - \lambda E\big[L_Q\big] \geq s$$

in which:

$$E\big[L_R\big] = \text{the expected value of the random lead-time } L_R$$

and $\lambda E[L_Q]$ must be included, since, if $R_m = 0$, the earliest expected time that a replenishment order of quantity Q might arrive is $\tau_m + L_R + L_Q$. Rearranging, the criterion becomes:

$$X_m + R_m \geq K$$

in which:

$$K = s + \lambda E[L_Q] + \lambda E[L_R].$$

Since $K \geq K_1$, the event of placing an order when the inventory position reaches K contains the event of placing an order when the inventory position reaches K_1. The ordering strategy which represents this stochastic MRP control can be summarized by:

$$\text{if} \quad K - X_m - R_m > 0, \qquad \text{order } Q$$
$$\text{if} \quad K - X_m - R_m \leq 0, \qquad \text{order } 0$$

When the inventory position at time τ_m is less than the prespecified value K, the quantity Q is ordered. Otherwise, no order is placed at time τ_m.

In a deterministic MRP environment, the system is observed at fixed time intervals and replenishment orders for required quantities are placed at these times. In the stochastic environment, the system is observed at random times $\tau_m : m \in N$. The decision setting the order quantity R_{m+1} at time τ_m is made considering the expected inventory level at time τ_{m+1}, and, thus, represents the look-ahead feature of an MRP system.

Modeling the production center as a server for which processing times are independent and generally distributed random variables, and treating demand for the end product as Poisson, Wilhelm and Som (1994) show that end-product inventory positions and order quantities jointly follow a Markov renewal/semi-regenerative structure. They find that the capacity of the buffer I must be bounded from above by (finite) K for any ergodic distribution of the end-product inventory position to exist. In the limit as $K \to \infty$, the long-term probability reduces to the trivial solution, zero. They obtain several performance measures of the inventory position process including the Palm probability vector $\mathbf{v}$, based on observations at replenishment order arrival times; the stationary probability vector η, based on observations at arbitrary times; and the stationary probability vector of the end-product inventory position f. The underlying probability structure affords flexibility in modeling, allowing generally distributed processing times for the production operations.

To model the effects of kitting on end-product inventory position in this MRP-controlled system, the results, which are based on G.I. processing times at the production center, must be modified to account for assembly as detailed in Fig. 6.6. Raw materials for each part type are withdrawn from an unlimited pool and a random amount of time is required to process part i at machine M_i. When all n parts required for assembly have been readied, a kit is said to be composed and is sent immediately to the assembly station M. When the MRP control system places a production order for quantity Q, raw materials are withdrawn, the part production processes start, and each produces a lot large enough to produce Q end products (an end product may require more than one part of each type). To model these assembly operations, let:

X_i = time to produce a lot of part i at machine M_i to assemble Q end products;

V = time required to assemble end products in lot size Q at machine M_A.

Then, the lead time to compose a kit is:

$$Z = \max_{i=1,\dots,n}\{X_i\},$$

and the total lead-time to produce a lot of Q end products, defined previously as L_Q, can be defined in more detail to incorporate kitting as:

$$L_Q = Z + V$$
$$= \max_i\{X_i\} + V$$

For the period when no order is placed (i.e. the order quantity is 0), the lead-time is L_R, since no kitting is associated with such an order.

Wilhelm and Som (1994) present an example based on the following assumptions:

1. X_i, the processing time at machine M_i, distribution function $\Theta_i(t) = 1 - e^{-\mu_i t}$ and density function $d\Theta_i(t) = \mu_i e^{-\mu_i t}$;
2. L_Q, the lead-time to assemble a lot size of Q end products at machine M_A, has distribution function $\rho(t) = 1 - e^{-\mu t}$ and density function $d\rho(t) = \mu e^{-\mu t}$;
3. L_R, the review duration during which no end products are assembled, has distribution function $\Psi(t) = 1 - e^{-vt}$ and density function $d\Psi(t) = v e^{-vt}$;
4. the distribution function of the time between demands for end products is $\xi(t) = 1 - e^{-\lambda t}$ and the density function is $d\xi(t) = \lambda e^{-\lambda t}$.

The distribution function of the kitting time Z, described as $\chi(t)$, can be obtained from:

$$\chi(t) = \prod_{i=1}^{n}\Theta_i(t)$$
$$= \prod_{i=1}^{n}\left(1 - e^{-\mu_i t}\right).$$

The distribution function $\Phi(t)$ of the lead-time required to produce Q end products from raw materials, including the kitting operation, is obtained using the convolution operator $*$:

$$\Phi(t) = \chi * \rho(t)$$

Part processing times and the assembly time on machine M are assumed to be exponentially distributed, since it is convenient to work with both the maximum and convolution operators when the related distributions are exponential. The approach, however, is not restricted to exponential distributions.

Wilhelm and Som (1994) present five example cases using $\mu = 8$, $\lambda = 10$, $\nu = 4$, $Q = 5$, $K = 6$ and five different values of n (1, 2, 3, 4 and 5) with corresponding $\mu_1 = 5$, $\mu_2 = 4$, $\mu_3 = 6$, $\mu_4 = 5.5$ and $\mu_5 = 4.5$. To demonstrate some of the analysis, consider the case with $n = 2$, so that the density function of the time to compose a kit is:

$$\chi(t) = \prod_{i=1}^{2} \Theta_i(t)$$

$$= \prod_{i=1}^{2} \left(1 - e^{-\mu_i t}\right)$$

$$= 1 - e^{-\mu_1 t} - e^{-\mu_2 t} + e^{-(\mu_1 + \mu_2)t},$$

the distribution function of kitting plus assembly time is:

$$\Phi(t) = \chi * \rho(t)$$

$$= \int_0^y \left(1 - e^{\mu_1(y-t)} - e^{-\mu_2(y-t)} + e^{-(\mu_1+\mu_2)(y-t)}\right)\mu e^{-\mu t} dt$$

$$= 1 - e^{-\mu t} - \frac{\mu}{\mu - \mu_1}\left[e^{-\mu_1 t} - e^{-\mu t}\right] - \frac{\mu}{\mu - \mu_2}\left[e^{-\mu_2 t} - e^{-\mu t}\right]$$

$$- \frac{\mu}{\left(\mu - \mu_1 - \mu_2\right)}\left[e^{-(\mu_1+\mu_2)t} - e^{-\mu t}\right]$$

and the density function of kitting plus assembly time is:

$$d\Phi(t) = \mu e^{-\mu t} dt + \frac{\mu}{\left(\mu - \mu_1\right)}\left[\mu_1 e^{-\mu_1 t} - \mu e^{-\mu t}\right]$$

$$+ \frac{\mu}{\left(\mu - \mu_2\right)}\left[\mu_2 e^{-\mu_2 t} - \mu e^{-\mu t}\right] - \frac{\mu}{\left(\mu - \mu_1 - \mu_2\right)}\left[\left(\mu_1 + \mu_2\right)e^{-(\mu_1 - \mu_2)t} - \mu e^{-\mu t}\right].$$

By substituting $\Phi(t)$, $d\Phi(t)$, $\Psi(t)$ and $d\Psi(t)$ into the models of Wilhelm and Som (1994), it is possible to quantify the long-term average end-product inventory position. Average inventory positions for $n = 1, \ldots, 5$ are listed in Table 6.4.

From these results it is clear that as the number of components required to assemble an end product increases, the average end-product inventory posi-

Table 6.4 Long-run average end-product inventory position

Number of part types in a kit: n	1	2	3	4	5
Average inventory level: I	52	47	36	15	4

tion decreases, assuming that the same order size Q is used for each value of n. The difference in average inventory position for the cases $n = 1$ and 2 is relatively minor, indicating that when the end product is assembled from a kit consisting of either one or two parts, the effect of kitting is not significant. However, the inventory position decreases sharply for $n = 4$ and 5, indicating that the kitting of several types of parts does affect the average inventory position significantly by increasing the production lead-time required to compose the kit of required parts.

The optimal lot size, Q^*, may be different for each value of n (1, 2, 3, 4 and 5) so that corresponding values of the average inventory position may, thus, be different from those tabled. In addition, if the system anticipated future orders so that the processing of at least some parts could begin before a replenishment order is placed, average end-product inventory position would not decline so dramatically as the number of parts in a kit increases.

Models of several configurations that operate according to the push philosophy are discussed in this section; the next section addresses the pull philosophy of material management.

6.3 Pull systems

In a pull system (Monden, 1983; Joo and Wilhelm, 1993), an end product is withdrawn when a customer places a demand, 'pulling' materials from upstream stations. A kanban is attached to each lot produced by an upstream station and is detached when another kanban authorizes the movement of the parts from the downstream station. Production at an upstream station is authorized by accumulating an appropriate number of kanban cards that circulate between the pair of sequential stations. The number of kanban cards in circulation determines the level of work-in-process (WIP) inventory at each station, so the pull system has been the focus of much attention in the last decade as manufacturers have sought to reduce WIP.

A number of models are available to evaluate the performance of production systems that operate according to the pull philosophy. However, relatively few models of pull systems in assembly are available. This section describes several models of assembly systems that operate according to a specialized pull philosophy called CONWIP (CONstant Work-In-Process) (Spearman, Hopp and Woodruff, 1989; Spearman and Zazanis, 1992). By circulating a fixed number of kanbans between the **first** and **last** stations, CONWIP holds the total inventory in the system constant, so that new (matching) parts are released to fabrication lines only when an end product completes assembly. Jobs are pushed from one station to the next. These models relate to a configuration in which a single machine assembles J different part types, each of which is produced on an independent fabrication line consisting of n_j stations (for $j = 1, \ldots, J$).

Duenyas and Hopp (1993) propose a method for developing an upper bound on the throughput rate for this assembly system. They assume that the processing time at each station is exponentially distributed.

The throughput for a serial line consisting of n_j stations plus the assembly machine gives an upper bound on the throughput for the assembly system. In fact, the minimum throughput of J serial lines configured in this manner gives a valid upper bound on the throughput of the assembly system. Duenyas and Hopp (1993) improve this bound by approximating the amount of time a part of one type must wait for matching parts from the other flow-lines. Invoking the CONWIP and exponential assumptions, each flow-line may be modeled as a closed queuing network to approximate the required waiting times.

They compare the estimates derived from this approach with those from simulation in a variety of test cases, including balanced and unbalanced flow-lines, stations with single or multiple machines, and systems with either fast or slow assembly times. In these tests, their method gave a close approximation, typically within 4% of simulation estimates. Their analysis resulted in several observations that may be used as design guidelines: 'throughput is a non-decreasing function of machine speed', 'the throughput of an assembly system with exponential processing times is a lower bound on the throughput of an assembly system with IFR processing times with the same mean processing times', 'a bottleneck at assembly limits throughput more than an equivalent bottleneck in fabrication', and 'the optimal card count for an assembly system with exponential processing times is a lower bound for an assembly system with IFR processing times with the same mean processing time'.

Duenyas and Hopp (1992) study the case with deterministic processing times and random outages (DPRO), assuming that times between failures (repairs) are exponentially distributed with mean rate Π (μ) depending upon the station. Such a case may arise in automated systems, for example in the assembly of circuit cards.

Again, they show that the throughput for a serial line consisting of n_j stations plus the assembly machine gives an upper bound on the throughput for the assembly system and that the minimum throughput of J serial lines configured in this manner gives a valid upper bound on the throughput of the assembly system. They also approximate the throughput of the DPRO system using an approximation for the throughput of a tandem DPRO line (Hopp and Spearman, 1991).

Their tests show that the method gives good approximations (within 4% of simulation results) especially when the bottleneck is sharp and 'times between failures are long relative to processing times'. They conclude that 'throughput is an increasing function of processing rates' and 'throughput tends to be higher if the bottleneck is located in fabrication rather than assembly', providing design guidelines for the DPRO assembly system.

Duenyas and Keblis (1994) study the case with general processing times (Duenyas, 1994). Devising a heuristic based on a state space heuristic for closed queuing networks with blocking, they show that CONWIP achieves a given throughput level with less WIP than does a kanban (JIT) release system.

Chandrasekhar and Suri (1995) study the case with exponentially distributed processing times, developing methods to approximate throughput and mean queue length using mean value analysis. Tests show that these methods give good estimates, typically within 4% of simulation estimates and competitive with the estimates from the approximations described above.

The first two sections of this chapter have classified models according to the material flow management philosophy they reflect. The next section focuses on flexible assembly.

6.4 Flexible assembly systems

For some time, the trend in manufacturing has been to produce a variety of products, each in low volume (Groover, 1987). Flexible assembly systems (FASs) have been developed for this production environment.

Models of an FAS configured as a closed loop are presented in section 6.4.1. Systems of this type are found frequently in automotive and other industries. Since robots are programmable and can quickly change over from one model to the next, they are ideal components of FASs. Section 6.4.2 presents models of robotic assembly operations.

6.4.1 Closed loop, flexible assembly systems

The productivity of the FAS designer can be greatly enhanced by a model which can quickly evaluate the performance of each alternative under consideration. For example, the number of pallets must be determined, types of equipment at each workstation must be selected and buffer capacities must be determined. Kamath, Suri and Sanders (1988) present such an analytical model, which is described in this section.

The model applies to an asynchronous FAS in which workstations are arranged in a circular configuration as shown in Fig. 6.7, so that, upon completion of an operation, a material handling device (e.g. conveyor or automated guided vehicle) transports the workpiece on a pallet from one station to the next. A fixed number of pallets circulate in the system, limiting the total

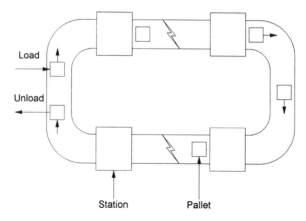

Fig. 6.7 Closed loop, flexible assembly system (adapted from Kamath, Suri and Sanders (1988) with permission from Kluwer Academic Publishers).

number of workpieces that can be in process at any time. Workpieces enter the system at a load station where each is mounted on a pallet. Upon completing assembly, the end product is removed from the pallet at the unload station and leaves the FAS.

A workstation typically consists of part feeders (e.g. vibratory bowl feeders) and a suitable mechanism to perform assembly operations, and can be hard-automated or programmable depending upon the flexibility needed. Inspection stations may be incorporated, and repair/rework loops may also be provided. Even though workstations may all have the same cycle time, variability may be introduced by parts jamming as difficulties are encountered randomly. Each jam stops the station for a random time duration while the operator clears it and may thus cause blocking at upstream stations as well as starving at downstream stations.

The mean and variance of the total time at each station may be expressed using:

C = deterministic cycle time at the station;

α = portion of assemblies for which the station jams;

$E[J]$ = expected value of the time required to clear a jam;

$V[J]$ = variance of the time required to clear a jam;

so that:

$$P = C + \delta J,$$

where δ is an indicator random variable that has value 1 if a jam occurs and 0 if not. The mean $E[P]$ and variance $V[P]$ of assembly time are thus:

$$E[P] = C + E[\delta]E[J]$$
$$= C + \alpha E[J]$$

$$V[P] = V[\delta J]$$
$$= V[\delta]V[J] + V[\delta]E[J]^2 + V[J]E[\delta]^2$$
$$= \alpha \{ V[J] + E[J]^2 (1-\alpha) \}$$

The model applies to the small FAS and invokes the following assumptions:

1. at any given time, the FAS is producing assemblies for a single end product;
2. since the system is small, no transport delays are incurred as a workpiece is moved from one station to the next; and
3. adequate buffer capacities are provided so that no blocking occurs between workstations.

Kamath, Suri and Sanders (1988) view the system as a closed, circular network of tandem, single-server queues with unlimited queuing space and first-come first-served (FCFS) discipline at each queue. The network is first decomposed so that each workstation is treated as a *GI/G/*1 queue whose arrival stream is the output process of the preceding station (queue). The arrival process must be characterized to approximate the performance of the FAS.

The variability of the arrival, departure and service processes in a *GI/G/*1 queue are related (approximately) according to:

$$C^D \approx (1 - U^2)C^A + U^2 C^S,$$

in which:

C^A = squared coefficient of variation of interarrival times;
C^D = squared coefficient of variation of interdeparture times;
C^S = squared coefficient of variation of service times;
U = utilization of the server.

The expected waiting time in a *GI/G/*1 queue may be closely approximated by $E[W]$:

$$E[W] \approx F(U, C^A, C^S) \left(\frac{C^A + C^S}{2} \right) \left(\frac{E[P]U}{1-U} \right),$$

in which:

$$F(U, C^A, C^S) = \begin{cases} \exp\left(\dfrac{-2(1-U)}{3U} \dfrac{(1-C^A)^2}{(C^A + C^S)} \right) & \text{if } C^A < 1 \\[4mm] \exp\left(-(1-U)\dfrac{(C^A - 1)}{(C^A + 4C^S)} \right) & \text{if } C^A > 1 \end{cases}$$

If U, C^A and C^S are relatively small, factor $F(U, C^A, C^S)$ is generally significantly less than 1.

In the network of queues, we are given:

M = number of stations;

N = number of pallets;

$E[P_i]$ = expected processing time at station i;

C_i^S = squared coefficient of variation of service time at station i;

and must find:

λ = steady-state throughput rate (arrival rate to each queue)

and C_i^A, C_i^D, U_i and $E[W_i]$, which specialize the previous notation for each station i.

For the case in which all stations have the same parameters:

$$E[P_i] = E[P] \qquad \text{for } i = 1, \dots, M$$
$$C_i^S = C^S \qquad \text{for } i = 1, \dots, M;$$

it is true that:

$$C_i^A = C^A \qquad \text{for } i = 1, \dots, M$$
$$C_i^D = C^D \qquad \text{for } i = 1, \dots, M$$
$$E[W_i] = E[W] \qquad \text{for } i = 1, \dots, M.$$

Furthermore, it is possible to show that:

$$C^A = C^S = C^D,$$

so that the approximation for $E[W]$ becomes:

$$E[W] = F(U, C^A, C^S)\left(\frac{E[P]UC^S}{1-U}\right).$$

This approximation can be improved by multiplying by the factor f where:

$$f = \left(\frac{N-1}{N}\right)\left(\frac{1}{1+(E[W]/E[P]N)}\right).$$

Now, applying Little's law to the FAS:

$$N = \lambda M(E[P] + fE[W])$$

or, since $U = \lambda E[P]$:

$$ME[P]U + MfE[W]U - NE[P] = 0 \tag{6.2}$$

Kamath, Suri and Sanders (1988) propose a bisection method to solve this nonlinear equation (note that $E[W]$ and f are functions of U) in the single unknown, U. The left-hand side of equation (6.2) is a function of U, $g(U)$, and we need the value of U, say U^*, such that $g(U^*) = 0$. Nine iterations of the bisection algorithm itemized in Table 6.5 are sufficient to guarantee an estimate U that is within 0.1% of U^*, assuring sufficient accuracy for the approximation.

The resulting U and λ give approximations to the steady-state utilization of each server and the throughput rate. Nine iterations are sufficient to assure this level of accuracy regardless of the system size and values of other parameters.

Kamath, Suri and Sanders (1988) evaluate the accuracy of their method in a set of test problems involving 4, 6 or 10 stations, geometrically or uniformly distributed clear times, and a selected set of other parameter values. Model approximations were shown to be generally within 5% of simulation estimates for most practical cases (e.g. $M > 15$ and about two pallets per station). However, errors of approximation are somewhat larger than 15% in cases that represent extreme conditions: for example, small systems ($M \leq 10$), only about one pallet per station, long clear times ($E[J] > 5\ D$) and low jam rate ($\alpha < 0.01$).

Kamath, Suri and Sanders (1988) presented a similar approximation method for the unbalanced case in which stations have different parameter values. Subsequently, Kamath and Sanders (1991) developed models to predict the effect of operator interference on the steady-state mean throughput rate. Since jams are relatively rare events, fewer than M operators would usually be provided in a FAS. If too few operators are provided, jammed

Table 6.5 The bisection solution method of Kamath, Suri and Sanders

---

Input: N, M, E[P], C^s
Initialize: L = 0.0, H = 1.0, U = 0.5
For I = 1 to 9 do
    Calculate g(U) using the current U and equation (6.2)
    If g(U) < 0
        Set L = U
    Else
        Set h = U
    endif
    Set U = (L + H)/2
End
Set λ = U/E[P].

Source: Kamath and Sanders (1988) with permission from Kluwer Academic Publishers.

stations will wait to be cleared, reducing productivity. If too many operators are provided, they will be idle much of the time. Thus, a model to evaluate the effect of operator interference can be used to determine how many operators should be assigned to the FAS. Building on this work, Bulgak and Sanders (1989) developed methods for the stochastic design optimization of an asynchronous FAS with statistical process control and repair loops.

6.4.2 *Robotic assembly*

The technological capabilities of robots and their control systems have evolved to the point at which robots are now capable of performing many assembly operations. Since they are programmable, robots are especially well suited for the flexible assembly environment. This subsection presents selected models of robotic assembly.

Consider the unitary cell which produces only one product type at a time (Seidmann and Nof, 1985a, 1985b, 1989; Seidmann, Schweitzer and Nof, 1985) (see also section 9.3). The unitary cell might be used to avoid frequent tooling changeovers, for example in electronics assembly. Parts are delivered to the cell where the robot performs material handling and assembly functions, completing assembly at a set of stations. The assembly is tested and leaves the cell if it passes the test. However, an assembly that fails the test is recycled: it is disassembled, assembled again at the stations, and retested. Cycles involving assembly and rework are continued until the assembly is accepted at test. It is assumed that any necessary replacement parts are on hand.

The performance of this unitary cell can be modeled using the notation in Table 6.6. Here, A, N and T are random variables and p, q and Q are parameters that describe a particular cell design. The number of cycles until acceptance is described by the geometric distribution:

$$\Pr\left[N = n\right] = qp^{n-1} \qquad \text{for } n = 1, 2, \ldots .$$

Table 6.6 Notation to describe performance of the unitary cell

A	= time required to complete an assembly cycle including the test
N	= total number of assembly cycles to obtain a good assembly
p	= probability that an assembly will fail the test
q	= probability that an assembly will pass the test = $1 - p$
Q	= lot size
T	= time required to prepare for reassembly (e.g. disassembly)
T_A	= total time in assembly and test
T_T	= total time in preparing for reassembly
θ	= total time to complete a lot of size Q accepted assemblies
CV	= coefficient of variation
COV	= covariance operator
ρ	= coefficient of correlation

Table 6.7 Moments for several measures of cell performance

Measure	Mean	Variance
N = Number of assembly cycles	Q/q	Qp/q^2
T_A = Total time in assembly	$QE[A]/q$	$Q\{pE^2[A]/q + V[A]\}/q$
T_T = Total time to prepare for rework	$QE[T]\,p/q$	$Qp\{E^2[T]/q + V[T]\}/q$
θ = Total time to complete Q accepted assemblies	$Q\{E[A] + pE[T]\}/q$	$Qp(E[A] + E[T])^2/q^2$ $+ Q\{V[A] + pV[T]\}/q$

Assuming that A and T are strictly bounded, moment-generating functions can be used to derive the mean and variance of several measures of cell performance as shown in Table 6.7. These moments can be used to evaluate the performance of alternative cell designs, for example comparing the total assembly time to the total time in the cell and evaluating the number of cycles as a function of the cell parameters.

Seidmann and Nof (1985a) identify a fundamental operating characteristic of the unitary cell: the distribution of time to complete an assembly may be multimodal. If each of L assembly tasks requires time that is exponentially distributed with parameter μ, if the robot's total travel time is a constant α, and if $Q = 1$, the distribution of A is a shifted gamma:

$$f_A(t) = \begin{cases} \exp\left\{-\mu(t-\alpha)\mu^L(t-\alpha)^{L-1}\big/(L-1)\right\} & \text{for } t > \alpha; \ L = 1, 2, \ldots \\ 0 \text{ otherwise} \end{cases}$$

and, in this case, $E[A] = \alpha + L/\mu$ and $V[A] = (L/\mu)^2$. Now, if T is a constant, the probability density function of the total time required to produce a good assembly is multimodal. The frequency of these modes is $1/(E[A] + E[T])$, representing completion of a good assembly after 0, 1, 2, . . . rework cycles. The height of each successive mode decreases, indicating that fewer cycles are more likely. The peakedness of each mode increases with the parameter p.

Seidmann, Schweitzer and Nof (1985) continue this analysis, indicating that the peaks become less pronounced as Q increases; for large Q (i.e. for $Q \geq 10$), the distribution of total time approaches a 'unimodal bell shape'. They also show that interleaving a number of product types offers no productivity improvement. However, the capability to interleaf may offer a flexibility that allows the number of cells to be minimized. These models allow the performance of the unitary cell to be evaluated as a function of fundamental parameters (e.g. p and Q).

Seidmann and Nof (1989) present a number of additional performance measures, including:

$$CV_N \quad = (p/Q)^{1/2}$$

$$CV_\theta \quad = \left\{ p \left(E[A] + E[T] \right)^2 + q \left(V[A] + pV[T] \right) \right\}^{1/2} \Big/ \left\{ Q^{1/2} \left(E[A] + pE[T] \right) \right\}$$

$$\text{COV}[\theta, N] = Qp \left(E[A] + E[T] \right) / q^2$$

$$\text{COV}[T_A, N] = QpE[A]/q^2$$

$$\text{COV}[T_T, N] = QpE[T]/q^2$$

$$\rho[\theta, N] \quad = \left\{ 1 + \left[\left(V[A] + pV[T] \right) q \right] \Big/ \left[p \left(E[A] + E[T] \right)^2 \right] \right\}^{-1/2}$$

$$\rho[T_A, N] \quad = E[A] \left\{ E^2[A] + qV[A]/p \right\}^{-1/2}$$

$$\rho[T_T, N] \quad = E[T] \left\{ E^2[T] + qV[T] \right\}^{-1/2}$$

and results show that correlations $\rho[\theta, N]$, $\rho[T_A, N]$ and $\rho[T_T, N]$ are independent of Q. They also show how these basic models may be adapted to reflect the characteristics of a particular cell. For example, the number of rework cycles may be limited with constant, decreasing or increasing probabilities that an assembly will fail the test. They present models to evaluate performance relative to task, product and systems levels as well as an extensive example showing application in the cell design process.

It is possible to devise analytical models of numerous assembly systems and approximations for even more configurations, and these models provide useful insights into system performance. However, simulation offers the capability of modeling a vast variety of configurations and drawing inferences about their performance. The next section focuses on simulation models of assembly systems.

6.5 Simulation analysis

This section addresses a variety of modeling issues and suggests how simulation might be used to test fundamental hypotheses about the operation of actual assembly systems. The primary advantages of simulation models are that they can represent most assembly systems and that estimates of performance can be improved by making more replications. However, simulation models may be time-consuming to construct and run, and they require an extensive background in statistics to design appropriate experiments and interpret results.

A simulation study which results in guidelines for system design and insights into the effects of buffers is described in section 6.5.1. Section 6.5.2 describes the use of simulation in evaluating the performance of cellular configurations for circuit card assembly, relating design decisions to those that prescribe operating policies such as sequencing rules.

6.5.1 Buffers in push and pull systems

Motivated by the difficulties that assembly systems present to analysis, Baker, Powell and Pyke (1990) designed a set of simulation experiments with the goal of identifying the underlying characteristics of assembly systems and developing guidelines for their design. They study a system that assembles two parts at one station. Each part is produced on a flow-line, which consists of 1, 2, 3, 5 or 7 stations, depending upon the test. The station farthest upstream from assembly is fed by an infinite pool of raw materials, but a buffer is provided ahead of each other station. Stations are assumed to process according to either the uniform, exponential or truncated normal distribution, depending upon the test.

They investigate both push and pull systems of material flow management. In particular, they note that, in an unbuffered system, the push mode will achieve a higher throughput rate because it allows a completed part to be passed along to an idle station rather than being blocked as a pull mode would require. Thus, the push mode allows each station to also serve as a buffer.

They seek to establish (1) the effects of providing a buffer at a single station, (2) the optimal stations at which to assign a fixed number of buffer locations, and (3) the relative improvements that a buffer makes in comparison with decreasing processing time.

Relative to their first objective, they find that if a buffer is to be located at just one station, the decomposition principle should be used to select the station (Conway *et al.*, 1988), which holds that the impact of a buffer at a single station can be assessed by determining the effect of a buffer of infinite capacity at that station, which would decompose the production system into two component lines. The expected throughput rate for an *n*-station serial line is (Muth, 1973):

$$E[T] = \left\{ E\left[\max\left(P_1, P_2, \ldots, P_n \right) \right] \right\}^{-1},$$

so that of a three-station assembly system as depicted in Fig. 6.8 is:

$$E[T_a] = \left\{ E\left[\max\left(P_1, P_2, P_3 \right) \right] \right\}^{-1},$$

in which:

$$P_i = \text{processing time on station } i.$$

Thus, when the fabrication lines that produce parts are not long, the best place to provide a buffer is ahead of the assembly station. However, if the fabrication lines are very long, it is optimal to provide a buffer at a station upstream.

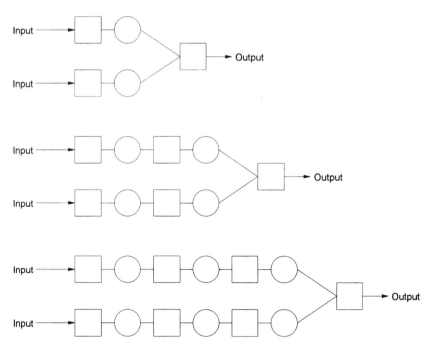

Fig. 6.8   Three-, five- and seven-station assembly systems (adapted from Baker, Powell and Pyke (1993)).

The material flow management philosophy – push or pull – also has an impact on where a buffer should be provided, since the push system can enlist a station to serve as a buffer at times. In a pull system, it is best that the first buffer be installed ahead of the assembly station, but this is not necessarily true for a push system.

Relative to their second objective, the authors note that Conway *et al.* (1988), who studied the allocation of a fixed number of buffers in a balanced serial line, found that buffers of equal capacity should be located at each station and that any leftover buffer locations should be provided evenly along the line. Baker, Powell and Pyke (1993) found that the same principle should be used in designing assembly systems. In addition, they found that their assembly systems exhibited the 'build up' property, whereas the flow-line does not. That is, the optimal provision of buffers for an n-stage assembly system is built up by augmenting the optimal solution for the $(n-1)$-stage system. They also found that push and pull systems perform similarly if buffers are provided.

Variability has a significant influence on the performance of assembly systems. For example, the variability in lead-times required to produce parts often causes some part(s) to wait for others in order to complete kits, detracting from overall performance. Buffers tend to ameliorate the impact of variability. As in other studies, Baker, Powell and Pyke (1993) find that a buffer of modest capacity is able to achieve a substantial portion of the efficiency that a buffer of infinite capacity would offer and that additional capacity makes only small, incremental improvement. However, systems with greater variability – for example, the case with exponential processing times – require larger buffer capacity to achieve an equivalent efficiency.

Relative to their third objective, Baker, Powell and Pike (1993) note that, without knowing the costs involved, it is not possible to determine whether it is best to increase buffer capacity or reduce processing time in order to increase the throughput rate. However, they find that the effects of buffers dominate those of assembly time, so buffers are an important means of improving throughput.

Bhatnagar and Chandra (1994) use a simulation model to study the effects of variability on processing times, workstation reliability and yield on a three-station assembly system. For the case with Poisson arrivals, reliable stations and perfect yield, they test normal, uniform and exponential processing times and conclude that throughput decreases as variability (i.e. the coefficient of variation) increases and that the pull mode is more adversely affected than is the push mode. Confirming the results of Baker, Powell and Pike (1993), they find that the addition of buffers allows this negative effect of variability to be reduced at a rate inversely proportional to the coefficient of variation and that the pull mode is more sensitive than is the push mode. The push mode operates differently because it allows stations to act as buffers. To evaluate the effects of variability in station reliability, they use unit processing times, perfect yield, exponentially distributed times between failures and both uniform and exponentially distributed repair times. They conclude that the throughput of such a system without buffers is a function of the ratio of the mean time to failure to the mean repair time. By adding buffers, throughput becomes a function of the mean repair time and losses due to variability can be recovered if buffer capacity is proportional to the mean repair time. They study yield variability on reliable stations using constant, uniform and exponential processing times. Poor yield at a part-processing station causes throughput to vary inversely with the coefficient of variation of processing time. Thus, synchronized stations with constant processing times were most adversely affected. For cases in which processing times are variable, the stations are not synchronized and a yield loss may occur at the same time as blocking or starving, partially offsetting negative effects. Yield losses at the assembly station have a direct and equal effect on throughput.

6.5.2 *Cellular configurations for circuit card assembly*

While simulation models may not lend the same depth of insight that analytical models do, they offer a versatile means for evaluating the performance of complex systems and are frequently used in the design process (Yano, Lee and Srinivasan, 1991). This section describes a simulation model that was devised to analyze the performance of cellular systems in the assembly of circuit cards. Rockwell and Wilhelm (1990) provide a structure by which to analyze the effects of cell design alternatives, the role of buffers, the influence of the material handling system and material flow management practices on system performance.

Cellular manufacturing is defined (Wemmerlov and Hyer, 1987) as the 'processing of collections of similar parts [i.e. families] on dedicated [i.e. cells] of dissimilar machines' and offers numerous advantages, including reducing work-in-process (WIP) inventory and cycle time, and increasing quality and flexibility (Rathmill and Leonard, 1977; Black, 1983; Green and Sadowski, 1983). Circuit cards are often produced in small-lot quantities because they are custom-designed for individual customers. Keen international competition places a premium on reducing WIP and cycle time and increasing quality and flexibility. However, circuit card assembly is well structured and highly standardized, so detailed analysis is required to integrate the design and operation of a cellular configuration and assure its competitive performance.

Each circuit card may be populated with several hundred components, including dual-in-line pin (DIP) (integrated circuits), axial (resistors), radial (capacitors), surface mount (SMT) and oddform shapes. Family membership may be based on criteria such as required processes (e.g. pin-through-hole or SMT), type of inserted components (e.g. DIP or other types), size and shape (identifying fixturing needs) and production volume (leading to automatic, semi-automatic, robotic or manual operations).

Machines are highly standardized and may assemble only one or two types of components (e.g. DIP or SMT), but at high rates. Manual assembly is slower but more flexible. Insertion machines usually form a generalized flow-line in which each lot flows downstream but may not be processed at each machine. DIP machines are usually located at the first stage of assembly because of tool clearances required, and manual assembly is usually done at the last stage. Pre-assembly operations consist of kitting parts and preparing components for assembly, and post-assembly processes include cleaning and fixturing, flow soldering, test and rework. Since they are quite expensive, cards that are rejected at test are typically reworked and retested until an acceptable product results. Insertion and soldering are not usually bottlenecks, but inspection, test and rework frequently are.

The simulated system consists of input and output buffers (for arrivals and departures, respectively) and a set of cells, each with input and output buffers. All of these buffers are served by the between-cell material handling system. Each cell consists of a set of stations, each with input and output buffers. Within-cell handling is accomplished by robots or by conveyors accompanied by load/unload devices. Each station includes input and output buffers and is tended by a robot. Nine types of station are used as shown in Table 6.8, which gives the number of machines, insertion times and setup times at each station. Test and rework are not included because they are product-specific operations. Handling devices are assumed to require the grasp, release and travel times given in Table 6.9. Buffers allow handoff of lots into and out of the system, each cell and each station. In addition, they provide some storage space so that blocking does not occur as soon as one operation is finished.

Table 6.8 Station descriptions

Station	Number of machines	Insertion time per component (sec)	Setup time per lot (sec)
DIP	2	1.8	20
Axial	2	0.9	30
Radial	1	1.1	30
Multiple	1	1.8	40
Pin	1	0.9	20
Robotic	2	5.8	20
Manual	2	5.8	60
Resolder	1	–	330
Flow solder	1	–	360

Source: Rockwell and Wilhelm, 1990.

Table 6.9 Service times of material handling devices

Domain	Grasp (sec)	Release (sec)	Travel (ft/sec)
Between cells	6.0	6.0	7.0
Within cells	6.0	6.0	3.0
Within station	6.0	6.0	2.0

Source: Rockwell and Wilhelm, 1990.

Circuit card families were defined (Table 6.10) to represent high-tech products (e.g. with most components being DIPs or axial types), those that contain a high percentage of components that can be assembled by robots, and those that consist predominantly of oddform components that require manual assembly. Production volume was assumed to be 240 cards per day; each arriving lot was randomly named to a family according to the fractions given in Table 6.10. The total number of components on each card was generated randomly from a discrete uniform distribution for each family and the mix of components on a card was generated randomly from a multinomial distribution with parameters given in Table 6.10. The processing time of a card at a station was calculated by multiplying the number of components by the time per component in Table 6.8. Processing times at non-insertion stations were generated from a normal distribution with mean unique to the station and coefficient of variation equal to 0.2.

While a large number of cellular configurations is possible, pragmatic considerations limit the number of feasible alternatives. For example, card families should be assigned to cells to minimize the amount of between-cell

Table 6.10   Composition and mix of card families

Station	Expected component mix			
	Family 1	Family 2	Family 3	Family 4
DIP	0.17	0.30	0.30	–
Axial	0.35	0.40	–	–
Radial	0.23	–	–	–
Multiple	–	0.20	0.20	–
Pin	0.17	–	–	–
Robotic	0.08	–	0.50	–
Manual	–	0.10	–	1.00
Total number of components (discrete uniform distribution)				
Lower bound	50	100	125	20
Upper bound	400	300	200	150
Mean	225	200	163	85
Production volume				
Fraction of 240 circuit cards/day	0.4	0.3	0.2	0.1

Source: Rockwell and Wilhelm, 1990.

material handling, and each cell should function as a flow-line to facilitate control. Slower manual operations should not be mixed with faster automated ones. Configurations 1 and 2, described in Table 6.11, resulted from this analysis; they represent feasible but inherently different designs (see also Rockwell and Wilhelm (1987)). Each offers certain advantages. Configuration 1 is based on processing card families in different cells and was effected by assigning automatic insertion machines and robots to different cells. The second is based on grouping processes and was effected by assigning automatic machines and robots in one cell, manual operations in a second cell and post-assembly operations in the third cell. Cell 1 in configuration 1 is large, encompassing both cells 1 and 2 of configuration 2, and can be expected to affect between-cell movement and workload balance. Table 6.11 describes each station, including the number of machines, whether it is attended by a handling device, the card families it processes, and its expected utilization based on a target production level of 240 cards in an eight-hour day.

A number of detailed assumptions were required to structure the system: lots of size one were used, each station had to be set up before it could process a lot, no breakdowns were considered, times for material handling tasks were considered to be deterministic, components were either prestaged at stations or were transported in the kit, one or more stations composed each cell, each handling device transports materials either within a cell or between cells, the

Table 6.11 Alternative configurations

Stations		Number of machines	MHD?	Part Families				UT*
				F1	F2	F3	F4	
Configuration 1								
Cell #1	Handling device ———————————— 84% max							
	1. DIP 1	1	No	X				30%
	2. Axial 1	1	No	X				33%
	3. Radial	1	No	X				29%
	4. Pin	1	No	X				18%
	5. Robotic 1	1	Yes	X				42%
Cell #2	Handling device ———————————— 69% max							
	1. DIP 2	1	No		X	X		50%
	2. Axial 2	1	No		X			26%
	3. Multiple	1	No		X	X		44%
	4. Robotic 2	1	Yes			X		81%
Cell #3	Handling device ———————————— 81% max							
	1. Manual	2	Yes		X		X	45%
	2. Presolder	4	Yes	X	X	X	X	69%
	3. Flow Solder	8	Yes	X	X	X	X	38%
Between-cell handling device——————————— 63% max								
Configuration 2								
Cell #1	Handling device ———————————— 100% max							
	1. DIP	2	Yes	X	X	X		40%
	2. Axial	2	Yes	X	X	X		32%
	3. Radial	1	No	X				29%
	4. Multiple	1	No		X	X		44%
	5. Pin	1	No	X				18%
	7. Robotic	2	Yes	X		X		62%
Cell #2	Handling device ———————————— 18% max							
	1. Manual	2	Yes		X		X	45%
Cell #3	Handling device ———————————— 75% max							
	1. Presolder	4	Yes	X	X	X	X	69%
	2. Flow solder	8	Yes	X	X	X	X	38%
Between-cell handling device ——————————— 72% max								

*UT = utilizations which are based on a throughput of 30 cards per hour.
MHD = material handling device.

Source: Rockwell and Wilhelm, 1990.

output buffers of cells have unlimited capacity so that they do not block stations, the input buffer of the system has unlimited capacity so that all lots can be accepted, and lots leave the system upon reaching the system output buffer.

Each kit held enough components for a lot. Kits were delivered to the system input buffer periodically in a group size that provided a daily input of 240 cards. Each kit was subsequently transported by the system handling device to the input buffer of the appropriate cell(s), and a within-cell handling device transported each kit to the series of stations on its routing. When completed in a cell, the kit was returned to the cell output buffer, and upon completion of an end product, the lot was transported to the system output buffer.

The discrete event simulation model employs a unique means of prioritizing activities. An entity representing a lot ready for transport is placed in a file representing the destination (e.g. a cell or station input buffer). A file representing each material handling device stores an entity associated with each idle destination 'calling' for work. Calls are prioritized with station output buffers first, idle machines next, and station input buffers last. Once a destination achieves the highest priority, it must be serviced next by the handling device. A transport can begin only when the applicable material handling device is idle, the destination 'calls' for a lot, and the lot is available. The filing system facilitates a smooth and orderly flow of work, maintaining utilizations as high as possible.

Statistical analysis of the simulation output representing system performance involved several unique aspects. Since kits were delivered periodically, the in-process inventory at any point in time was decreasing and the system operated in a periodic mode. To deal with a measure that attains a steady state,

Table 6.12 Factors and levels

#	Factor	Description	Levels (−)	(+)	Initial
1	MSR	Machine selection rule	FCFS	HUF	(−)
2	JSR	Job selection rule	FCFS	SPT	(−)
3	CIB	Cell input buffer	1 lot	unlimited	unlimited
4	MIB	Station input buffer	*	unlimited	unlimited
5	IAP	Interarrival period	60 min	2 min	(−)
6	PV	Production cards/day	240	264	(−)
7	TT	Device service time	1.3 (Table 6.7)	Table 6.7	(+)
8	MODE	Handling device mode	Conveyor	Robot	(−)
9	CNFG	Configuration	$CNFG_1$	$CNFG_2$	(−)

FCFS = first-come first-served.
HUF = highest utilization first.
SPT = shortest processing time first.
* station input buffer capacities = 0, except where stations have load/unload devices which require capacities of 1.

Source: Rockwell and Wilhelm, 1990.

the average inventory over each interarrival period was used. Run-in (transient) intervals were determined from preliminary analysis. Steady-state measures were collected over 512 interarrival periods and were analyzed using a batch means algorithm that eliminated serial correlations among observed values. The same 'random' arrivals and processing times were generated for each run as a variance-reduction tactic. The primary measures used to evaluate system performance were the average level of WIP per interarrival period for the system and for each cell, and the average utilization of material handling devices.

Experiments involved nine factors, each with two levels, as shown in Table 6.12. Machine and job selection rules prioritize material handling service. Different cell and station input buffer capacities were tested to assess their influence on workflow. The interarrival period and total number of cards input per day were tested to evaluate their effects on resource utilization and on the smoothness of workflow. The types and service times of handling devices were tested to assess their effect on workflow. Finally, the two configurations were compared.

A series of 2^2 and 2^3 factorial experiments were conducted to study the main effects of each factor and certain two-factor interactions. Specific numerical results may not apply in general cases, but some of the conclusions may be useful in forming hypotheses concerning the performance of actual systems. Tests analyzing configuration 1 led to the conclusions that, in contrast with other system configurations, job dispatching and machine selection rules did not have a significant effect on system performance. Buffers at system, cell and station levels distribute WIP and affect system performance in complex ways: the capacities of cell input buffers have a significant effect on WIP and smaller capacities tend to improve performance; the effects of station input buffer capacity is less dramatic, although the bottleneck station is a prime determinant of system WIP; and restricting cell and station input buffer capacities tends to increase WIP held 'outside' of the cells. Smoothing inputs by changing to the shorter interarrival duration reduced WIP significantly and rendered buffer capacities less important to workflow, and the higher production volume amplified the effects of station input buffer capacity and improved resource utilization. In contrast with other studies, the type of material handling device and its speed had significant effects on performance; WIP increased as material handling service time increased, and, in comparison with robots, conveyors reduced system WIP but had mixed effects at the cell level. In general, configuration 2 performed similarly, but the effects of station input buffer capacities contrasted due largely to inefficiencies caused by more inter-family blocking in configuration 1.

6.6 Summary

This chapter describes some of the approaches taken to model the steady-state operation of stochastic assembly systems, providing a summary of models that might be used to evaluate the performance of certain assembly configurations that are used commonly in industry. Various insights gained from analytical approaches are related, including the design factors that typically determine whether an assembly system will reach a steady state, guidelines for locating buffers in assembly systems and the effects of the kitting process on assembly operations.

In an application involving the design of an actual assembly system, models can be used to evaluate system performance and, thus, provide insights that can be used to make trade-offs in selecting design alternatives, leading to the most effective combination of station configuration, machine selection, tooling provision, buffer capacities, material handling devices and operating procedures. A fundamental aspect that makes assembly systems unique is that part arrivals must be coordinated so that kits may be composed to support assembly schedules effectively. Stochastic models provide a means of understanding this important process and its effects on overall system performance. The number of configurations that might be used for assembly is virtually unlimited. However, models are currently available for relatively few configurations, since this area has just begun to attract the interests of researchers.

Most models based on queuing analysis invoke the assumption that parts arrive according to independent processes. This assumption facilitates analysis and may lead to useful results, but may not reflect the manner in which many assembly systems operate. For example, in a push system, when an end product is scheduled for assembly, due dates for its parts are established, so that deliveries of parts are coordinated, improving control over the kitting process. One analytical model that coordinates part deliveries in a stochastic MRP environment is presented in section 6.4.2. Other means of controlling part deliveries to the kitting process are described in Chapter 8. The next chapter deals with operating decisions: sequencing and scheduling.

6.7   Review questions

1. Describe the bowl phenomenon and the reversibility property of flow-lines.
2. What roles do buffers play in enhancing productivity?
3. Discuss criteria for an assembly system to attain a steady state.
4. Develop an array relating the bounds available to estimate assembly system performance to the applicable system configuration.
5. Describe the most general assembly system configuration for which approximate measures of performance are available.
6. Describe a double-ended queue and how it arises in the kitting process.
7. Under what conditions can the assembly system in section 6.2.3 be decoupled for analysis and what is the output process describing the kitting process under these conditions?
8. Describe the influences of buffers in assembly systems.

9. Describe the stochastic MRP system modeled by Wilhelm and Som (1994) and the effects of kitting on system operation.
10. How can JIT and CONWIP philosophies be implemented in an assembly system?
11. Summarize design guidelines (rules of thumb) that models have been able to establish for the operation of pull systems.
12. Explain the need for flexible assembly systems.
13. What roles do buffers play in pull and push systems?
14. Summarize design guidelines (rules of thumb) for placing buffers in assembly systems.
15. What are the 'decomposition principle' and the 'build up property'?
16. What effects do variability have on system performance and how can they be ameliorated?
17. Describe how cellular concepts can be used in assembly.
18. Design a simulation experiment to analyze the operation of a cellular configuration used to assemble circuit cards.
19. Describe how buffers (at system, cell and station levels), material handling devices and interarrival periods can affect WIP and assembly system performance.

References

Altiok, T. and Shiue, G.A. (1994) 'Single-stage, Multi-product production/inventory systems with back-orders', *IIE Transactions*, **26** (2), 52–61.

Baker, K.R., Powell, S.G. and Pyke, D.F. (1993) *'Buffered and Unbuffered Assembly Systems with Variable Processing Times'*, *Management Science*, **39** (1), 101–6.

Bhat, U.N. (1970) 'A controlled transportation queueing process', *Management Science*, **16**, 446–52.

Bhat, U.N. (1986) 'Finite capacity assembly queues', *Queueing Systems: Theory and Applications*, **1**, 85.

Bhatnagar, B. and Chandra, P. (1994) 'Variability in assembly and competing systems: effect on performance and recovery', *IIE Transactions*, **26** (5), 18–31.

Bitran, G.R. and Dasu, S. (1992) 'A review of open queueing network models of manufacturing systems', *Queueing Systems*, **12**, 95–134.

Black, J.T. (1983) 'Cellular manufacturing systems reduce set-up time, make small lot production economical', *Industrial Engineering*, **15** (11), 36–48.

Bonomy, F. (1987) 'An approximate analysis for a class of assembly-like queues', *Queueing Systems: Theory and Applications*, **1**, 289.

Bulgak, A.A. and Sanders, J.L. (1989) *Modeling and Design Optimization of Asynchronous Flexible Assembly Systems with Statistical Process Control and Repair*, Working Paper, School of Industrial Engineering, University of Wisconsin, Madison, WI.

Buxey, G.M., Slack N.D. and Wild, R. (1973) 'Production flow line system design – a review', *AIIE Transactions*, **5** (1), 37–48.

Buzacott, J.A. (1967) 'Automatic transfer lines with buffer stocks', *International Journal of Production Research*, **6**, 183–200.

Buzacott, J. and Hanifin, L.E. (1978) 'Models of automatic transfer lines with inventory banks: a review and comparison', *AIIE Transactions*, **10**, 197–207.

Buzacott, J.A. and Shanthikumar, J.G. (1992) 'Design of manufacturing systems using queueing models', *Queueing Systems*, **1**, 135–214.

Buzacott, J.A. and Shanthikumar, J.G. (1994) 'Safety stock versus safety lead time in MRP-controlled production systems', *Management Science*, **40** (12), 1678–89.

Chandrasekhar, P. and Suri, R. (1995) 'Approximate queueing network models for closed fabrication/assembly systems: Part I: Single level systems', *Production and Operations Management*.

Conway, R.W., Maxwell, W.L., McClain, J.O. and Thomas, L.J. (1988) 'The role of work-in-process inventory in serial production lines', *Operations Research*, **36**, 229–41.

Dallery, Y., David, R. and Xie, X.L. (1988) 'An efficient algorithm for analysis of transfer lines with unreliable machines and finite buffers', *IIE Transactions*, **20** (3), 280–3.

Dallery, Y., David, R. and Xie, X.L. (1989) 'Approximate analysis of transfer lines with unreliable machines and finite buffers', *IEEE Transactions on Automatic Control*, **34** (9), 943–53.

Dallery, Y. and Gershwin, S.B. (1992) 'Manufacturing flow line systems: a review of models and analytical results', *Queueing Systems*, **12**, 3–94.

David, R., Xie, X.L. and Dallery, Y. (1990) 'Properties of continuous models of transfer lines with unreliable machines and finite buffers', *IMA Journal of Mathematics in Business and Industry*, **6**, 281–308.

de Koster, M.B.M. (1988) 'Capacity Oriented Analysis and Design of Production Systems', Eindhoven University of Technology, PhD thesis.

Di Mascolo, M., David, R. and Dallery, Y. (1991) 'Modeling and analysis of assembly systems with unreliable machines and finite buffers', *IIE Transactions*, **23** (4), 315–30.

Disney, R.L. and Kiessler, P.C. (1987) *Traffic Processes in Queueing Networks: A Markov Renewal Approach*, John Hopkins University Press.

Disney, R.L. and Konig, D. (1985) 'Queueing networks: a survey of their random processes', *SIAM Review*, **27**, 335–403.

Dobbie, J.M. (1961) 'A double-ended queueing problem of Kendall', *Operations Research*, 9755–7.

Duenyas, I. (1994) 'Estimating the throughput of a cyclic assembly system', *International Journal of Production Research*, **32**, 1403–19.

Duenyas, I. and Hopp, W.J. (1992) 'CONWIP assembly with deterministic processing and random outages', *IIE Transactions*, **24** (4), 97–108.

Duenyas, I. and Hopp, W.J. (1993) 'Estimating the throughput of an exponential CONWIP assembly system', *Queueing Systems*, **14**, 135–57.

Duenyas, I. and Keblis, M.F. (1994) *Release Policies for Assembly Systems*, Technical Report 93-15, Dept. of Ind. and Ops. Eng., Univ. of MI.

Gavish, B. and Graves, S.C. (1980) 'A one-product production/inventory problem under continuous review policy', *Operations Research*, **28** (5), 1228–36.

Gavish, B. and Graves, S.C. (1981) 'Production/inventory systems with a stochastic production rate under a continuous review policy', *Computers and Operations Research*, **8** (3), 169–83.

Gershwin, S.B. (1987) 'An efficient decomposition method for the approximate evaluation of tandem queues with finite storage space and blocking', *Operations Research*, **35** (2), 291–305.

Gershwin, S.B. (1991) 'Assembly/disassembly systems: an efficient decomposition algorithm for tree-structured networks', *IIE Transactions*, **23** (4), 302–14.

Gershwin, S.B. and Schick, I.C. (1983) 'Modeling and analysis of three-stage transfer lines with unreliable machines and finite buffers', *Operations Research*, **31** (2), 354–80.

Green, T. and Sadowski, R. (1983) 'Cellular manufacturing control', *Journal of Manufacturing Systems*, **2** (2), 137–45.

Groover, M.P. (1987) *Automation, Production Systems, and Computer Integrated Manufacturing*, Prentice-Hall, Englewood Cliffs, NJ.

Harrison, J.M. (1973) 'Assembly-like queues', *Journal of Applied Probability*, **10**, 354–67.

Hillier, F.S. and Co, K.C. (1991) 'The effect of the coefficient of variation of operation times on the allocation of storage space in production line systems', *IIE Transactions*, **23** (2), 198–206.

Hopp, W.J. and Simon, J.T. (1989) 'Bounds and heuristics for assembly-like queues', *Queueing Systems*, **4**, 137–55.

Hopp, W.J. and Spearman, M.L. (1991) 'Throughput of a constant work in process manufacturing line subject to failures', *International Journal of Production Research*, **29**, 635.

Hunt, G.C. (1956) 'Sequential arrays of waiting lines', *Operations Research*, **4**, 674–83.

Joo, S.H. and Wilhelm, W.E. (1993) 'A review of quantitative approaches in just-in-time manufacturing', *Production Planning & Control*, **4**, 207–22.

Kala, R. and Hitchings, G.G. (1973) 'The effects of performance time variance on a balanced, four-station manual assembly line', *International Journal of Production Research*, **11** (4), 311–53.

Kamath, M. and Sanders, J.L. (1991) 'Modeling operator/workstation interference in asynchronous automatic assembly systems', *Discrete Event Dynamic Systems: Theory and Applications*, **1**, 93–124.

Kamath, M., Suri, R. and Sanders, J.L. (1988) 'Analytical performance models for closed-loop flexible assembly systems', *The International Journal of Flexible Manufacturing Systems*, **1**, 51–84.

Kashyap, B.R.K. (1965) 'A double-ended queueing system with limited waiting space', *Proceedings of the National Institute of Science of India*, **31**, 559–70.

Kashyap, B.R.K. and Chaudhury, M.L. (1988) *An Introduction to Queueing Theory*, A&A Publications, Kingston, Ontario, Canada.

Latouche, G. (1981) 'Queues with paired customers', *Journal of Applied Probability*, **18**, 684–96.

Lipper, E.H. and Sengupta, B. (1986) 'Assembly-like queues with finite capacity: bounds, asymptotics and approximations', *Queueing Systems*, **1**, 67–83.

Magazine, M.J. and Silver G.L. (1978) 'Heuristics for determining output and work allocation in series flow lines', *International Journal of Production Research*, **16** (3), 169–81.

Monden, Y. (1983) *Toyota Production System*, Industrial Engineering and Management Press, Atlanta, GA.

Muth, E.J. (1973) 'The production rate of a series of work stations with variable service times', *International Journal of Production Research*, **11**, 155–69.

Muth, E.J. (1979) 'The reversibility property of production lines', *Management Science*, **25**, 152–8.

Muth, E.J. and Alkalf, A. (1987) 'The bowl phenomenon revisited', *International Journal of Production Research*, **25**, 161–73.

Nahmias, S. (1989) *Production and Operations Analysis*, Irwin, Homewood, IL.

Neuts, M.F. (1981) *Matrix-Geometric Solutions in Stochastic Models: An Algorithmic Approach*, The Johns Hopkins University Press.

Orlicky, J.A. (1975) *Materials Requirements Planning*, McGraw-Hill, New York.

Rao, N.P. (1976) 'A generalization of the bowl phenomenon in series production systems', *International Journal of Production Research*, **14**, 437–43.

Rathmill, K. and Leonard, R. (1977) 'The fundamental limitations of cellular manufacture when contrasted with efficient functional layout', *Proceedings of the Fourth International Conference of Production Research*, 777.

Rockwell, T.H. and Wilhelm, W.E. (1987) 'Applying cellular structures in small-lot assembly of circuit cards', *Proceedings, ICPR Conference*, Cincinnati, OH.

Rockwell, T.H. and Wilhelm, W.E. (1990) 'Material flow management in cellular configurations for small-lot circuit card assembly', *International Journal of Production Research*, **28** (3), 573–93.

Sarker, B.R. (1984) 'Some comparative and design aspects of series production systems', *IIE Transactions*, **16** (3), 229–39.

Seidmann, A. and Nof, S.Y. (1985a) 'Unitary manufacturing cell design with random product feedback flow', *IIE Transactions*, **17** (2), 188–93 (see also **17** (5), 403 (1985b)).

Seidmann, A. and Nof, S.Y. (1989) 'Operational analysis of an autonomous assembly robotic station', *IEEE Transactions on Robotics and Automation*, **5** (1), 4–15.

Seidmann, A., Schweitzer, P.J. and Nof, S.Y. (1985) 'Performance evaluation of a flexible manufacturing cell with random multiproduct feedback flow', *International Journal of Production Research*, **23** (6), 1171–84.

Simon, J.T. and Hopp, W.J. (1991) 'Availability and average inventory of balanced assembly-like flow systems', *IIE Transactions*, **23** (2), 161–8.

Som, P., Wilhelm, W.E. and Disney, R.L. (1994) 'Kitting process in a stochastic assembly system', *Queueing Systems: Theory and Applications*, **17** (III–IV), 471–90.

Spearman, M.L. and Zazanis, M.A. (1992) 'Push and pull production systems: issues and comparisons', *Operations Research*, **28**, 521.

Spearman, M.L., Hopp, W.J. and Woodruff, D.L. (1989) 'A hierarchical control architecture for constant work-in-process (CONWIP) production systems', *Journal of Manufacturing and Operations Management*, **2**, 147.

Suri, R., Sanders, J. and Kamath, M. (1993) 'Performance evaluation of production networks', *Handbooks in OR and MS*, Vol. 4, Elsevier Science (eds S.C. Graves *et al.*), 199–286.

Vollman, T.E., Berry, W.L. and Whybark, D.C. (1984) *Manufacturing Planning and Control Systems*, Irwin, Homewood, IL.

Wemmerlov, U. and Hyer, N. (1987) 'Research issues in cellular manufacturing', *International Journal of Production Research*, **25** (3), 413–31.

Wilhelm, W.E. and Som, P. (1993) *Analysis of Stochastic Assembly with GI Distributed Assembly Time*, Working Paper INEN/MS/WP/-93, Department of Industrial Engineering, Texas A&M University.

Wilhelm, W.E. and Som, P. (1994) *Analysis of an MRP Controlled Stochastic*

Assembly/Inventory System: A Markov Renewal Approach, Working Paper, Department of Industrial Engineering, Texas A&M University.

Yano, C.A., Lee, H.F. and Srinivasan, M.M. (1991) 'Design and scheduling of flexible assembly systems for large products: a simulation study', *Journal of Manufacturing Systems*, **10**, 54–5.

Yazamaki, G. and Sakasegawa, H. (1975) 'Properties of duality in tandem queueing systems', *Ann. Inst. Statist. Math.*, **27**, 201–12.

Zipkin, P. (1989) *A Kanban-like Production Control System: Analysis of Simple Models*, Working Paper No. 89-1.

7

Sequencing and scheduling of assembly operations

7.1 Introduction

Design, planning and performance evaluation – the topics of Chapters 4, 5 and 6 – establish the system in which assembly operations are performed. Decisions based on design, planning and performance evaluation establish system capacity by fixing the number of resources (e.g. machines, tools and workers) and by determining fundamentals of the assembly process. Scheduling – and, more generally, production control – must operate within this constrained framework to ensure the timely flow of materials. Thus scheduling decisions are made in a dynamic, time-dependent environment. Even though scheduling decisions are made in a highly constrained environment, they can exert significant influence on the operation of an assembly system. In particular, scheduling is crucial to controlling in-process inventories and cycle times, and to providing competitive levels of customer (i.e. due date) performance. Thus scheduling is an important component in world-class assembly systems.

A sequence for a set of operations prescribes the order in which the operations are performed. A schedule prescribes the time at which each operation in a sequence is performed. In most production configurations, a schedule can easily be calculated once a sequence is determined. For example, if all operations are performed by a single machine with no idle time inserted between operations or pre-emption of operations, the starting time of an operation is determined by the sum of the processing times of all preceding operations. However, a schedule for a more complex production configuration may incorporate a great deal of information in addition to the sequence of operations. For example, delays waiting for parts and blockage due to insufficient buffer space both influence schedules.

The purpose of this chapter is to describe selected methods for sequencing and scheduling assembly operations. Since scheduling is a broad field, this chapter focuses on issues associated uniquely with assembly. For example, even though some products (e.g. circuit cards) are commonly assembled in a flexible flow-line (Wittrock, 1985, 1988), this chapter does not discuss that configuration because it is not uniquely associated with assembly. Configurations that deal with these more generic types of operations have been treated in other, readily available sources.

Assembly can be performed in a variety of configurations, but it seems that no comprehensive classification scheme for these configurations exists. A classification scheme has, however, been proposed for assembly lines (i.e. flow-lines) that produce single, multi- or mixed-models in high volumes (Buxey, Slack and Wild, 1973), and a scheme has been developed for flow-lines that deal with low volumes (Aneke and Carrie, 1984).

This chapter is organized into four sections. The first section deals with scheduling in push systems, the second with scheduling in pull systems. Methods for sequencing models in mixed-model assembly lines are described in the third section. Finally, conclusions are presented.

7.2 Scheduling in push systems

This section considers a broad range of scheduling problems in environments that might be classified as traditional push systems. Solution methods for three different types of problems are described. The first is a basic configuration consisting of two stations where parts are processed at the first station and assembled at the second. The second section reviews scheduling rules in large-scale job shop assembly systems. Scheduling kitting operations by assigning parts to kits in small-lot, multi-echelon assembly systems is the topic of the third section.

Table 7.1 Notation for sequencing in two-station assembly

Π	= a schedule
O_{ij}	= operation ij (i.e. part i of job J_j is processed on machine M_i)
$S_{ij}(\Pi)$	= start time of operation O_{ij} in schedule Π
p_{ij}	= processing time of operation O_{ij}
$C_{ij}(\Pi)$	= completion time of operation ij in schedule Π
$R_{Aj}(\Pi)$	= time at which all m parts are ready to begin the assembly operation for job J_j in schedule Π
$S_{Aj}(\Pi)$	= start time of the assembly operation for job J_j in schedule Π
p_{Aj}	= processing time of the assembly operation for job J_j
$C_{Aj}(\Pi)$	= completion time of the assembly operation for job J_j in schedule Π
π	= $(j_1, j_2, \ldots, j_n)$, the permutation of the jobs in schedule Π

Source: Potts *et al.*, 1995.

7.2.1 Scheduling in a two-station assembly process

Consider a two-station system in which the first station processes parts and consists of m machines in parallel ($M_1, \ldots, M_m$), and the second station assembles the parts and consists of a single machine, M_A, as shown in Fig. 7.1. Each job J_j ($j = 1, \ldots, n$) consists of m parts; part i must be processed on machine M_i ($i = 1, \ldots, m$) at station 1, and all m parts are assembled in one operation on the single machine at station 2. The objective is to minimize makespan, C_{max}. Notation used to formulate this problem is defined in Table 7.1.

Each machine processes jobs in the same order, machine M_i processing part i. Each part may have a unique processing time, so that a kit of parts required to assemble job J_j is completed at time $R_{Aj}(\Pi)$, and assembly can start whenever the assembly machine becomes ready at or after that time. Fundamental relationships which describe the operation of the system are given by:

$$C_{ij}(\Pi) = S_{ij}(\Pi) + p_{ij} \quad \text{for } i = 1, \ldots, m; j = 1, \ldots, n$$
$$R_{Aj}(\Pi) = \max_{i=1,\ldots,m} C_{ij}(\Pi) \quad \text{for } j = 1, \ldots, n$$
$$R_{Aj}(\Pi) \leq S_{Aj}(\Pi) \quad \text{for } j = 1, \ldots, n$$
$$C_{Aj}(\Pi) = S_{Aj}(\Pi) + p_{Aj} \quad \text{for } j = 1, \ldots, n.$$

Potts *et al.* (1995) study this two-stage problem, observing that it is a generalization of the two-machine flow-line scheduling problem. They establish several fundamental characteristics of the assembly problem proving that 'there exists an optimal solution which is a permutation schedule' and that it is NP-hard in the strong sense so that an effective heuristic would be useful in solving large-scale, industrial cases. They also show that makespan can be expressed as:

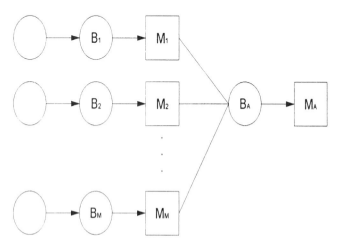

Fig. 7.1 A two-stage assembly system (Potts *et al.*, 1995).

$$C_{max}(\Pi) = \max_{u=1,\dots,n} \left\{ \max_{i=1,\dots,m} \left[\sum_{k=1}^{u} p_{i,j_k} \right] + \sum_{k=u}^{n} p_{A,j_k} \right\}.$$

Intuitively, one might interpret job u as the one which initiates the last busy period of the assembly station so that it begins assembly as soon as its last part is processed by the first station (i.e. the kit does not wait in queue ahead of the second station). Maximizing over u then establishes the makespan for the schedule.

Potts *et al.* (1995) present a three-step heuristic for this problem:

1. Define a two-machine flowshop problem with:

$$a_j = \left(\frac{1}{m} \right) \sum_{i=1}^{m} p_{ij} \quad = \text{processing time of job } J_j \text{ on machine 1}$$

$$b_j = p_{Aj} \quad = \text{processing time of job } J_j \text{ on machine 2.}$$

2. Apply Johnson's rule (Pinedo, 1995) to solve the two-machine flowshop problem with makespan as the objective (i.e. $n/2/F/C_{max}$) using processing times:

$$a_j \text{ and } b_j \quad \text{for } j = 1, \dots, n.$$

Then, let $\pi^* = (j_1, \dots, j_n)$ be the permutation prescribed by Johnson's rule.

3. Construct the assembly schedule Π using permutation π^*. **STOP**.

Potts *et al.* (1995) show that this heuristic has worst case performance bound given by:

$$\frac{\text{Value of heuristic solution for instance } I}{\text{Value of optimal solution for instance } I} \leq 2 - 1/m$$

Example 7.1 Consider the $n = 6$ job sequencing problem described in Table 7.2.

1. a_j and b_j values are determined for job j as noted in Table 7.2.
2. Johnson's rule assigns job j to set U if $a_j < b_j$ and to set V if $a_j \geq b_j$ (Table 7.2);

   - arranges jobs in set U with smallest a_j first: {4, 3, 5};
   - arranges jobs in set V with smallest b_j last: {2, 6, 1};
   - prescribes π^* with the ordered jobs from U followed by the ordered jobs from V.

3. The resulting job sequence π^* is {4, 3, 5, 2, 6, 1}.

Thus, we have a heuristic to solve the two-stage assembly scheduling problem. Now consider a system of larger scale in which parts are produced in a job shop.

Table 7.2 Sequencing in a two-station assembly process

Job j	Part processing machines M_i				Assembly machine, M_A	a_j values	b_j values	Assign to set
	M_1	M_2	M_3	M_4				
1	4	8	6	2	3	5	3	V
2	9	7	5	7	6	7	6	V
3	10	4	8	6	8	7	8	U
4	2	4	6	4	9	4	9	U
5	12	8	9	11	12	10	12	U
6	3	5	7	9	4	6	4	V

Source: Potts *et al.*, 1995.

7.2.2   Sequencing rules for job shop assembly

A push philosophy is often used in low volume environments in which a production facility, perhaps a job shop, produces parts that are subsequently assembled to produce end products. Numerous simulation studies have investigated the influence of sequencing rules and other production control techniques in managing job shops (Wilhelm, 1977; Blackstone, Phillips and Hogg, 1982), but it is important to understand how these factors influence the timely production of matched sets of parts that are used in assemblies.

In general, simulation studies must be analyzed in detail, reflecting on the adequacy of statistical analysis techniques, appropriateness of the shop configuration, and practicality of factors tested. Simulation studies can provide insights into underlying relationships and guidelines for studying actual systems. One must take care in assuming that the results of a particular simulation study apply directly to all actual production systems; fortunately, current technologies allow actual systems to be simulated in order to test hypotheses based on insights derived from the relatively simple systems on which research studies are typically based.

In one early study (Conway, Maxwell and Miller, 1967), a simulation model produces parts in a job shop then assembles each matched set of parts in a single operation to complete an end product. The bill of materials for each product is thus an assembly tree and all parts required to assemble each product must be kitted before assembly can begin. Thus scheduling the completion of each matched set of parts is crucial to the performance of the assembly system. It is noteworthy that the shortest processing time rule, which optimizes average flow-time on a single machine and which has been shown to perform well in the job shop environment (Blackstone, Phillips and Hogg, 1982), completes individual parts with a low average flow-time, but does poorly in coordinating the completion times of matched sets of parts so that lengthy delays are incurred by some parts waiting for others to complete each kit. In fact, delays increase with the number of parts required in a kit. Rules that use measures of the assembly tree to represent job structure in combination with using the shortest processing time to break ties outperformed rules based purely on due dates.

In an ambitious study, Russell and Taylor (1985) analyze the effects of four factors on assembly system performance, including product structure, dispatching rules for sequencing, methods for setting due dates and rules for assigning workers. Their test facility consists of four machine centers, each with two machines, and a group of three workers to man each pair of machine centers. Jobs arrive according to a Poisson process, processing times are exponentially distributed, and the bottleneck resource has an 84% utilization. Based on their statistical analysis of simulation results, they identify a strong interaction between rules for sequencing and labor assignment. They conclude that jobs with 'tall' structures should be scheduled and managed differently from those with 'flat' structures; in particular, certain rules should be avoided, depending upon product structure. Overall, the best performing combination involves a sequencing rule based on the branch slack plus the square of the number of operations remaining rule that sets due dates according to the length of the longest path in the structure of a product, and a rule that assigns workers to the longest queue. Fundamental results were confirmed and extended by a more recent study which is described next.

Fry *et al.* (1989) study the effects of both the sequencing rule and bill of materials structure on assembly system performance. Their hypothetical test facility consists of a six-machine job shop and a four-machine assembly facility in which machine utilization is 85% and processing times are normally distributed, reflecting the authors' opinions that the distributions of actual processing times can be modeled adequately by that distribution. Job arrivals are modeled as Poisson, due dates are set to reflect the planned lead-time of a product, and each part is used in only one product. They investigate 14 sequencing rules, including rules based on processing characteristics, product structure and due dates. In particular, they investigate rules based on using MRP lead-times to set due dates for individual parts. They consider 10 bills of materials categorized as flat, tall and combined.

For the flat bill of materials, four of the sequencing rules consistently outperform others relative to average (job) flow-time, percentage of jobs tardy and average tardiness. In order of their ranking for all three flat structures tested, these rules are:

1. closest level to the end product first (with ties broken by shortest processing time);
2. number of subsequent levels in the branch;
3. due date for the branch except the estimated finish time (with no queuing) of the branch is used if it is larger; and
4. earliest job due date first.

(The order of the first two rules was reversed in one test case.) While rules (1) and (2) performed well relative to these three measures, they did poorly relative to average absolute lateness.

Again, the shortest processing time rule performed rather poorly. One reason for this result may be that processing times were normally distributed. Most studies of job shop sequencing rules assume that processing times are exponentially distributed, a case which may magnify the effect of the shortest processing time rule. Other studies of the job shop (Dar-El and Wysk, 1982) have also found that the shortest processing time rule is less effective if processing times are normally distributed. Results also show that the percentage of jobs tardy was quite sensitive to the sequencing rule used.

For the tall bills of materials, sequencing rules based on due dates performed best, and rules based on processing times performed worse than they did in the flat structures. Rules based on the number of levels remaining and on shortest processing time performed poorly.

For the combined structure, rule (1) gave best values of average flow-time, percentage of jobs tardy and average tardiness on bills of materials with three levels. However, it again did poorly relative to average absolute lateness, and its performance degraded on the more complex structures tested. Rule (4), which is also based on due dates, seemed to be the best selection for combined structures.

Fry *et al.* (1989) conclude that, in their study, tardiness tended to be greater for tall structures, the shortest processing time rule did not perform well overall, the performance of rule (4) was consistently good in all cases, and rules based on due dates performed better for taller structures (although the rather simple rule (3) performed much better than more detailed rules that set due dates for each operation). Should these results prove to hold in actual production, they bear important implications for production control and should be considered carefully.

The kitting process in which required parts are accumulated before production can be initiated is unique to assembly. The next section addresses the problem of scheduling kitting operations in assembly systems.

7.2.3 Scheduling kitting operations in small-lot, multi-echelon assembly

The kitting problem in multi-echelon assembly is to allocate on-hand stock and anticipated future deliveries to kits in order to minimize total cost, consisting of job earliness, job tardiness and in-process holding costs. This section, which is taken from Chen and Wilhelm (1993), describes the kitting problem and compares the performance of three solution heuristics – two that are commonly used in industry and a new one. Computational experience shows that the new heuristic outperforms the others so it is expected to find application in the large-scale problems encountered in industry. Solutions will facilitate time-managed flow control, prescribing kitting decisions that promote cost-effective performance to schedule.

Problem description   All parts required by an assembly are commonly gathered in a kit, which is 'launched' (i.e. 'released') to initiate production. Operating according to the logic of MRP, each subassembly is assigned a date to be kitted (i.e. a due date), allowing adequate time to complete the end product by its due date. The entire assembly schedule thus relies upon the ability to compose each kit on its due date. This is often impossible, however, because there may not be sufficient stock on hand to meet scheduled requirements. Even though MRP sets vendor due dates to support the assembly schedule, unexpected events affect deliveries, disrupting kitting operations and, therefore, the assembly schedule.

This problem is especially prevalent in the electronics industry. For example, each circuit card may require several hundred components, so that a kit for a lot of circuit cards may require several thousand components of various types. Some components, such as integrated circuits, may be subject to yield loss, others to transportation delays from distant vendors. Safety stocks might be used to hedge against these problems, but the high cost of electronic components makes this an unattractive alternative.

In multi-echelon assembly systems, kits may require components from vendors as well as subassemblies produced upstream. Thus a disruption at one echelon may readily propagate to other echelons, affecting the coordinated flow of materials and the assembly schedule. For example, consider the three-echelon system of subassemblies depicted in Fig. 7.2. At the top echelon, circuit cards are assembled from vendor-supplied components. The circuit cards are then assembled into subassemblies which are, in turn, assembled into an end product such as a personal computer. A kitting delay at any echelon can affect material flow, increasing the cost of in-process inventories (e.g. some circuit cards may have to wait for the completion of others that are used in the same subassembly) and leading to poor schedule performance.

The kitting problem in multi-echelon assembly systems requires on-hand stock and promised deliveries to be allocated to kits to minimize total cost, including job earliness, job tardiness and the cost of holding in-process inventories, which may result from poor coordination of material flows. Part availabilities present important limitations that must be observed. In addition, shop capacities must be considered so that bottlenecks are not created by

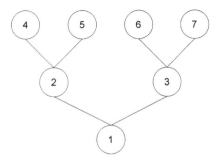

Fig. 7.2 A three-echelon assembly network (Chen and Wilhelm, 1993).

kitting an inappropriate set of subassemblies. The primary decision to be made is the day on which each kit is to be composed (each completed kit is immediately launched into production).

The kitting problem is prevalent in industry but has been the focus of little research. Three heuristic solution methods are compared in this section. Two of the heuristics reflect current industrial practice, while the third is a new method which, in fact, gives better test results.

This section is organized into the following subsections. Fundamental concepts are discussed first, including the context of kitting, the model formulated by Chen and Wilhelm (1993) and the two heuristics (H2 and H3) commonly used in industry. The new heuristic (H1) is then detailed and computational evaluation is discussed. Methods that have been devised to prescribe optimal solutions to the kitting problem are described and, finally, conclusions are related.

Fundamentals This section describes the kitting problem in the context of small-lot, multi-echelon, multi-product assembly systems. The presentation deals with information related to jobs, the structure of each job, subassemblies and part relationships, shop operations, the kitting schedule and part availabilities. A mathematical model is then presented to give a complete statement of the problem. Finally, the two heuristics used in industry are described. Notation is defined as it is presented and is summarized in Table 7.3 for convenience.

Jobs The goal is to assemble a set of jobs J. Each job, $j \in J$, represents a customer order that requires a single end product. The due date of job j is denoted by γ_j ($\gamma_j \geq 0$) and its lot size by Q_j ($Q_j \geq 1$). A kit for subassembly ij must, therefore, contain a sufficient number of components to assemble N_j final assemblies. A number of customer orders may require the same end product, but each is modeled individually to allow flexibility in the kitting schedule.

Job structure The bill of materials for each job may be described by an assembly tree in which each node has at most one successor (denoted by the set $K(i,j)$). Each node represents a subassembly. $PATH(m,i,j)$ is the set of nodes in the path from node m to node i (inclusive) for job j. Node ij is in echelon e_{ij} (where $e_{ij} = |PATH(i,1,j)|$). $M(i,j)$ $[A(i,j)]$ is the set of immediate (all) predecessors of node i. Job j is composed of the set of nodes $I(j)$, including the end product $1j$ and all of its predecessors.

Subassembly and part relationships The final assembly of each job is completed at echelon 1. $SE(e,j)$ denotes the set of subassemblies of job j in echelon e (for $e = 1, \ldots, E_j$). Subassemblies in echelon E_j are assumed to consist exclusively of parts supplied from vendors. Subassembly ij in echelon $e_{ij} < E_j$ is composed of subassemblies $m \in M(i,j)$ and requires quantity q_{ijp} of part p, where q_{ijp} reflects the total requirements of order quantity Q_j.

Table 7.3 Notation to model the kitting problem

Indices

d	$= 1, \ldots, D$ index for days in the planning horizon
i	$= 1, \ldots, I(j)$ index for subassemblies composing job j
j	$\in J$ a job (customer order)
e	$= 1, \ldots, E_j$ index of echelons in the structure of job j
s	$= 1, \ldots, S$ index for shops
p	$\in P$ an A^+ part

Cost parameters

$C_{ij}(d)$	= total cost incurred by kitting subassembly ij on day d
H_{ij}	= holding cost/day for subassembly ij [$i \in I(j)$] in lot size Q_j
T_j	= tardiness cost/day for job j

Decision variables

X_{ijd}	= 1 if subassembly ij is kitted on day d; 0 otherwise

Job information

$I(j)$	= set of subassemblies comprising job j
J	= set of jobs
Q_j	= lot size of job j
$SE(e,j)$	= set of subassemblies of job j in echelon e
γ_j	= due day of job j

Part information

n_{p0}	= number of part p on hand when the kitting policy is set
n_{pd}	= number of part p with delivery promised on day $d = 1, \ldots, D - 1$
N_{pd}	= cumulative number of part p available through day $d = \sum_{t=0}^{d-1} n_{pt}$ for $d = 1, \ldots, D$
P	= set of A^+ parts

Shop information

R_{sd}	= capacity of shop s on day d (i.e. number of kits that can be launched into shop s on day d)

Subassembly information

$A(i,j)$	= set of all predecessors of subassembly ij
e_{ij}	= echelon of subassembly ij
s_{ij}	= the shop in which subassembly ij is processed
$K(i,j)$	= set comprised of the immediate successor of subassembly ij
$M(i,j)$	= set of immediate predecessors of subassembly ij
$PATH(m'j,ij)$	= the set of nodes on the path from node $m'j$ to node ij (inclusive)
q_{ijp}	= quantity of part p required by subassembly ij to assemble lot size Q_j
$SS(s,j)$	= $\{i \mid s_{ij} = s\}$: set of subassemblies of job j processed in shop s
σ_{ij}	= planned lead-time to produce subassembly ij, an integer number of days

---

Source: Chen and Wilhelm, 1993.

Shop operations Subassembly *ij* is assembled in shop s_{ij}. The set of all subassemblies of job *j* that are assembled in shop *s* is denoted by $SS(s,j)$.

The capacity of shop *s* ($s = 1, \ldots, S$) is measured in terms of the number of kits that can be launched into it each day, R_{sd}. This measure of shop capacity appears to be acceptable, since small-lot production involves many lots, none of which dominate the processing capability of an entire shop. More detailed measures of capacity (involving, for example, the number of lots in process or the processing time requirements of each lot) could be incorporated at the cost of making the model somewhat more difficult to solve.

The planned lead-time of subassembly *ij* in shop s_{ij} is denoted σ_{ij}. This planned lead-time is an integer number of days that can be estimated by simulation, queuing models or any method used to determine planned lead-times for MRP systems.

The kitting schedule The kitting policy for each day over the horizon $d = 1, \ldots, D$ is to be prescribed. Using binary **decision variables**:

$X_{ijd} = 1$ if subassembly *i* on job *j* is kitted on day *d*; 0 otherwise,

the day on which subassembly *ij* is kitted is described by:

$$\sum_{d=1}^{D} d X_{ijd}.$$

Precedence relationships are invoked by:

$$\sum_{d=1}^{D} d X_{ijd} \geq \sum_{d=1}^{D} d X_{mjd} + \sigma_{mj} \quad \text{for } m \in M(i,j)$$

in which σ_{mj} is the planned lead-time for subassembly *mj*. Operation *mj* completes on day:

$$\sum_{d=1}^{D} d X_{mjd} + \sigma_{mj} - 1$$

so that subassembly *mj* could be kitted for subassembly *ij* on the following day, according to this convention for accounting time.

Part availability The traditional inventory classification scheme identifies *A*, *B* and *C* classes of items (Hax and Candea, 1984). A new class, A^+, which designates parts that are likely to disrupt planned schedules (e.g. integrated circuits), is important in the kitting problem. Representing technological capabilities, it is assumed that each A^+ part is assembled by only one of the *S* shops. Kitting schedules need be prescribed only for the set *P* of A^+ parts, since only A^+ parts require this additional management attention. However, the cost of holding a subassembly, say *ij*, as in-process inventory should include the cost of constituent A^+, *A*, *B* and *C* parts as well as the total contributed value built into all subassemblies $m \in M(i,j)$.

The quantity of part p $\{p \mid p \in P\}$ on hand, n_{p0}, is known when the kitting policy is prescribed. Vendors have promised to deliver quantity n_{pd} of part p on day d, and these parts may be kitted on or after day $d + 1$. Thus the cumulative number of part p available for use through day d is:

$$N_{pd} = \sum_{t=0}^{d-1} n_{pt} \quad \text{for } d = 1, \ldots, D$$

and the total available over the horizon, N_{Pd}, must be sufficient to complete all scheduled jobs.

The model The objective is to minimize total cost, which consists of job earliness, job tardiness and subassembly holding cost. Cost parameters H_{1j}, job earliness cost/day and T_j, job tardiness cost/day, penalize performance that deviates from schedule, while H_{ij}, the cost/day of holding subassembly ij, penalizes poor material flow coordination from one echelon to the next. H_{ij} reflects the contributed value added by each subassembly operation and the value of the parts required by subassembly ij. This holding cost is assessed for each day that subassembly ij must wait to be kitted to produce subassembly $k \in K(i,j)$. H_{1j} is the cost of holding a finished end product until its due date.

If assembly $1j$ is kitted on day $d = \gamma_j - \sigma_{1j} + 1$, job j would be completed on its due day γ_j. Thus, earliness is incurred if $1j$ is kitted on day $d < \gamma_j - \sigma_{ij} + 1$, and the cost of earliness is:

$$H_{1j} \operatorname{Max}\{0, \gamma_j - \sigma_{ij} + 1 - d\}. \tag{7.1}$$

Tardiness is incurred if assembly $1j$ is kitted on day $d > \gamma_j - \sigma_{ij} + 1$, and the cost of tardiness is:

$$T_j \operatorname{Max}\{0, d - (\gamma_j - \sigma_{ij} + 1)\}. \tag{7.2}$$

The cost of holding subassembly $mj \in M(i,j)$, if subassembly ij is kitted on day d, is:

$$H_{mj}\left[d - \sum_{t=1}^{d-\sigma_{mj}} (t + \sigma_{mj}) X_{mjt}\right] \tag{7.3}$$

and the cost of holding subassembly $ij \in A(1,j)$ if it is kitted on day d is:

$$H_{ij}\left[\sum_{t=d+\sigma_{ij}}^{D} t X_{kjt} - (d + \sigma_{ij})\right] \quad \text{for } k \in K(i,j) \tag{7.4}$$

in which the bracketed term in (7.3) ((7.4)) defines the number of days over which subassembly mj (ij) is held. The cost of holding subassembly ij over the planned lead-time, σ_{ij}, is not included, since this cost is incurred regardless of the kitting schedule. Objective function coefficients $C_{ij}(d)$, which incorporate job earliness, job tardiness and in-process holding costs, may be determined using equations (7.1)–(7.4) (Chen and Wilhelm, 1993).

The kitting problem may be formulated as

Problem P_0:

Minimize: $Z_0 = \sum\limits_{j \in J} \sum\limits_{i \in I(j)} \sum\limits_{d=1}^{D} C_{ij}(d) X_{ijd}$ (7.5)

Subject to:

$\sum\limits_{j \in J} \sum\limits_{i \in SS(s,j)} X_{ijd} \leq R_{sd}$ for $d = 1, \ldots, D$; $s = 1, \ldots, S$ (7.6)

$\sum\limits_{j \in J} \sum\limits_{i \in I(j)} \sum\limits_{t=1}^{d} q_{ijp} X_{ijt} \leq N_{pd}$ for $d = 1, \ldots, D-1$; $p \in P$ (7.7)

$\sum\limits_{j} \sum\limits_{i \in I(j)} \sum\limits_{t=1}^{D} q_{ijp} X_{ijt} = N_{pd}$ for $p \in P$ (7.8)

$\sum\limits_{d=1}^{D} (d + \sigma_{mj}) X_{mjd} \leq \sum\limits_{d=1}^{D} d X_{ijd}$ for $j \in J$; $i \in I(j)$; $m \in M(i,j)$ (7.9)

$\sum\limits_{d=1}^{D} X_{ijd} = 1$ for $j \in J$; $i \in I(j)$ (7.10)

$X_{ijd} = 0 \text{ or } 1$ for $j \in J$; $i \in I(j)$; $d = 1, \ldots, D$ (7.11)

The objective as noted in (7.5) is to minimize the total cost. Shop capacities, which limit the number of kits that can be launched each day, are imposed by constraint (7.6). Inequality (7.7) ensures that the kitting schedule observes part availabilities, so that no more than N_{Pd} parts, the cumulative number of parts on hand initially and planned for delivery on days $1, \ldots, (d-1)$, can be used in kits composed through day d. Constraint (7.8) assures that part deliveries satisfy all requirements over the horizon of D days. Inequality (7.9) invokes precedence relationships so that all predecessors $m \in M(i,j)$ must be assembled before subassembly ij can be kitted. Constraint (7.10) ensures that each subassembly is kitted exactly once during the planning horizon, and (7.11) requires decision variables to be binary integers.

Two heuristics used in industry   One heuristic commonly used in industry 'dedicates' on-hand stock to specific kits; the other requires kits to 'compete' daily until all required parts are available. In both heuristics, each kit is assigned a 'due day' on which it becomes a candidate for the allocation of parts. In general, the due day of a kit is determined by the due day of its associated end product. Thus each kit has a priority based on its due day and, perhaps, other business considerations according to which it is allocated parts.

The 'dedicate' heuristic, H2, allocates on-hand parts (i.e. dedicates them) to the priority kit. A partially completed kit (i.e. the dedicated parts that are on-hand) must wait for future deliveries of missing components from vendors or upstream production/assembly operations.

The 'compete' heuristic, H3, allocates parts to a kit only when **all** the parts required by the kit are on hand, so that a kit must compete daily until it finds all required components on hand. When two kits compete for the same parts, the priority kit has the competitive advantage.

In actual operations, partially completed kits may be launched into production to initiate assembly operations. However, this practice incurs additional costs (future deliveries must be specially handled to 'catch up' with the kit) and should be discouraged, particularly in the automated assembly environment. Thus we assume that no partial kits are launched into assembly.

The new heuristic proposed by Chen and Wilhelm (1993) can now be described.

Heuristic H1   Chen and Wilhelm (1993) argue that the kitting problem is NP-hard, so a heuristic that prescribes 'good' solutions in reasonable time is most useful in solving large-scale practical problems. A good heuristic must seek four goals simultaneously: (1) a good kitting sequence for jobs, (2) low job tardiness costs, (3) low job earliness costs and (4) low subassembly holding costs. Since each job may have unique cost parameters and planned lead-times, kitting jobs according to earliest-due-date-first may not result in a solution of high quality. Heuristic H1 uses a special procedure to determine job kitting sequence. In the first $|J|$ passes, jobs are scheduled in increasing order of slack. The slack of job j is defined as:

$$SLA_j = \gamma_j + 1 - \max_{mj \in SE(E_i, j)} \sum_{aj \in PATH(mj, 1j)} \sigma_{aj}$$

in which the maximum operator defines the longest series of planned lead-times to complete the end product among all subassemblies in echelon E_j. This term gives the MRP planned lead-time for job j. Subassemblies of a selected job are then sequenced according to earliest due day, $IDUE(i,j)$, first:

$$IDUE(i,j) = \gamma_j + 1 - \sum_{aj \in PATH(ij, 1j)} \sigma_{aj}$$

in which the summation gives the planned lead-time for subassembly ij. Each pass determines $ORDER(j)$, the sequence position in the final schedule, for the job j with the lowest potential tardiness cost ($ORDER(j)$, = 1, . . . , $|J|$).

To achieve goal (2), subassemblies should be kitted as early as possible. However, this may incur a large amount of job earliness and/or subassembly holding costs when a subassembly is completed and there are not enough parts to kit its immediate successor, and thus conflicts with goals (3) and (4). So, each time all subassemblies of a job are scheduled, the kitting times for all subassemblies of that job are right shifted in time (i.e. postponed) to avoid holding costs.

To summarize, heuristic H1 schedules jobs one at a time. In each of the first $|J|$ passes, jobs are scheduled in increasing order of slack, SLA_j, and the $ORDER(j)$ of one job, j, in the final sequence is determined. Each subassembly of each job is assigned a due day, $IDUE(i,j)$. Subassemblies of a selected job are then sequenced with the earliest $IDUE(i,j)$ first. After all subassemblies of a selected job are scheduled, the kitting days for all subassemblies of that selected job are right-shifted in time to reduce holding costs. The remaining jobs are rescheduled again and $ORDER(j)$ of another job, j, is fixed at the next iteration. The procedure iterates until $ORDER(j)$ is determined for each job $j \in J$. Then the kitting days of all subassemblies are determined in the $(|J| + 1)$th pass, scheduling job j according to $ORDER(j)$.

The heuristic determines $KIT(i,j)$, the scheduled kitting day for subassembly ij. Before presenting H1, several additional elements of notation must be defined (Table 7.4). $POSITION (= |J|, \ldots, 1)$ is a counter that identifies how many sequence positions remain to be considered. $FINAL$ is a code set to 0 for the first $|J|$ passes, then to 1 for the final pass which prescribes the kitting schedule (i.e. $KIT(i,j)$ for all i and j). Π_1 and Π_2 are subsets of J, the set of jobs. TC_j, the tardiness cost of sequencing job j to complete on day D, is defined as:

$$TC_j = T_j \max\left[0,\ D + \sigma_{1j} - \gamma_j - 1\right].$$

Table 7.5 presents heuristic H1 in succinct, algorithmic form; each step is now described from an intuitive perspective. To begin, Π_1, Π_2, $POSITION$ and $FINAL$ are initialized (Step 0-A). Subassembly due day and job slack are calculated in Step 0-B. In Step 1-A, a job is selected to determine its kitting schedule. In each of the first $|J|$ passes, the unscheduled job with the smallest slack SLA_j is selected first. In the $(|J| + 1)$th pass, job j is selected according to $ORDER(j)$, which is determined during the first $|J|$ passes.

After selecting a job $[j^* \in \Pi_1]$ to schedule in Step 1-A, the procedure

Table 7.4 Notation for heuristic H1

$FINAL$	= 0 for the first $|J|$ passes, then 1 for pass $(|J| + 1)$, which prescribes the schedule
$IDUE(i,j)$	$= \gamma_j + 1 - \sum_{aj \in PATH(ij,1j)} s_{aj}$ = due day for kitting subassembly ij
$KIT(i,j)$	= the kitting day for subassembly ij scheduled by heuristic H1
$POSITION$	= counter of the number of sequence positions remaining to be considered = $|J|, \ldots, 1$
SLA_j	$= \gamma_j + 1 - \max_{mj \in SE(E_j)}, \sum_{aj \in PATH(mj,1j)} s_{aj}$ = slack of job j
TC_j	= tardiness cost of sequencing job j to complete on day $D = T_j \max[0, D + \sigma_{1j} - \gamma_j - 1]$
Π_1 and Π_2	= subsets of J

Source: Chen and Wilhelm, 1993.

Table 7.5 The heuristic of Chen and Wilhelm

Step 0. Initialization

    0-A. Set $\Pi_1 = \Pi_2 = J$, POSITION = $|J|$, FINAL = 0

    0-B. For each $j \in J$, set $IDUE(1,j) = \gamma_j + 1 - \sigma_{1j}$,
        $SLA_j = \gamma_j + 1 - \max_{mj \in SS(E_j, j)} \sum_{aj \in PATH(mj.1j)} \sigma_{aj}$, and for each $ij \in A(1, j)$ set
        $IDUE(i,j) = IDUE(k,j) - \sigma_{ij}$, in which $kj \in K(i,j)$.

Step 1. Schedule all jobs in set Π_1

    1-A. If FINAL = 0, select $j^* = \operatorname{argmin}_{j \in \Pi_1} \{SLA_j\}$,
        breaking ties by selecting the one with the smallest order quantity
        first. Otherwise, select $j^* = \operatorname{argmin}_{j \in \Pi_1} \{ORDER(j)\}$.

    1-B. Select subassembly i^* as the one with the earliest $IDUE(i, j^*)$
        (i.e. $i^* = \operatorname{argmin}_{i \in I(j^*)} \{IDUE(i,j^*)\}$).
        Break ties by selecting subassembly $i^* = \operatorname{argmin}_{i \in I(j^*)} \{H_{ij^*}\}$.
        Set $KIT(i^*,j^*)$ as early as possible without violating constraints (7.6)–
        (7.9). (Update the RHSs of constraints (7.6)–(7.9) every time a kitting
        schedule for a subassembly is determined.)

    1-C. Select subassembly i^* as the one with the latest $KIT(i,j^*)$
        (i.e. $i^* = \operatorname{argmax}_{i \in I(j^*)} \{KIT(i,j^*)\}$).
        Break ties by selecting i^* as the subassembly $i^* = \operatorname{argmax}_{i \in I(j^*)} \{H_{ij^*}\}$.

        1-C-1. For $i^* = 1$
            If $KIT(i^*,j^*) < IDUE(i^*,j^*)$, set $KIT(i^*,j^*)$ as close to (but
            not later than) $IDUE(i^*,j^*)$ as possible, observing (7.6)–(7.9).
            Otherwise, $KIT(i^*,j^*)$ is not changed. Go to Step 1-D.

        1-C-2. For $i^* > 1$
            If $KIT(i^*,j^*) < KIT(k,j^*) - \sigma_{i^*,j^*}$, in which $k \in K(i^*,j)$ set
            $KIT(i^*,j^*]$ as close to (but not later than) $[KIT(k,j^*) - \sigma_{i^*,j^*}]$,
            as possible, observing (7.6)–(7.9). Otherwise, $KIT(i^*,j^*)$ is not
            changed. Go to Step 1-D.

    1-D. Set $\Pi_1 \leftarrow \Pi_1 - \{j^*\}$. If $\Pi_1 \neq \varnothing$, go to Step 1.
        Otherwise, $(\Pi_1 = \varnothing)$, set $D = \max_j KIT(1,j)\}$.

    1-E. If FINAL = 0, go to Step 2.
        Otherwise, (FINAL = 1), stop.

Step 2. Determine the ORDER of job j + for final scheduling

    2-A. For $j \in \Pi_2$, calculate $TC_j = T_j \max[0, D + \sigma_{1j} - \gamma_j - 1]$.

    2-B. Let $j+ = \operatorname{argmin}_{j \in \Pi_2} \{TC_j\}$.
        Break ties by setting $j+ = \operatorname{argmax}_{j \in \Pi_2} \{IDUE(1, j)\}$.
        Break any second tie by designating $j+ = \operatorname{argmin}_{j \in \Pi_2} \{Q_j\}$.
        Update $\Pi_2 \leftarrow \Pi_2 - \{j+\}$.
        Set $ORDER(j+) = $ POSITION, and decrement
        POSITION $\leftarrow$ POSITION $- 1$.

    2-C. If POSITION > 1, set $\Pi_1 = \Pi_2$ and go to Step 1.
        Otherwise, (POSITION = 1), go to Step 3.

Step 3. Determine final schedule

    Set $j+ = $ the index of the remaining job in set Π_2, $ORDER(j+) = 1$, $\Pi_1 = J$,
    FINAL = 1, re-initialize constraints (7.6)–(7.9), and go to Step 1.

Source: Chen and Wilhelm, 1993.

proposes a feasible kitting schedule (i.e. relative to constraints (7.6)–(7.9)) for all subassemblies of job j^* (Step 1-B). The subassembly i^* with the earliest $IDUE(i,j)$ amongst all subassemblies of job j^* is scheduled first. A tie is broken by giving priority to the subassembly with the smallest in-process holding cost/day. The kitting schedule of i^* is set on the earliest day feasible with respect to constraints (7.6)–(7.9). In Step 1-C, kitting schedules for all subassemblies of job j^* are refined, if possible, to reduce in-process holding cost. The subassembly i^* with the latest kitting day, $KIT(i^*,j^*)$, is rescheduled first. $KIT(1^*,j^*)$ is rescheduled to be as close as possible to γ_j (Step 1-C-1) and $KIT(1^*,j^*)$, $i^* = 1$, is rescheduled to be as close as possible to the day when its immediate successor requires it (i.e. $KIT(k,j^*) - \sigma_{i^*,j^*}$) (Step 1-C-2). In Step 1-D, Π_1 is updated. If all jobs have been scheduled, the kitting horizon D relative to all jobs in Π_1 is determined. Otherwise, another job is selected to schedule. If the final schedule has been obtained (i.e. in the $(|J| + 1)$th pass), H1 terminates in Step 1-E. Otherwise, the procedure progresses to Step 2 to determine $ORDER(j)$ of another job $j \in J$.

In Step 2-A, the total tardiness cost for each job is calculated based on the assumption that it is kitted on day D. The job $j+$, which will incur the smallest tardiness cost if it is kitted on day D, is selected to determine $ORDER(j+)$ (Step 2-B). The $ORDER(j+)$ of job $j+$ is set to the current value of $POSITION$. If $ORDER(j)$ of all jobs (except one job) has not been determined, Π_1 is updated and H1 returns to Step 1. Otherwise, $POSITION$ is set to 1 and the procedure progresses to Step 3.

In Step 3, $ORDER(j+)$ of job $j+$, the one that has not been determined in the first $|J| - 1$ passes, is set to 1. $FINAL$ is set to 1 and H1 returns to Step 1 for the $(|J| + 1)$th pass to determine the final schedule. After the final schedules for all subassemblies have been determined, H1 terminates in Step 1-D. The upper bound on the optimal objective value of Problem P_0, Z_0, may then be obtained according to $KIT(i,j)$ for all i and j.

Computational evaluation The runtime and solution quality of H1 are compared with those of H2 and H3 in this section using a set of 24 test problems. All tests were performed on an IBM 3090 running the MVS-ESA operating system. The matrix generators and heuristics were coded in FORTRAN.

Although generated randomly, test problems were designed to study the effects of five factors: (1) the number of jobs, $|J|$; (2) the number of echelons, E; (3) the number of parts in class A^+, $|P|$; (4) the number of different part types in class A^+ required by all subassemblies; and (5) the part delivery time distribution. The magnitudes of $|J|$ and $|P|$ affect the size of the problem, and the magnitude of E reflects the assembly structure. The number of each different part type in the A^+ class required by a subassembly dictates the density of constraints (7.7) and (7.8). The part delivery time distribution influences the kitting horizon D, affecting the number of constraints and variables.

Based on these factors, the 24 test problems were defined as noted in Table 7.6. The pattern by which levels of the five factors were set for each problem is apparent so the interested reader can easily identify a pair of problems that differ by a change of one factor to assess its impact on runtime and solution quality.

Given levels of the five factors that define these 24 problems, Table 7.6 indicates the number of binary decision variables and constraints involved in each problem P_0 formulation. These problems are not large in terms of the parameters that describe their practical contexts but they do represent rather large binary integer problems.

Table 7.6 Test problem parameters and sizes

Problem no.			Factors			Problem size	
	$|J|$	E	$|P|$	DS	PDP dist. no.	NC	NV
1	6	3	56	1	1	890	588
2	10	3	56	1	1	897	910
3	6	3	56	3	1	890	588
4	10	3	56	3	1	897	910
5	6	3	112	2	1	1660	588
6	10	3	112	2	1	1625	910
7	6	3	112	4	1	1688	588
8	10	3	112	4	1	1625	910
9	13	2	56	1	1	681	429
10	20	2	56	1	1	727	660
11	13	2	56	3	1	703	429
12	20	2	56	3	1	738	660
13	13	2	112	2	1	1231	429
14	20	2	112	2	1	1332	660
15	13	2	112	4	1	1308	429
16	20	2	112	4	1	1343	660
17	6	3	56	1	2	1064	714
18	6	3	56	3	2	1064	714
19	6	3	112	2	2	1999	714
20	6	3	112	4	2	2033	714
21	13	2	56	1	2	849	546
22	13	2	56	3	2	877	546
23	13	2	112	2	2	1549	546
24	13	2	112	4	2	1647	546

DS = Distribution set to generate the number of different A^+ part types required by each subassembly.
PDP dist. no. = Part delivery perturbation distribution number.
NC = Number of constraints in P_0.
NV = Number of binary variables in P_0.

Source: Chen and Wilhelm, 1993.

Each test problem involved a binary assembly tree in which each node (other than those in echelon E) had two immediate predecessors. This structure does not limit interpretation of test results, since any assembly tree can be recast into the binary form. Jobs are described either by an $E = 2$ or an $E = 3$ echelon binary network so that $|I(j)|$ is 3 or 7, respectively. For simplicity, it was assumed that subassemblies in level e would be assembled in shop $s = e$ so that $S = E$.

The number of jobs was either 6 or 10 in the three-echelon problems and either 13 or 20 in the two-echelon problems. The number of parts in class A^+, $|P|$, was either 56 or 112. The interested reader can refer to Chen and Wilhelm (1993) for complete details, including shop capacities, sojourn times, job due dates and the number of A^+ parts assembled in each echelon, and the probability distributions used to generate the number of A^+, A, B and C parts required by each subassembly, the part delivery times and customer order quantities.

Test results, runtimes and solution values for heuristics H1, H2 and H3 are given in Table 7.7. In general, runtime increases with the number of jobs (e.g. compare problems 3 and 4) and the delays of part deliveries (e.g. compare problems 1 and 17), since they extend the kitting horizon and thus increase the size of the problem. This indicates, as expected, that problems of larger size are more difficult to solve. Runtime is also influenced by the number of A^+ parts required by subassemblies. Roughly speaking, the CPU time required to solve the problem increases with the number of A^+ parts required (e.g. compare problems 5 and 6), indicating that problems with denser constraints are more 'difficult' to solve, as expected.

H1 obtained optimal solution values for 17 out of the 24 test problems with all solution times under 0.231 seconds. H2 provided a better solution value than H1 in only one problem (problem 21), as did H3 (problem 22). In all test problems, H3 required less runtime than did H1 and H2. However, overall, H3 did not prescribe solutions of the quality of those prescribed by H1 and H2.

The reason that H1 did not perform best in problems 21 and 22 may be because parts that were supposed to be assigned to subassemblies of two (or more) jobs with small order quantities were used by subassemblies of a job with a large order quantity. According to Step 2 of H1, a job with a large order quantity may be scheduled earlier than a job with a small order quantity, provided they have the same due day. Kitting subassemblies of a job with a large order quantity may result in kitting delays of the subassemblies of other jobs with small order quantities. Furthermore, this situation may result in higher tardiness costs. In a sense, H1 allows subassemblies of a job with a large order quantity to 'steal' parts from the subassemblies of other jobs with small order quantities. Since this situation results in higher tardiness costs, H1 could be refined to explicitly deal with it if the application warranted.

On average, H1 provided solutions that were within 0.6% of the optimal solutions. The average value of a solution from H1 was 3.4% better than that from H2 and 10.7% better than that from H3.

Table 7.7 Comparison of heuristic procedures

Problem no.	Optimal solution value $v(OPT)$	Heuristic H1		Heuristic H2		Heuristic H3	
		Solution value $v(H1)$	Runtime (secs)	Solution value $v(H2)$	Runtime (secs)	Solution value $v(H3)$	Runtime (secs)
1	894.5	*894.5	0.038	929.7	0.050	1198.0	0.023
2	1567.0	*1567.0	0.079	1617.3	0.078	1954.3	0.030
3	979.0	*979.0	0.043	1001.7	0.052	1004.2	0.023
4	1567.0	*1567.0	0.093	1621.6	0.079	1771.7	0.031
5	979.0	*979.0	0.060	986.4	0.085	1232.7	0.026
6	1482.5	*1482.5	0.121	1501.4	0.132	1719.9	0.036
7	979.0	*979.0	0.070	996.5	0.092	998.6	0.028
8	1401.0	*1401.0	0.145	1453.5	0.144	1653.1	0.040
9	690.0	696.0	0.060	713.9	0.049	714.9	0.022
10	984.0	*984.0	0.120	1014.2	0.070	1015.2	0.028
11	567.0	573.0	0.066	634.9	0.052	634.9	0.023
12	981.0	1011.0	0.147	1078.0	0.076	1085.8	0.030
13	676.5	*676.5	0.102	692.8	0.083	692.8	0.027
14	1143.0	*1143.0	0.215	1166.4	0.129	1169.6	0.036
15	696.0	*696.0	0.120	725.8	0.093	752.8	0.028
16	1047.0	*1047.0	0.231	1092.8	0.141	1070.4	0.039
17	2278.0	*2378.5	0.056	2379.5	0.077	2639.3	0.028
18	2319.5	*2319.5	0.065	2485.7	0.095	2671.9	0.031
19	2447.5	*2447.5	0.100	2468.2	0.158	2720.8	0.042
20	2404.0	2447.5	0.117	2466.6	0.180	2920.3	0.043
21	1740.0	1791.0	0.076	**1786.2	0.053	2009.8	0.024
22	1548.0	1590.0	0.084	1698.0	0.059	***1588.8	0.025
23	1684.5	*1684.5	0.132	1742.2	0.097	1758.1	0.031
24	1746.0	1789.5	0.156	1790.8	0.104	1810.0	0.033

*   H1 obtained the optimal solution value.
**   H2 obtained better solution value than H1.
***   H3 obtained better solution value than H1.
$v(*)$ objective function value for solution procedure*.
Source: Chen and Wilhelm, 1993.

Optimizing methods   In a related study, Chen and Wilhelm (1994) devised an optimizing approach for the kitting problem employing Lagrangian relaxation. Their Lagrangian problem decomposes into subproblems related to independent jobs. They incorporated several preprocessing steps, an efficient dynamic programming algorithm to solve the subproblems, dominance properties to enhance efficiency and a specialized branching rule. They showed that this approach outperforms OSL, a standard mathematical programming package, on the set of 24 test problems.

Subsequently, Wilhelm, Chen and Parija (1994) devised a cutting plane method to prescribe optimal solutions to the kitting problem. They identified facets related to precedence relationships and derived valid inequalities related to shop capacities and part availabilities. In addition, they gave a polynomial time algorithm to solve the separation problem associated with each family of valid inequalities. The cutting plane method posted encouraging computational results, requiring less runtime than did OSL on all 24 test problems and less time than did their specialized Lagrangian relaxation approach in all 14 problems on which the latter approach required more than 2.1 seconds.

While these optimizing approaches were successful in application to a set of problems of modest size, it is expected that heuristic methods are needed to deal with truly large-scale problems found in some industries. Heuristic H1 addresses that need.

Conclusions   The kitting problem is a crucial aspect of managing material flow in small-lot, multi-product, multi-echelon assembly systems. The problem requires the materials manager to allocate parts to kits to minimize total cost – including job earliness, job tardiness and in-process holding cost – while observing shop capacity, part availability and subassembly precedence relationships.

A new heuristic, H1, is described in this section and compared empirically with two heuristics commonly applied in industry, namely 'dedicate' (H2) and 'compete' (H3). Since H1 has been shown to run in polynomial time in the worst case (Chen and Wilhelm, 1993) and since it outperforms heuristics H2 and H3 on average, it could be used by any optimizing approach which requires initial upper bounds on the horizon D and on the optimal value of the objective function of problem P_0. In addition, it appears to be well suited for direct application to large-scale industrial problems.

A set of 24 test problems was used to evaluate the computational characteristics of the heuristics. Heuristic H1 obtained optimal solution values for 17 out of the 24 test problems with solution times under 0.231 seconds (on average). H2 provided a better solution value than did H1 in only one problem, as did H3. On average, H1 prescribed solutions that were within 0.6% of the optimal solutions. The value of a solution from H1 was 3.4% better than that from H2 and 10.7% better than that of H3 (on average). These percentages are not large, but they would represent significant cost savings in actual industrial applications.

The new heuristic would promote material flow coordination throughout a multi-echelon assembly system, providing a balance between the cost of in-process inventory and schedule performance, and prescribing the allocation of parts to achieve cost-effective schedule performance while promoting shop efficiency. Time-managed flow control would result, coordinating material flow throughout the multi-echelon system to achieve cost-effective due date performance.

7.3 Level scheduling in just-in-time pull systems

The fundamental precept of just-in-time (JIT) is that it demands the production of only the necessary items in the necessary quantities at the necessary time. Any surplus that is not consistent with this precept may be considered waste (Hay, 1988). Excessive inventories represent the 'waste' associated with poor scheduling practices. For example, in mixed-model assembly, the sequence in which end products are assembled determines the rate at which components parts are withdrawn and, consequently, affects their in-process inventories. Thus, a mixed-model assembly system must produce a mix of end products every day to meet customer demand and level the requirements for components.

Consider the case in which n different part types are manufactured and used to assemble a mix of end products, each of which may require a unique set of parts (similar to the case in Fig. 7.1). In a JIT system, each process withdraws the necessary items in the necessary quantities from its preceding processes; however, if the number of parts withdrawn by the process is different at every withdrawal, the preceding process must maintain sufficient capacity (e.g. inventory, equipment, labor) to meet the highest possible demand. The impact of these fluctuations can propagate to stations further upstream (Kimura and Terade, 1981). Hence, it is desirable to keep the usage rate of each part as constant as possible to help each fabrication station to avoid overloads during some time periods and underloads during other time periods and to maintain the necessary part flow with a minimum inventory. A large fluctuation in the part usage rate may make it impossible for fabrication stations to provide parts just-in-time. This viewpoint leads to the requirement for level scheduling, one of the fundamental concepts of JIT production (Sugimori *et al.*, 1977; Monden, 1983; Hall, 1984).

A level schedule is defined as a sequence for **end products** which minimizes the deviation of the quantity of **each part** withdrawn from its target level over a series of **time slots**. The target level can be set by the production control manager, and a time slot is the duration required to assemble a product.

For production **planning** purposes, the approximate capacity of each station can be set to the average daily production quantity determined by the master schedule for production over several months (Monden, 1983; Hall, 1984; Hay, 1988; Groeflin *et al.*, 1989; Miltenburg, 1989). During planning, the processes that precede the final assembly line are given only estimates of the production quantities that will be required. At the end of every scheduling horizon (typically a day according to Hay (1988) and Monden (1983) or a week according to Miltenburg and Sinnamon (1992)), actual demand for each product is received for the next scheduling horizon, we assume the next day. A day's demand must be satisfied by the end of that day. The estimated production capacity can be adjusted to meet actual demand, for example by changing workforce level or utilizing multi-function workers (Monden, 1983; Hay,

1988), unless a **significant** difference between forecast and actual demand occurs. With this information on the quantity of each product demanded for a day, an end-product assembly schedule must be determined to level the production of parts for the day.

According to the JIT system, processes that precede final assembly do not need a production schedule in advance because they are driven by final assembly, which follows a planned production sequence and uses kanbans (cards authorizing production or part withdrawal) to order the production of parts. In this sense, the schedule for final assembly is crucial to production in the entire plant.

A number of quantitative approaches have been proposed for various problems associated with JIT production, including level scheduling, determining the number of kanbans (Askin, Mitwasi and Goldberg, 1993) and lot sizing. An extensive literature review of these quantitative approaches is presented by Joo and Wilhelm (1993a). Monden (1983) reports the first analytical, level scheduling model which minimizes the total variation of part usage from a target quantity in each time slot. He defined a time slot as the time during which an end product is assembled and assumed that all types of products require the same assembly time. Monden (1983) introduced a myopic heuristic algorithm called the Goal-Chasing Method to resolve the problem. Groeflin *et al.* (1989) also assumed an identical assembly time for each product and devised a means of prescribing the lexicographical minimum variability of part usage rates. They developed a heuristic to prescribe a sequence for final assembly.

Miltenburg (1989) investigated the use of level schedules in multi-level production systems. He assumed that each product would require approximately the same number of each part. With this assumption, a constant rate of part usage is achieved, independent of the assembly sequence. Consequently, his objective was to prescribe a schedule that keeps the proportion of each product relative to the total produced as close as possible to the proportion of total demand represented by that product. Miltenburg and Sinnamon (1989) presented a similar model but relaxed the assumption that each end product requires approximately the same number of each part. Their objective was to schedule the system so as to maintain a constant rate of part usage at all echelons. They proposed a heuristic to resolve the problem.

Kubiak and Sethi (1991) formulated an assignment problem which is equivalent to the problem studied by Miltenburg (1989). They showed that an optimal solution for Miltenburg's problem can be constructed in polynomial time from an optimal solution to the assignment problem. Sumichrast, Russell and Taylor (1992) conducted a simulation experiment to compare the performance of several heuristics. One of their findings is that Miltenburg's method was superior to other available heuristics in minimizing the fluctuation of part usage rates. Inman and Bulfin (1992) proposed an ad hoc heuristic for multi-echelon production systems. Their heuristic considers only the sequence of end products, ignoring the production of parts.

Steiner and Yeomans (1993) developed a polynomial time algorithm to solve a particular version of the level scheduling problem, which involves using a **min-max** objective function (rather than minimizing a total sum) to minimize the maximum deviation of **end-product** production from the desired level rather than the total deviation of **part** production. In addition, they offer no means of dealing with additional side constraints.

Modeling, developing solution algorithms and effectively minimizing part usage variations are the main issues in these scheduling problems. Only the work of Kubiak and Sethi (1991) and Steiner and Yeomans (1993) has led to optimizing procedures, since runtime may prohibit optimizing these problems which are NP-hard. Most of these studies focused on developing heuristics that can be used effectively on large-scale industrial problems.

This section presents two methods for prescribing a level schedule in just-in-time assembly systems. The first method is a heuristic, and the second is an optimizing approach.

7.3.1 A heuristic method

Ding and Cheng (1993) present a heuristic for level scheduling with the objective of minimizing the squared deviation of part usage rate from target (i.e. a uniform rate) for each time slot, then adding over all slots:

$$\text{Minimize}: \sum_{t=1}^{T} \sum_{m \in M} \left(Y_{m,t} - \frac{t d_m}{T} \right)^2$$

in which the decision variable is:

Table 7.8 The heuristic of Ding and Cheng

---

1. Set $t = 1$ and $Y_{m0} = 0$ for $m \in M$

2. Determine the model $p \in M$ where $p = \text{argmin}_{m \in M} \left[Y_{m,t-1} - \frac{(t+0.5)d_m}{T} \right]$

3. Determine the model $q \in M$ giving

$$q = \text{argmin} \left[\min_{m \neq p} \left(Y_{m,t-1} - \frac{(t+1)d_m}{T} \right), \left(Y_{p,t-1} + 1 - \frac{(t+1)d_p}{T} \right) \right]$$

4. IF $p \neq q$ and $\left(Y_{p,t-1} - \frac{(t+0.5)d_p}{T} \right) - \left(Y_{q,t-1} + 1 - \frac{(t+0.5)d_q}{T} \right) > \left(\frac{0.5}{T} \right)(d_q - d_p)$

    then schedule model q at time t. Otherwise, schedule model p at time t.

5. Update Y_{mt} for all $m \in M$.
    IF $t = T$, **STOP**.
    ELSE, increment $t \leftarrow t + 1$, and GOTO Step 2.

Source: Ding and Cheng, 1993.

$$Y_{mt} = \quad \text{number of units of model } m \text{ produced}$$
$$\text{in first } t \text{ time periods } \left(t = 1, \ldots, T\right)$$

and parameters are:

$$m = \text{model index } m \in M$$
$$d_m = \text{demand for model } m$$
$$T = \text{total demand} = \sum_{m \in M} d_m.$$

As shown in Table 7.8, the heuristic sequences one model in each time period $t = 1, \ldots, T$ to meet the total demand, T.

Ding and Cheng (1993) show that Steps 2–4 minimize the squared deviation over the next two time periods t and $t + 1$, and that this myopic, greedy heuristic has complexity $O(|M| T)$.

They report computational experience comparing this heuristic with 'Mittenburg's algorithm 3 using heuristic 1'. On the set of problems tested, their heuristic prescribes the same solutions but takes much less runtime.

7.3.2   An optimizing approach

This section describes an optimizing approach developed by Joo and Wilhelm (1993b) to resolve an extension of the level scheduling problem studied by Miltenburg (1989) and Monden (1983). The model is based on the assumption (e.g. see Monden (1983)) that all types of end products require the same assembly time, which is defined as the duration of a time slot. However, it extends earlier versions of the level scheduling problem by incorporating additional, practical restrictions that disallow one product type, say q, from following another, say p, in the assembly sequence. For example, assembly of product p may require a special adhesive that should not contact product q. The remainder of this section relates the model developed by Joo and Wilhelm (1993b), describes their cutting plane approach and reviews some of their computational experience.

As in previous research (Monden, 1983; Miltenburg and Sinnamon, 1989), the target usage rate of part j is defined as the total number of part j required, divided by the time required to produce all products. This objective can be represented in various forms as indicated by Miltenburg (1989). Unfortunately, none of the forms can be easily handled mathematically because they lead to non-linear objective functions. However, Joo and Wilhelm (1993b) use a form that can be transformed into a linear objective function.

Only parts which directly feed the final assembly line are considered. The level scheduling problem is complicated by the number of products and the variety of parts required by different products. If each product required the same number of each part, the part usage rate would remain nearly constant, regardless of the product assembly sequence.

Because many companies serve demands for a variety of end products, each in low volume, the mixed-model assembly line is becoming a predominant configuration in industry. For example, even an assembly line for a particular type of car assembles an enormous number of distinct cars, depending on the optional features ordered by customers. Many types of parts are used by more than one product in such a case. The typical company assembles many end products, each in low volume. Each product is distinguished by the set of parts it requires.

A special set of sequence limitations is considered. For two specified products, say p and q, assembly of product p cannot be followed by assembly of product q. For example, a product requiring white paint cannot follow one requiring black paint. Final assembly can make a mix of products in any sequence that is feasible with respect to these limitations.

No significant setups should be required, since setup time is reduced to a negligible level in a JIT production system. It is assumed that an end product is launched (i.e. assembly is initiated) every time slot. The total number of positions in the sequence is equal to the total demand for all products over the horizon (e.g. a day).

A set of parts must be withdrawn from the production stations that feed the final assembly line whenever a product is launched. A weight is given to each part in order to indicate its importance relative to other parts. A model of this level scheduling problem is presented next using the notation in Table 7.9.

Problem P_0: Minimize: $$\sum_{k=1}^{T}\sum_{j\in J} w_j \left| kR_j - \sum_{t=1}^{k}\sum_{i\in I} r_{ij}x_{it} \right| \qquad (7.12)$$

Subject to: $$\sum_{i\in I} x_{it} = 1 \qquad t = 1, \ldots, T \qquad (7.13)$$

$$\sum_{t=1}^{T} x_{it} = d_i \qquad i = 1, \ldots, |I| \qquad (7.14)$$

$$x_{pt} + x_{q, t+1} \le 1 \qquad t = 1, \ldots, T-1 \; (p, q) \in Q \qquad (7.15)$$

$$x_{it} \in \{0, 1\} \qquad i = 1, \ldots, |I| \; t = 1, \ldots, T. \qquad (7.16)$$

In objective function (7.12), kR_j represents the target quantity of part j to support production in the first k sequence positions. The objective is to minimize the absolute deviation of the quantity used from the target usage level, summed over all time slots for all parts. This objective function penalizes any deviation from the target level so that the actual number of parts used is as close as possible to the target level at all times. Schedules that allow large deviations for 'important' parts may be avoided by putting relatively large weights on the deviations associated with them. Constraint (7.13) ensures that one product is assigned to each sequence position, and (7.14) ensures that the demand for each product is satisfied. Inequality (7.15) ensures that product q will not follow product p in the assembly sequence for $(p, q) \in Q$.

Table 7.9 Notation to model the level scheduling problem in JIT assembly

Sets

I = set of products

J = set of parts

Q = set of pairs of products in which one cannot follow the other

Indices

i	= product i	$(i = 1, \ldots, |I|)$
j	= part j	$(j = 1, \ldots, |J|)$
t	= sequence position	$(t = 1, \ldots, T)$
k	= sequence position	$(k = 1, \ldots, T)$

(p, q) = pairs of products such that $(p, q) \in Q$

Parameters

a_{ij} = $w_j(R_j - r_{ij})$

d_i = demand for product i over the horizon

T = total demand for all products

= $\sum_{i \in I} d_i$ = total number of positions in a sequence

r_{ij} = the number of part j required to assemble one product i

R_j = target usage for part $j = \dfrac{\sum_{i \in I} r_{ij} d_i}{\sum_{i \in I} d_i}$

w_j = weight associated with part j (i.e. relative importance of a deviation in the usage of part j from its target level)

Computed values

R_j = target usage for part $j = \dfrac{\sum_{i \in I} r_{ij} d_i}{\sum_{i \in I} d_i}$

$\sum_{i \in I} d_i$ = total demand for all products = T

$\sum_{i \in I} r_{ij} d_i$ = total number of part j required to assemble all end products demanded

Decision variables

x_{it} = 1 if product i is assigned to sequence position t, 0 otherwise

---

Linearizing the objective function (7.12), and refining the model by eliminating certain variables and constraints, the problem can be expressed as the 0/1 mixed integer program:

Problem P_1: Minimize: $\displaystyle\sum_{j \in J}\sum_{i \in I}\left|a_{ij}\right|\left(x_{i1} + x_{iT}\right) + \sum_{j \in J}\sum_{k=2}^{T-2}\left(2z_{jk}^{-} + \sum_{t=1}^{k}\sum_{i \in I}a_{ij}x_{it}\right)$

Subject to: $(7.13) - (7.16)$ and:

$$z_{jk}^- + \sum_{t=1}^{k} \sum_{i \in I} a_{ij} x_{it} \geq 0 \quad k = 2, \ldots, T-2; \quad j = 1, \ldots, |J|$$

$$z_{jk}^- \qquad\qquad \geq 0 \quad k = 1, \ldots, T; \qquad j = 1, \ldots, |J|$$

in which:

$$a_{ij} = w_j \left(R_j - r_{ij} \right).$$

Joo and Wilhelm (1993b) describe several preprocessing methods that can further reduce the problem size and strengthen the formulation by reducing values of a_{ij} coefficients. By studying the polyhedral structure of problem P_1, they are able to devise several families of valid inequalities and a polynomial time separation heuristic for each. Their optimizing approach uses these valid inequalities in a branch-and-bound framework to prescribe optimal level schedules.

Joo and Wilhelm (1993b) evaluated this solution approach on a set of 40 test problems designed to assess the effects of six factors: (1) the number of different products, $|I|$; (2) the distribution of product demand, d_i; (3) the number of elements in set Q, $|Q|$; (4) the number of different parts, $|J|$; (5) the number of part j required to assemble product i, r_{ij}; and (6) the portion of products which require part j, f_j. The values of $|I|$, d_i and $|J|$ affect the size of a test problem, while the distributions of r_{ij} and f_j reflect the assembly structure.

The cutting plane method required less runtime than did the commercially available Optimization Subroutine Library (OSL) in 37 problems solved to optimality within four hours. The percentage of the gap between the optimal values of the linear programming relaxation and the integer program is, in some cases, a measure of the efficacy of the cuts added in solving a problem. Gap reductions ranged from 19.7% to 81.4% and averaged 52.0% for the 37 problems solved within four hours. Eighteen cases were defined, each with two levels of the f_j factor (6). In 12 out of these 18 cases, the valid inequalities seem to be more effective for problems with smaller f_j values. According to the computational results, the average gap reduction for the problems with smaller $|J|$ values (52.9%) was a little higher, but not significantly, than for those with larger $|J|$ values (51.1%). As problem size becomes larger, the problems with smaller $|Q|$ values required more runtime than did those with larger $|Q|$ values.

These computational tests demonstrate the effectiveness of the valid inequalities, showing that they make meaningful gap reductions, facilitating optimization. The effectiveness of each family of valid inequalities depends on the problem structure and the specific parameter values that were generated randomly.

Following these discussions of methods for scheduling both push and pull systems, the next section describes methods to sequence mixed-model assembly lines.

7.4 Sequencing in mixed-model assembly lines

Section 5.2.6 describes methods for formulating and solving mixed-model line balancing problems. It cautions that, even though a line is balanced with respect to all models, severe imbalances may exist for certain individual models. The negative effects of such imbalances can be ameliorated by the sequence in which models are launched and assembled.

This section describes models that can be used to prescribe sequences that promote the efficiency of mixed-model assembly lines. In addition, it describes a method for dealing with models that each require certain options, entailing extra workload at certain stations.

7.4.1 Sequencing models for efficiency

Bard, Dar-El and Shtub (1992) propose six mixed 0/1 integer programming models for sequencing models in a mixed-model assembly line. The sequencing models are designed for application after a line balancing model has assigned tasks to stations; they are based on the assumption that models are spaced equally on the conveyor line (i.e. that fixed rate launching is in use), and they are formulated relative to the minimal part set (MPS). An MPS is the smallest set for which the proportion of each model is the same as the portion of demand for each model. The notion of an MPS can be expressed more precisely using:

d_m = demand for model m
q = the smallest common divisor of d_m for all $m \in M$
M = set of models.

The portion of demand for each model is $d_m/\sum_{m \in M} d_m$, and the number of each model in the $MPS = \delta_m = d_m/q$ for $m \in M$.

Models reflect the activities of a worker in detail. The worker engages the workpiece and performs assigned tasks while traveling downstream with it. Upon completion of a workpiece, the worker travels upstream to engage the next workpiece. If the station downstream is **open**, the worker may follow the workpiece somewhat beyond the boundary of his station to finish it. If the station downstream is **closed**, the worker cannot go beyond the boundary of his station. An **early start** occurs when the worker engages the first workpiece at the initial boundary of his station, and a **late start** occurs if the worker engages the first workpiece after it has entered his station.

The modeling objective is either to maximize the total idle time of all workers (if the length of the line has already been determined) or to minimize throughput time. Four of the models maximize idle time using pairwise combinations of early or late start along with closed or open stations. Two of the models minimize throughput time using either early or late start along with closed stations.

As an example of these models, consider the case in which total idle time is to be maximized (by minimizing the length of the line used by workers) for a line with early start and closed stations. According to the early start rule, a worker travels upstream to the beginning boundary of his station upon completing a unit. The conveyor brings each new unit into the station and it is engaged by the worker as soon as possible. By minimizing the length of line used, workers are more likely to reach the station boundary and incur idle time waiting for the next unit to engage. The model uses the notation defined in Table 7.10 and is formulated below.

Minimize: $\displaystyle\sum_s Y_s$ (7.20)

Subject to:

$$\sum_m X_{mt} = 1 \qquad t = 1, \ldots, T \tag{7.21}$$

$$\sum_t X_{mt} = \delta_m \qquad m \in M \tag{7.22}$$

$$Z_{s,t+1} \geq Z_{st} + v\sum_m p_{ms}X_{mt} - w \qquad s = 1, \ldots, S; \quad t = 1, \ldots, T \tag{7.23}$$

$$Y_s \geq Z_{st} + v\sum_m p_{ms}X_{mt} \qquad s = 1, \ldots, S; \quad t = 1, \ldots, T \tag{7.24}$$

$$X_{mt} = \{0, 1\} \qquad m \in M; \qquad t = 1, \ldots, T \tag{7.25}$$

$$Y_s \geq 0 \qquad s = 1, \ldots, S \tag{7.26}$$

$$Z_{st} \geq 0 \qquad s = 1, \ldots, S; \quad t = 1, \ldots, T. \tag{7.27}$$

Table 7.10   Notation for Bard, Dar-El and Shtub sequencing models

Indices
m = index for models $m \in M$
s = index for stations
t = index for sequence position

Parameters
δ_m = number of model m in an MPS
p_{ms} = total processing time for model m at station s
S = number of stations on the line
T = total number of units to be sequenced = $\displaystyle\sum_{m\in M}\delta_m$
v = velocity of the conveyor
w = the distance which the worker walks upstream (assuming no waiting)

Sets
M = set of models

Decision variables
X_{mt} = 1 if model m is in sequence position t, 0 otherwise
Z_{st} = position of the worker at station s upon engaging the tth unit
Y_s = length of the line used by the worker at station s

The objective (7.20) minimizes the total line length used by all workers, maximizing total idle time. Constraint (7.21) ensures that a model is assigned to each sequence position and (7.22) ensures that MPS demand for model m is satisfied. Inequality (7.23) defines the trajectory of the worker. The term $\sum_m p_{ms} X_{mt}$ describes the processing time of the tth unit at station s, so that worker displacement is obtained by multiplying by v. Inequality (7.24) defines the length of station s as the largest displacement over the MPS. Constraints (7.25)–(7.27) impose non-negativity and integer requirements.

This model can be adapted to the open station case using an additional decision variable:

$$O_s = \text{overlap from station } s \text{ to station } s + 1.$$

The problem of maximizing total idle time (by minimizing the length of the line used by workers) for a line with early start and **open** stations can be formulated as:

Minimize: $\displaystyle\sum_{s=1}^{S} Y_s - \sum_{s=1}^{S-1} O_s$

Subject to: $(7.21)-(7.27)$ and:

$$O_s \leq Y_s - \left(Z_{st} + v\sum_m p_{ms} X_{mt} \right) + Z_{s,t+1} \qquad s = 1, \ldots, S-1; \qquad t = 1, \ldots, T-1$$

$$O_s \geq 0 \qquad\qquad\qquad\qquad\qquad s = 1, \ldots, S.$$

The authors report computational experience using GAMS to solve over 150 problems with 3–6 models, 10–25 units in an MPS and 3–12 stations. They report the interesting result that all formulations are strong; that is, solutions to the linear programming relaxations were typically within 2 or 3% of the integer optimal solutions. Such strong formulations give tight lower bounds and facilitate solution by branch-and-bound.

Test results indicate that the two objectives appear to be nearly equivalent, since the optimal solution relative to one was typically within 5% of optimum relative to the other. However, using an early start improved solutions relative to the objective of minimizing idle time and using an open station improved solutions relative to the objective of minimizing throughput time. These results recommend using the minimum throughput time objective when the length of the line is fixed and production runs are small, and the idle time objective when processing times are uncertain.

Bard, Shtub and Joshi (1994) formulate a mixed integer non-linear model of a just-in-time sequencing problem in which the objective is to maintain the assembly rate for each model equal to its demand rate. They found that branch-and-bound was effective in solving problems with up to 20 units. To deal with larger problems, they devised an adjacent pairwise exchange heuristic which obtained good solutions for up to 25 units. They also devised a tabu search heuristic which prescribed good solutions for larger problems with up to 50 units.

7.4.2 Sequencing with product options on a paced assembly line

In a paced assembly line, jobs are moved from one station to the next at the end of each cycle time interval. The worker at a station must complete processing a job while it is within a window that does not interfere with the window of an adjacent station. Upon completing a job, the worker must walk upstream to be in a position to engage the next job.

Should a job entail more work than can be accommodated in the window, the worker could follow the job downstream or work faster to finish within the envelope. Neither of these alternatives is desirable: the former may entail unproductive conflicts with the next station, and the latter is likely to lead to poor product quality. More acceptable alternatives involve manning the line for peak loads, employing utility workers who move from station to station dealing with overloads, or completing unfinished operations at the end of the line. These alternatives increase cost so that it is important to minimize the work overloads at stations.

Yano and Rachamadugu (1991) consider a mixed-model, paced assembly line in which each station may incorporate some optional feature in a portion of the jobs. They give an example in which sunroofs must be incorporated in 20% of all automobiles assembled. Thus if manpower levels can deal with this additional work on every fifth vehicle, it is important that the sequence of jobs does not deviate far from fifth-vehicle sequencing. However, if each station deals with a different optional feature, it is not an easy task to prescribe the sequence of jobs that would minimize the work overload for the entire assembly line. Using the notation defined in Table 7.11, their model for this multiple station problem is formulated below.

Table 7.11   Notation for the Yano and Rachamadugu model

Indices

i = station index $i = 1, \ldots, I$
j = job index $j = 1, \ldots, J$
k = sequence position index $k = 1, \ldots, J$

Parameters

O_i = number of workers at station i
F_{ki} = finish time of work at station i on the job in position k
P_{ki} = processing time of the job in sequence position k at station i
R = rate of work = $O_i \times$ (efficiency factor)
W_i = length of the window at station i

Decision variables

S_{ki} = starting time of the job in sequence position k at station i
t_{ki} = workload associated with the job in position k at station i
x_{jk} = 1 if job j is assigned to sequence position k, 0 otherwise

---

Maximize: $\displaystyle\sum_i O_i \sum_k t_{ki}$ (7.28)

Subject to:

$$\sum_k x_{jk} = 1 \qquad \text{for all } j \qquad (7.29)$$

$$\sum_j x_{jk} = 1 \qquad \text{for all } k \qquad (7.30)$$

$$t_{ki} \le \sum_j P_{ki} x_{ki} \qquad \text{for all } k \text{ and } i \qquad (7.31)$$

$$S_{ki} \ge k-1 \qquad \text{for all } k \text{ and } i \qquad (7.32)$$

$$S_{ki} \ge S_{k-1,i} + t_{k-1,i} \qquad \text{for all } k \text{ and } i \qquad (7.33)$$

$$S_{ki} + t_{ki} \le k-1 + W_i \qquad \text{for all } k \text{ and } i \qquad (7.34)$$

$$t_{ki} \ge 0 \qquad \text{for all } k \text{ and } i \qquad (7.35)$$

$$x_{jk} = \{0, 1\} \qquad \text{for all } j \text{ and } k. \qquad (7.36)$$

The objective (7.28) maximizes the amount of work effort required at each station, which is equivalent to minimizing work overload. Equation (7.29) ensures that each job is assigned to one sequence position and (7.30) ensures that some job is assigned to each position. Work input to each job cannot exceed its processing time according to (7.31). Work starts at station i on the job in position k when it is indexed into the station (inequality (7.32)) or when the job in sequence position $k-1$ is completed, whichever is later. Inequality (7.33) ensures that the work done at station i on the job in position k will not exceed the time duration afforded by the window at station i, W_i. Work at station i on the job in position k will finish at time:

$$F_{ki} = \min\{S_{ki} + P_{ki}, k-1 + W_i\}$$

so that the associated work overload is given by:

$$R \text{ maximum}\{0, P_{ki} + [S_{ki} - (k-1)] - W_i\}.$$

In a feasible solution, x_{jk} variables prescribe a sequence of jobs and S_{ki} variables give a time schedule of operations.

For the case of a single station, Yano and Rachamadugu (1991) prove a key property that 'For any given sequence, a non-preemptive FCFS (first-come first-served) work schedule provides an optimal solution.' They note that this result implies that the optimal schedule for the jobs in the first k positions is not changed when additional jobs are added, so the additional jobs cannot reduce overloads incurred by the earlier jobs. Then, focusing on the case in which each station deals with (at most) a single optional feature, they describe a regeneration point as an epoch at which the worker returns to the beginning of the window to start the next job. The worker will move toward the end of the window if several optional jobs must be processed in series. If too many optional jobs are sequenced consecutively, an overload will occur. Using:

P = processing time for a job with the optional feature
p = processing time for a basic job
m_o = number of consecutive jobs with the optional feature in a cycle
m_b = number of consecutive basic jobs in cycle

each repetitive cycle would consist of m_o jobs with optional features followed by m_b basic jobs; and, since it is assumed that $p < 1 < P$, labor would be fully utilized if:

$$m_o P + m_b p = m_0 + m_b \qquad (7.37)$$

where m_o is a given integer, the portion of jobs that would have the optional feature is:

$$m_o / (m_o + m_b)$$

and it is assumed that the value of m_b in (7.37) is an integer. Such a sequence offers several advantages: tools need be changed only twice each cycle, the system would regenerate at the beginning of each cycle and the schedule is easy to follow. If the number of jobs requiring the optional feature is not an integer multiple of m_o, minor corrections must be implemented at the end of the sequence. If fewer than m_o jobs require the optional feature in the last cycle, there will be some idle time but no work overload. However, if there are more than m_o, some overload cannot be avoided. Yano and Rachamadugu (1991) prove a proposition that gives an optimal sequence for a single station:

> Given n_o jobs that require the optional feature and n_b basic jobs, and a repetitive sequence that has m_o jobs that require the optional feature followed by m_b basic jobs in each cycle, an optimal sequence has (a) $C = \min \{ \lfloor n_o / m_o \rfloor, \lfloor n_b / m_b \rfloor \}$ repetitive cycles followed by (b) $\min \{ n_o - C m_o, m_o \}$ jobs requiring the optional feature, followed by (c) $n_b - C m_b$ basic jobs, followed by (d) $\max \{ 0, n_o - (C + 1) m_o \}$ jobs that require the optional feature.

For the multiple station problem, Yano and Rachamadugu (1991) propose a greedy heuristic that determines a lower bound for the work overload associated with an entire sequence to determine which job to sequence in the next available position. The bound is calculated by solving a series of single-station problems and adding their objective function values. This procedure gives a valid lower bound, since each individual station problem gives the best sequence for a particular station, since it may not coincide with the globally optimal sequence and since the actual work overload cannot be smaller than the value calculated in this manner. Given the finishing time of the last job in a partial sequence, the actual work overload associated with sequencing a candidate job next can be computed using the equations given above, and a lower bound associated with the remaining jobs can be easily computed. The candidate job will be sequenced in position k, and the solution for a single station involving positions $k, \ldots, J$ can have only one of two forms, depending upon whether the candidate job is basic or requires the option. The complexity of this heuristic is $O(IJ^2)$.

If the value of m_b that satisfies (7.37) is not an integer, Yano and Rachamadugu recommend rounding to an integer before applying their heuristic. They also give a dynamic programming procedure for solving the single-station problem and note that it could be used as an alternative means of computing lower bounds, especially if m_b from (7.37) is not an integer.

To evaluate their heuristic, they devised a set of 20 test problems based on actual data, each involving 1000 jobs and 12 stations. In these problems, the heuristic reduced work overload by an average of 55% in comparison with the procedure that had been used in industry.

7.5   Summary

This chapter describes methods for resolving scheduling and control problems unique to assembly. These problems arise in both push and pull environments, typically resulting from the conjoining of parts which compose each assembly. Often, parts are accumulated in a kit before assembly can be initiated, although parts can be supplied to assembly stations in a variety of ways. This chapter presents effective methods for scheduling assembly operations in push environments including two-stage systems, large-scale job shops and small-lot, multi-echelon systems. It also describes effective methods for level scheduling in just-in-time assembly.

Other types of scheduling problems arise in mixed-model assembly lines. Even though the line is well balanced with respect to the collection of models, severe imbalances may occur as individual models progress down the line. The negative effects of such imbalances can be ameliorated by the sequence in which models are launched and assembled, and this chapter presents effective methods for doing so. These problem formulations do not address the availability of parts needed for assembly. Rather, they are assumed to be made available by the conveyor or by stock held at individual stations.

Even though scheduling decisions are made within the environment established by design, planning and performance evaluation, they have significant effects on in-process inventories and cycle times as well as on customer (i.e. due date) performance. Thus effective scheduling methods are crucial to establishing world-class assembly operations.

Formulations of scheduling problems are most frequently based on the assumption that the system operates deterministically. This assumption permits resolution by a variety of prescriptive solution methods that provide effective guidance for scheduling decisions. However, many systems operate in a dynamic, stochastic environment. The solution strategy for such a stochastic environment is to apply the deterministic, prescriptive model periodically (e.g. daily) to compensate for random events such as job arrivals and machine breakdowns.

The next chapter introduces a new approach to time-manage material flow in stochastic environments, such as those in which small-lot assembly systems typically operate. This approach provides the ability to 'look ahead' in a stochastic environment, allowing schedule improvements to be devised with the goal of improving customer performance.

7.6   Review questions

1. How does the scheduling function relate to design and planning and to performance evaluation?
2. Construct data for a two-station, $n = 10$ job sequencing problem and apply the heuristic of Potts *et al.* (1995) to prescribe a sequence.
3. List factors that are important in the performance of job shop assembly operations.
4. List rules that might be used to sequence job shop assembly systems and describe the conditions under which you expect each to be effective.
5. What is kitting and why is it important in assembly operations?
6. Describe the problem of scheduling kitting operations in small-lot, multi-product, multi-echelon assembly systems.
7. Devise data for a small-lot, multi-product, multi-echelon kitting problem and apply three heuristics ('dedicate', 'compete' and the Chen and Wilhelm method) to solve it.
8. Describe the level scheduling problem in just-in-time assembly systems.
9. Devise data for a level scheduling problem and apply the Ding and Chen heuristic to propose a schedule.
10. Why is it important to sequence models in a mixed-model line balancing problem?
11. Describe the effects of open and closed stations and of early and late starts.
12. What is a minimal part set?
13. Describe the operation of a mixed-model, paced assembly line and some difficulties that might be ameliorated by sequencing appropriately.
14. Devise data for a single-machine, mixed-model, paced assembly line and use the proposition of Yano and Rachamadugu to prescribe an optimal solution.

References

Aneke, N.A.G. and Carrie, A.S. (1984) 'A comprehensive flowline classification scheme', *International Journal of Production Research*, **22** (2), 281–97.

Askin, R.G., Mitwasi, M.G. and Goldberg, J.B. (1993) 'Determining the number of kanbans in multi-item just-in-time systems', *IIE Transactions*, **25** (1), 89–98.

Bard, J., Dar-El, E.M. and Shtub, A. (1992) 'An analytic framework for sequencing mixed model assembly lines', *International Journal of Production Research*, **30**, 35–48.

Bard, J., Shtub, A. and Joshi, S.B. (1994) 'Sequencing mixed-model assembly lines to level parts usage and minimize line length', *International Journal of Production Research*, **32** (10), 2431–54.

Blackstone, J.H., Phillips, D.T. and Hogg, G.L. (1982) 'A state-of-the-art survey of dispatching rules for manufacturing job shop operations', *International Journal of Production Research*, **20** (1), 27–45.

Buxey, G.M., Slack, N.D. and Wild, R. (1973) 'Production flow line system design – a review', *AIIE Transactions*, **5** (1), 37–48.

Chen, J.F. and Wilhelm, W.E. (1993) 'An evaluation of heuristics for allocating components to kits in small-lot, multi-echelon assembly systems', *International Journal of Production Research*, **31** (12), 2835–56.

Chen, J.F. and Wilhelm, W.E. (1994) 'An approach for optimizing the allocation of components to kits in multi-echelon assembly systems', *Naval Logistics Research*, **41** (2), 229–56.

Conway, R.W., Maxwell, W.L. and Miller, L.W. (1967) *Theory of Scheduling*, Addison-Wesley, Reading, Mass.

Dar-El, E.M. and Wysk, R.A. (1982) 'Job shop sequencing research', *Journal of Manufacturing Systems*, **1** (1).

Ding, F.Y. and Cheng, L. (1993) 'An effective mixed-model assembly line sequencing heuristic for just-in-time production systems', *Journal of Operations Management*, **11**, 45–50.

Fry, T.D., Oliff, M.D., Minor, E.D. and Leong, G.K. (1989) 'The effects of product structure and sequencing rule on assembly shop performance', *International Journal of Production Research*, **27** (4), 671–86.

Groeflin, H., Luss, H., Rosenwein, M.B. and Wahls, E.T. (1989) 'Final assembly sequencing for just-in-time manufacturing', *International Journal of Production Research*, **27** (2), 199–214.

Hall, R. (1984) 'Leveling the schedule', *Zero Inventory Philosophy & Practices APICS Seminar Proceedings*, American Production and Inventory Control Society, Falls Church, VA, 111–19.

Hax, A.C. and Candea, D. (1984) *Production and Inventory Management*, Prentice-Hall, Englewood Cliffs, NJ.

Hay, J.E. (1988) *The Just-in-Time Breakthrough Implementing the New Manufacturing Basics*, John Wiley & Sons, New York.

Inman, R.R. and Bulfin, R.L. (1992) 'Quick and dirty sequencing for mixed-model multi-level JIT systems', *International Journal of Production Research*, **30**, 2011–18.

Joo, S.H. and Wilhelm, W.E. (1993a). 'A review of quantitative approaches in just-in-time manufacturing', *Production Planning & Control*, **4**, 207–22.

Joo, S.H. and Wilhelm, W.E. (1993b) *A Cutting Plane Approach to the Level Scheduling Problem for Assembly in a Just-in-Time System*, Working Paper INEN/MS/WP/-93, Department of Industrial Engineering, Texas A&M University.

Kimura, O. and Terade, H. (1981) 'Design and analysis of pull system, a method of multistage production control', *International Journal of Production Research*, **19**, 241–53.

Kubiak, W. and Sethi, S. (1991) 'Level schedules for mixed-model assembly lines in JIT production systems', *Management Science*, **37**, 121–2.

Miltenburg, J. (1989) 'Level schedules for mixed-model assembly lines in just-in-time production systems', *Management Science*, **35**, 192–207.

Miltenburg, J. and Sinnamon, G. (1989) 'Scheduling mixed-model multi-level just-in-time production systems', *International Journal of Production Research*, **27**, 1487–509.

Miltenburg, J. and Sinnamon, G. (1992) 'Algorithms for scheduling multi-level just-in-time production systems', *IIE Transactions*, **24**, 121–30.

Monden, Y. (1983) *Toyota Production System*, Industrial Engineering and Management Press, Atlanta, GA.

Potts, C.M., Sevastjanov, S.V., Strusevich, V.A., Van Wassenhove, L.N. and Zwaneveld, C.M. (1995) 'The two-stage assembly scheduling problem: complexity and approximation', *Operations Research*, **43** (2), 346–55.

Russell, R.S. and Taylor, B.W. (1985) 'An evaluation of scheduling policies in a dual resource constrained assembly shop', *IIE Transactions*, **17** (3), 219–32.

Steiner, G. and Yeomans, S. (1993) 'Level schedules for mixed-model, just-in-time processes', *Management Science*, **39**, 728–35.

Sugimori, Y., Kusunoki, K., Cho, F. and Uchikawa, S. (1977) 'Toyota production system and kanban system materialization of JIT and respect-for-human system', *International Journal of Production Research*, **5**, 553–64.

Sumichrast, R.T., Russell, R.S. and Taylor, III, B.W. (1992) 'A comparative analysis of sequencing procedures for mixed-model assembly lines in a just-in-time production system', *International Journal of Production Research*, **30**, 199–214.

Wilhelm, W.E. (1977) 'Recent progress in job shop materials management research', *Production and Inventory Management*, **17** (4), 108–19.

Wilhelm, W.E., Chen, J.F. and Parija, G.R. (1994) *Cutting Planes for Kitting in Small-Lot, Multi-Echelon Assembly Systems*, Working Paper, Department of Industrial Engineering, Texas A&M University.

Wittrock, R.J. (1985) 'Scheduling algorithms for flexible flow lines', *IBM Journal of Research and Development*, **29** (4), 401–12.

Wittrock, R.J. (1988) 'An adaptable scheduling algorithm for flexible flow lines', *Operations Research*, **36** (3), 445–53.

Yano, C.A. and Rachamadugu, R. (1991) 'Sequencing to minimize work overload in assembly lines with product options', *Management Science*, **37** (5), 572–86.

8

Time-managed material flow control

8.1 Introduction

One unique aspect of assembly is that it involves the merging of part flows. Assembly logistics thus require thoughtful coordination of material flows and present significant challenges to material flow managers. Poor coordination of material flows leads to excessive in-process inventories as well as inability to achieve assembly schedules.

This chapter describes a relatively new modeling approach that can be applied to time-manage material flow in assembly systems. Time management is a control philosophy that schedules each operation over time and involves looking ahead to coordinate material flows to assembly operations, ensuring schedule performance. In contrast, Material Requirements Planning (MRP) deals with large time buckets and is insensitive to the timing of specific operations and job completions. Just-in-time (JIT) systems that pull production in response to demand were originated to address high-volume production, not small-lot assembly. The look-ahead capability required to time-manage flow is provided by the new modeling approach. Descriptive models allow the manager to assess the implications of materials management policies, and associated prescriptive models give mathematically optimal solutions, minimizing total cost composed of inventory costs and penalties for poor performance according to schedule. In addition, this chapter explores some of the fundamental characteristics of material flow in assembly, developing insight that can be applied to improve system performance, whether these specific models are used or not.

Models for both one-time and repetitive assembly are described in this chapter. One-time assembly systems never reach a steady state. Rather, production planners and schedulers must estimate the likelihood of events that would affect assembly operations and devise control tactics to ensure timely completion of the end product. Few principles are available to facilitate this activity in the stochastic environment, because which characterizes actual assembly operations operation times and part delivery times are typically random variables.

To yield a mathematically tractable model, it is often assumed that the system operates deterministically. This assumption has enabled a number of approaches such as project planning and MRP, which can make substantial contributions in appropriate environments. However, the deterministic assumption leads to schedules that cannot be achieved in the actual stochastic assembly system.

By presenting a new approach for modeling time-dependent material flow, this chapter describes a number of principles that might be used by managers in their intuitive approaches to managing flow in stochastic assembly systems. However, the models can be applied directly to estimate forthcoming schedule performance, evaluating proposed flow management tactics and helping to identify improvements. In addition, some of the models may be used to prescribe mathematically optimal flow-control tactics.

The body of the chapter is organized into four sections. Section 8.2 focuses on the stochastic kitting process, and section 8.3 presents a method for modeling the transient performance of repetitive assembly operations. Section 8.4 uses the descriptive models from the first two sections to optimize the time management of material flow in kitting and repetitive assembly. Section 8.5 presents a summary and relates conclusions.

8.2   The kitting process

Before an assembly operation can be initiated, all required parts must be kitted. Thus kitting plays a central role in assembly and must be thoroughly understood to facilitate effective flow management.

Most prior research related to kitting is fragmented and has not led to a comprehensive understanding of the process. According to Orlicky (1975), MRP is an approach to time-phase material requirements, but it does not offer the ability to develop a schedule that anticipates random influences and hedges appropriately against them. The JIT system was designed to coordinate material flow for high-volume, level production in close proximity to the vendors who supply components. None of these conditions are met in the typical small-lot production environment.

McGinnis and Bozer (1984) introduce a generic descriptive model of the kitting process. They develop kit requirement planning techniques that provide a common database for material requirements, production control and material handling needs, but they do not consider time-phasing requirements. Deterministic approaches for prescribing component reorder intervals are developed by Maxwell and Muckstadt (1983), but these, too, find limited application to the probabilistic, special-order conditions which typify small-lot assembly. Lambrecht, Muckstadt and Luyten (1982) demonstrate the applica-

tion of (s, S) inventory policies to establish buffer stocks in multi-stage production systems. However, they consider only the case in which demand is the
source of variability. Sellers and Nof (1986) report results of a survey identifying nationwide trends in robotic kitting. Sellers and Nof (1989) investigate the
performance potential of several robotic kitting facilities, and Tamaki and Nof
(1991) describe methods for designing robot kitting systems.

This section describes fundamental aspects of the recursion modeling approach that was devised (Wilhelm and Wang, 1986) to model time-
dependent operations in a stochastic assembly system. The approach is first
used to describe some of the fundamental characteristics of the kitting
process. A model for prescribing the cost-optimal, time-managed control of
kitting is presented in section 8.4.1.

8.2.1 A modeling approach

Consider a simple case in which a product requires two parts that are special-
ordered from different vendors, and assume that both vendors can deliver
parts exactly at specified times to support the assembly schedule. In this
scenario, the due times for both parts should be equal to the time at which the
schedule calls for the assembly to be initiated. If vendors delivered at different
times, the assembly operation could not begin until the later arrival. This is a
fundamental characteristic that governs all production and assembly operations: an operation cannot begin until all required resources are ready.

This relationship may be expressed mathematically using the maximum
operator, max[. , .]. Letting

$$R_j = \text{the delivery time (i.e. the ready time) of part } j \quad \text{for } j = 1, 2$$

the kit of required parts will be ready for assembly at time:

$$A = \max[R_1, R_2]. \tag{8.1}$$

The later of the two deliveries determines when the assembly operation can be
initiated. Let:

$$H = \text{scheduled time to initiate the assembly operation,}$$

then due times $R_1 = H$ and $R_2 = H$ are appropriate so that $A = \max[R_1, R_2] = H$
for this scenario. Due times should not be set so that $A = \max[R_1, R_2] > H$,
since this would force the assembly operation to start after its scheduled time
and lead to the end product being tardy, giving poor customer service. On the
other hand, the due time of one part should not be earlier than that of the other,
since that part would incur in-process holding costs while it waited for delivery
of the second part. Thus, if the delivery times of parts are deterministic, it is
easy to set due times so that no in-process holding costs are incurred and the
assembly schedule is achieved, avoiding tardy delivery to the customer.

Recursion model fundamentals   If delivery times are probabilistic, the situation changes dramatically. Unfortunately, most actual systems involve random events that lead to uncertainty in delivery times. A probabilistic environment is more costly than a deterministic one because some parts will be delivered early and others will be late, leading to both in-process holding and tardiness costs that can be avoided in the deterministic system. Costs will be even higher if managers do not deal explicitly with randomness. For example, if managers set assembly schedules and due times assuming that a probabilistic system will operate like a deterministic one, both in-process holding and tardiness costs will be increased.

As in the deterministic case, the assembly operation cannot begin until the last part arrives. Thus, relationship (8.1) is still correct, but A is now a random variable. A is neither R_1 nor R_2 as it would be in the deterministic case; it is another random variable. Figure 8.1 depicts that, even if R_1 and R_2 are inde-

$$f_R(t) = \mu e^{-\mu t},\ t \geq 0$$

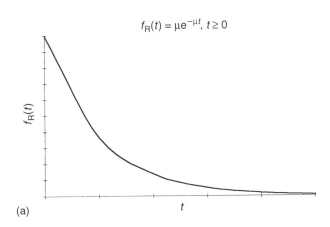

(a)

$$f_A(t) = 2\mu e^{-\mu t} - 2\mu e^{-2\mu t},\ t \geq 0$$

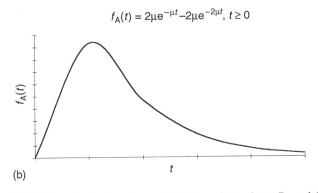

(b)

Fig. 8.1   Probability density functions of R_1, R_2 and A, where R_1 and R_2 are i.i.d random variables with the exponential distribution: (a) probability density function of R_1 and R_2; (b) probability density function of $A = \max[R_1, R_2]$.

pendent and **identically** distributed (i.i.d), A will have a third distribution, which is typically somewhat skewed to the right.

Describing the distribution of a random variable that is the maximum of several other random variables is a difficult problem. For the simple case in which R_1 and R_2 are independent random variables, the distribution function of the maximum is the product of the distribution functions of the individual random variables. Thus, A has distribution function

$$F_A(t) = F_{R_1}(t)F_{R_2}(t) \qquad \text{for } t \geq 0.$$

More generally, if A is the maximum of J i.i.d. random variables, its distribution function is the J-fold product of the distribution function which describes the individual random variables. Mathematically, if $A = \max[R_1, R_2, \ldots, R_J]$ and if the distribution function of R_j is $F_R(t)$ for $j = 1, \ldots, J$, the distribution function of A is:

$$F_A(t) = \prod_{j=1}^{J} F_R(t) \qquad \text{for } t \geq 0.$$

Example 8.1   Suppose that R_1 and R_2 are i.i.d. exponential random variables with probability density function:

$$f_R(t) = \mu e^{-\mu t} \qquad \text{for } t \geq 0$$

and distribution function:

$$F_R(t) = 1 - e^{-\mu t} \qquad \text{for } t \geq 0.$$

If $A = \max[R_1, R_2]$, its distribution function is thus:

$$F_A(t) = \left(1 - e^{-\mu t}\right)\left(1 - e^{-\mu t}\right) \qquad \text{for } t \geq 0$$

and its probability density function, the derivative of its distribution function, is:

$$f_A(t) = 2\mu e^{-\mu t} - 2\mu e^{-\mu t} \qquad \text{for } t \geq 0.$$

This special case is depicted in Fig. 8.1(b).

Example 8.1 demonstrates the principles involved in defining the distribution of A, but the results do not provide sufficient capability to model actual operations, since the mathematics require R_1 and R_2 to be independent. In practice, ready times would be independent only if the parts were supplied by independent sources. For cases in which parts are produced in the same factory, R_1 and R_2 are likely to be dependent (or correlated) random variables, since the production of one part may affect that of the other.

Fortunately, one case is both mathematically tractable and represents practical situations with some accuracy. This case is presented using a more general notation, since it is referenced in subsequent sections. Clark (1961) showed that if (η_1, η_2) is a bivariate normal (i.e. $\eta_1 \sim N(\mu_1, \sigma_1)$ and $\eta_2 \sim N(\mu_2, \sigma_2)$) with parameters:

$$\mu_j = \text{marginal mean} \qquad \text{for } j = 1, 2$$
$$\sigma_j = \text{marginal variance} \qquad \text{for } j = 1, 2$$
$$\rho = r[\eta_1, \eta_2] = \text{coefficient of correlation} \left(\text{with } \rho \neq 1\right)$$

then $Z = \max[\eta_1, \eta_2]$ has mean described **exactly** by:

$$E[Z] = \mu_1 \Phi(\alpha) + \mu_2 \Phi(-\alpha) + a\phi(\alpha) \qquad (8.2)$$

in which:

$$a^2 = \sigma_1^2 + \sigma_2^2 - 2\rho\sigma_1\sigma_2$$
$$\alpha = (\mu_1 - \mu_2)/a$$
$$\phi(\alpha) = \text{standard normal density function evaluated at } \alpha$$
$$\Phi(\alpha) = \text{standard normal distribution function evaluated at } \alpha.$$

Furthermore, the second moment of Z is described **exactly** by:

$$E[Z^2] = (\mu_1^2 + \sigma_1^2)\Phi(\alpha) + (\mu_2^2 + \sigma_2^2)\Phi(-\alpha) + (\mu_1 + \mu_2)a\phi(\alpha) \qquad (8.3)$$

so that, using these standard relationships, the **exact** variance of Z is:

$$V[Z] = E[Z^2] - E^2[Z] \qquad (8.4)$$

Example 8.2 To demonstrate equations (8.2)–(8.4), suppose $Z = \max[\eta_1, \eta_2]$, where (η_1, η_2) is a bivariate normal with parameters $\mu_1 = 100$, $\mu_2 = 100$, $\sigma_1 = 20$, $\sigma_2 = 20$, and coefficient of correlation $\rho = r[\eta_1, \eta_2] = 0.1$. Then $a^2 = 20^2 + 20^2 - 2(0.1)(20)(20) = 720$, $\alpha = 0$, $\phi(\alpha) = 0.399$, and $\Phi(\alpha) = 0.5$. Using these values, the random variable $Z = \max[\eta_1, \eta_2]$ has:

$$E[Z] = 100(0.5) + 100(0.5) + (26.8328)(0.399) = 110.706$$
$$E[Z^2] = (100^2 + 20^2)(0.5) + (100^2 + 20^2)(0.5)$$
$$+ (100 + 100)(26.8328)(0.399) = 12\,541.2587$$
$$V[Z] = 12\,541.2587 - 110.706^2 = 285.44 \text{ so that } V^{1/2}[Z]$$
$$= 16.895 < \sigma_1 = \sigma_2 = 20.$$

This example demonstrates an important principle that is fundamental to assembly operations. In general, the mean of the maximum is larger than the means of the individual variables within the operator, i.e. $E[Z] \geq E[\eta_1]$ and $E[Z] \geq E[\eta_2]$, where equalities hold only if η_1 and η_2 are deterministic with $V[\eta_1] = V[\eta_2] = 0$. This relationship occurs because the right-hand tails of the distributions of the individual variables tend to 'pull' the mean of the maximum to the right. This phenomenon has a profound effect on assembly operations, because it affects every kitting operation. It also explains why schedules based on the assumption that the system operates deterministically are overly optimistic and cannot be achieved in the stochastic environments in which assembly systems typically operate. Such a deterministic assumption incorrectly leads to setting $E[Z] = \max\{E[\eta_1], E[\eta_2]\}$ when, in fact, $E[Z] > \max\{E[\eta_1], E[\eta_2]\}$ if η_1 and η_2 are random variables.

Equations (8.2)–(8.4) specify the moments of $Z = \max[\eta_1, \eta_2]$ exactly, but not the distribution of the random variable Z. Clark proposed that the distribution of Z be **approximated** by a normal with the mean and variance given in equations (8.2) and (8.4), respectively. This approximation enhances modeling capability appreciably. For example, if a kit requires three parts, $A = \max[R_1, R_2, R_3]$, the maximum of three random variables. The maximum operator allows A to be expressed in different ways, including:

$$A = \max\left[R_1, R_2, R_3\right] = \max\left\{R_3, \max\left[R_1, R_2\right]\right\}$$

This allows the maximum to be approximated by dealing with variables in a pairwise fashion, first forming $A(12) = \max[R_1, R_2]$, then $\max[R_3, A(12)]$ for which the correlation $r[R_3, A(12)]$ is needed. Clark (1961) derived an **exact** expression for $r\{R_3, \max[R_1, R_2]\}$:

$$r\left[\eta_3, \max(\eta_1, \eta_2)\right] = \left[\sigma_1 \rho_1 \Phi(\alpha) + \sigma_2 \rho_2 \Phi(-\alpha)\right] / V^{1/2}[Z] \qquad (8.5)$$

where:

$$\rho_1 = r\left[\eta_3, \eta_1\right] = \text{pairwise coefficient of correlation between } \eta_3 \text{ and } \eta_1$$
$$\rho_2 = r\left[\eta_3, \eta_2\right] = \text{pairwise coefficient of correlation between } \eta_3 \text{ and } \eta_2.$$

Equation (8.5) allows moments of the maximum of three or more normally distributed random variables to be approximated, dealing with variables pairwise and approximating the distribution of each maximum by the normal.

Example 8.3   To demonstrate the use of equation (8.5) in conjunction with equations (8.2)–(8.4), suppose that $A = \max[\eta_1, \eta_2, \eta_3]$, where (η_1, η_2, η_3) is a trivariate normal with parameters given in Example 8.2, $\mu_3 = 100$, $\sigma_3 = 20$, and coefficients of correlation $\rho_1 = r[\eta_3, \eta_1] = 0.1$ and $\rho_2 = r[\eta_3, \eta_2] = 0.1$. First, calculate the moments of $Z = \max[\eta_1, \eta_2]$ as in Example 8.2 and approximate the distribution of Z using the normal. Then, view (η_3, Z) as a bivariate normal with means μ_3 and $E[Z]$, variances σ_3 and $V[Z]$ and coefficient of correlation as determined by equation (8.5):

$$r\big[\eta_3,\ \max(\eta_1,\ \eta_2)\big] = r\big[\eta_3,\ Z\big] = \big[(20)(0.1)(0.5)$$
$$+(20)(0.1)(0.5)\big]\big/(16.895) = 0.11837.$$

Applying equations (8.2)–(8.4) again, $a^2 = 20^2 + 16.895^2 - 2(0.11837)(20)$ $(16.895) = 605.4466$, $\alpha = (100 - 110.706)/(24.6058) = -0.435101$, $\phi(\alpha) = 0.363086$, and $\Phi(\alpha) = 0.331823$ so that the distribution of A is approximated by the normal with moments:

$$E[A] = (100)(0.331823) + (110.706)(0.668177)$$
$$+ (24.6058)(0.363086) = 116.08752$$
$$E[A^2] = (100^2 + 20^2)(0.331823) + (110.706^2 + 16.895^2)(0.688177)$$
$$+ (100 + 110.706)(24.6058)(0.363086) = 13713.19226$$
$$V[A] = 13713.19226 - 116.08752^2 = 236.87996 \text{ so that } V^{1/2}[A] = 15.391.$$

Some developments in subsequent sections require the correlation $r[\tau, \max(\tau, \eta)]$, a special case that may be evaluated by the following procedure (Saboo and Wilhelm, 1986):

1. Scale τ to be a unit normal and apply the same scaling to η. The scaled variable η has mean $(\mu_\eta - \mu_\tau)/\sigma_\tau$ and variance $\sigma_\eta^2/\sigma_\tau^2$.
2. Apply equations (8.2)–(8.4) to the moments for $\max(\tau, \eta)$ using scaled τ and η.
3. Apply equation (8.5) using these scaled results.

The resolution of one issue raises three new questions. Is it appropriate to use the normal to describe actual distributions in assembly systems? How well does the normal approximate the distribution of the maximum? How can the pairwise coefficients of correlation be determined? These questions are addressed in subsequent sections.

Safety lead-time   Commonly used to manage material flow in small-lot assembly, MRP determines the due time for a kit, H, by starting with the due time of the end product and 'backing off' a predetermined 'lead-time' for each operation. A safety lead-time may be incorporated for each component, hedging against the risk of late delivery.

Safety lead-times are typically determined by combining an intuitive assessment of vendor delivery performance with a knowledge of the urgency of the assembly schedule. An approach like this, in which it is assumed that the system will operate deterministically, may lead to undesirable results in the actual probabilistic environment. Safety lead-times that are too long lead to costly work-in-process inventories; those that are too short result in poor schedule performance and encourage inefficient practices such as cannibalizing other kits to obtain required components.

Even though component ready time is a random variable, the materials manager can influence delivery by setting the safety lead-time, shifting the ready-time distribution to be earlier or later in time. Suppose that assembly has been scheduled to begin at time H, and that the safety lead-time for component j must be set to give an acceptable level of assurance that the component will be available by time H. Let:

L_j = lead-time allowed for production/delivery of component j
PD_j = actual time that is required to produce/deliver component j.

The lead-time allowed for production/delivery, L_j, is a fixed duration (i.e. a constant), but PD_j is a random variable because random events influence the production/delivery process.

Component ready time, R_j, is a function of the two constants H and L_j and the random variable PD_j:

$$R_j = (H - L_j) + PD_j. \tag{8.6}$$

Part fabrication should begin at time $(H - L_j)$, and the component will be ready at time R_j. Since PD_j is a random variable, R_j will also be a random variable.

This relationship is depicted in Fig. 8.2. The area under the density function of R_j to the right of time H is the probability that the component will be late. If safety lead-time L_j is increased, the ready-time distribution shifts to the left so the chance of early delivery increases and the area to the right of H becomes smaller so that the chance of a late delivery reduces. Unfortunately, the cost of in-process inventory increases as the chance of early delivery increases. Thus,

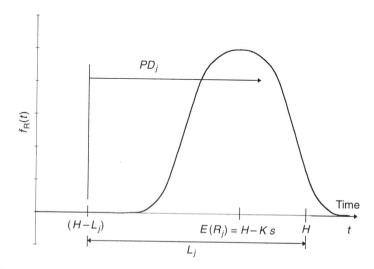

Fig. 8.2 Component ready times R_j relative to scheduled assembly time H.

a lead-time must be selected to balance the costs of holding in-process inventory and delaying assembly beyond H, the scheduled time.

Defining lead-time as:

$$L_j = E[PD_j] + K_j s_j \tag{8.7}$$

in which:

$E[PD_j]$ = mean of production/delivery time

K_j = management-determined parameter that determines the safety lead-time

$V[PD_j]$ = variance of production/delivery time

s_j = $V^{1/2}[PD_j]$ = standard deviation of production/delivery time

provides a safety lead-time expressed as a multiple K_j of the standard deviation of production/delivery time. The mean and variance of PD_j may be estimated from the historical performance of the vendor, but K_j can be set to manage the kitting process.

Substituting equation (8.7) into (8.6), the mean and variance of ready time R_j are:

$$E[R_j] = H - K_j s_j$$
$$V[R_j] = V[PD_j]$$

and $E[R_j]$ can be used as the due time of the component.

Ready time may be controlled by managers by setting parameter K_j, determining the expected safety lead-time $(H - E[R_j]) = K_j s_j$. A large K_j value yields a large safety lead-time and gives a high probability that the component will be delivered ahead of schedule (i.e. before H) as well as a high level of work-in-process. Reducing K_j reduces work-in-process but entails a greater risk of late delivery and, consequently, poor schedule performance. The optimal value of K_j gives the best possible trade-off between early and late delivery.

If PD_j is normally distributed, then R_j will also be normally distributed. In this case, a K_j value of 0.0 shifts the component delivery distribution so that $E[R_j] = H$ and late delivery occurs with probability 0.5. A K_j of 1.0 yields a 0.1587 probability of late delivery, and a K_j of 2.0 yields a 0.0228 probability of late delivery. K_j may thus be used by materials managers as a fundamental means of controlling the kitting process. A method to optimize this time-managed material flow control is described in section 8.4.1.

Kitting time The kitting time for a set of J components may be defined as:

$$A_J = \max[R_1, R_2, \ldots, R_J]$$

Ready times are random variables that are dependent if components are produced in a common facility. A_J is also a random variable, but its distribution cannot be defined exactly by existing mathematical theory. It is known only that, if component ready times are i.i.d. normal random variables, the distribution of A_J will approach the extreme value distribution as $J \to \infty$ (Johnson and Kotz, 1972).

A means of approximating the distribution of kit ready time using equations (8.2)–(8.5) is given by the following four-step procedure.

1. Define $A_1 = R_1$ and set $j = 2$.
2. Apply equations (8.2)–(8.4) to estimate the moments of $A_j = \max[A_{j-1}, R_j]$.
3. For $n = j + 1, \ldots, J$ apply equation (8.5) to compute $r[R_n, \max(A_{j-1}, R_j)]$ using:

$$\rho = r\big[A_{j-1}, R_j\big], \quad \rho_1 = r\big[R_n, A_{j-1}\big] \quad \text{and} \quad \rho_2 = r\big[R_n, R_j\big].$$

4. IF $j = J$ **STOP**. Otherwise, increment j to $(j + 1)$ and GOTO Step (2).

In Step (3), the required values of ρ and ρ_1 are computed at earlier steps in the recursion, and ρ_2 is assumed to result from an analysis of the component production process (e.g. see section 8.4). The correlation $r[A_{j-1}, R_j]$ required at Step (2) is computed during an earlier application of Step (3). Kitting time A_J is then approximated by a normal distribution with the calculated moments.

Example 8.4 Consider organizing the calculations in Examples 8.2 and 8.3 according to this algorithm, letting $R_1 = \eta_1$, $R_2 = \eta_2$ and $R_3 = \eta_3$:

1. Set $A_1 = R_1 = \eta_1$ and $j = 2$.
2. Compute the moments of $A_2 = \max[A_1, R_2] = \max[\eta_1, \eta_2]$ as in Example 8.2.
3. Calculate $r[R_3, \max(A_1, R_1)] = r[\eta_3, \max(\eta_1, \eta_2)]$ as in Example 8.3.
4. Set $j = 3$.
2. Compute moments of $A_3 = \max[A_2, R_3] = \max[(\eta_1, \eta_2), \eta_3]$ as in Example 8.3.

Model accuracy Now, consider the question of how accurately the normal approximates the distribution of kitting time. Wilhelm and Wang (1986) report numerical tests obtained using hypothetical test cases in which component ready times were assumed to be normally distributed with equal means, equal variances and equal pairwise correlations (see also Greer and La Cava (1979)). In a sense, this represents a 'worst case' analysis, since the errors of approximation can be expected to be largest if parameters are equal. If one ready time had a mean that is much larger than that of other components, it would dominate the definition of kitting time and thereby enhance the apparent accuracy of the procedure.

Test cases utilized the following parameters:

1. $E[R_j] = 100.0$ for all components ($j = 1, 2, \ldots, J$).
2. Coefficients of variation for component ready times: $C_j = 0.2, 0.3, 0.6$ and 0.9.
3. Pairwise correlations among component ready times: $\rho = 0.0, 0.3, 0.6$ and 0.9.
4. Twenty different kit sizes (i.e. number of components/kit): $J = 1, 2, \ldots,$ 20.

If all pairwise coefficients of correlation ρ were 1.0, all components would be ready at the same time, and the true distribution of A_J would, in fact, be normal. The largest error occurs if all $\rho = 0.0$; in this case, the distribution of A_J converges to the extreme value distribution as J increases.

Simulations of the 'worst case' (i.e. with $\rho = 0.0$) described the relative frequency distributions of A_J shown in Fig. 8.3 ($C_j = 0.2$ and $J = 1, 3$ and 10). Note that the titles of some diagrams presented in this section use the term 'accumulate' as a synonym of 'kit'. The distribution of A_J appears to converge to the extreme value distribution relatively slowly. Frequency distributions appear to be 'normal-like' (i.e. symmetrical and bell-shaped) for the values of J tested. Chi-square statistics for goodness of fit tests are able to discern the tendency of the distribution of kitting time to depart from the normal for values of J larger than 6 (Wilhelm and Wang, 1986). Nevertheless, the normal distribution provides a convenient conceptual model of the true distribution and a reasonable level of accuracy.

Wilhelm and Wang (1986) tested 160 cases (four values of C_j, four of ρ and 20 of J) and used 5000 simulation replications to derive 95% confidence intervals for values of $E[A_J]$. In all 160 cases, the recursion model estimate of $E[A_J]$ fell within the confidence intervals. The computer cost to obtain the simulation estimates was 700 times that required by the recursion model, so it

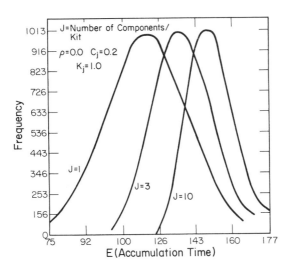

Fig. 8.3 Distribution of kitting time (Wilhelm and Wang, 1986).

appears that the recursion approach gives good estimates of kitting time at a favorable accuracy/runtime trade-off.

8.2.2   Characteristics of the kitting process

Some products such as electronic circuit cards are assembled from a single kit. Others, such as automobiles, have bills of materials with multiple levels and require kits to be composed at several echelons. This section deals with the single kit; subsequent sections address multi-echelon assembly.

This section identifies basic characteristics of the kitting process. The first subsection describes some performance measures, and the second demonstrates some relationships between these performance measures and the parameters that describe the kitting process.

Performance measures for the kitting process   Several measures of the kitting process are required to implement time-managed material flow control: kit earliness, kit tardiness, component shelf waiting time and component in-process time. This subsection describes how these performance measures can be calculated using the recursion modeling approach (see Wilhelm and Wang (1986) for details).

A kit completed ahead of its due time may be held in inventory for duration $(H - A_J)$ until the time H at which it is needed for assembly. Assuming that kit completion time is normally distributed, there is a probability that it will be completed early and a probability that it will be completed tardily. The expected value of kit earliness, $E[E]$, is:

$$E[E] = \left(H - E[A_J]\right)\Phi(Z) + V^{1/2}[A_J]\phi(Z) \qquad (8.8)$$

in which:

$\Phi(Z)$ = standard normal CDF evaluated at $Z = \left(H - E[A_J]\right)/V^{1/2}[A_J]$

$\phi(Z)$ = standard normal PDF evaluated at $Z = \left(H - E[A_J]\right)/V^{1/2}[A_J]$.

If a kit is completed after its due time H, tardiness results, and the assembly schedule may be jeopardized. Assuming that kit completion time is normally distributed, the expected value of kit tardiness, $E[T]$, is:

$$E[T] = \left(E[A_J] - H\right)\left(1 - \Phi(Z)\right) + V^{1/2}[A_J]. \qquad (8.9)$$

A component incurs shelf waiting time S_j from the time it is ready until kitting time:

$$S_j = A_J - R_j.$$

The mean of S_j may be computed using:

$$E[S_j] = E[A_J] - E[R_j]. \tag{8.10}$$

If the kit is early, the total time that component j spends as work-in-process before assembly begins, W_j, is the sum of the shelf waiting time of the component and kit earliness:

$$W_j = S_j + E.$$

If the kit is tardy, W_j is simply the shelf waiting time for component j. The expected total in-process time for component j, $E[W_j]$, is:

$$E[W_j] = E[S_j] + E[T]\Phi(Z). \tag{8.11}$$

These measures are now related to the parameters which define the kitting process.

Relationships of kitting performance to parameters Using the recursion model, Wilhelm and Wang (1986) demonstrate some relationships between the expected kitting time $E[A_J]$, the number of components/kit J, the pairwise correlations among component ready times ρ and the coefficients of variation of ready times C_j. Some of these relationships are described in this section.

If $J = 1$, kitting time is the same as the component ready time. As J increases, $E[A_J]$ increases but, as shown in Fig. 8.4 (for $\rho = 0.0$, $C_j = 0.2$, $K = 1$), at a decreasing rate.

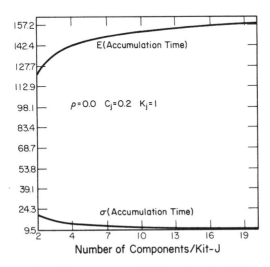

Fig. 8.4 Relationship of the moments of kitting time to kit size, J (Wilhelm and Wang, 1986).

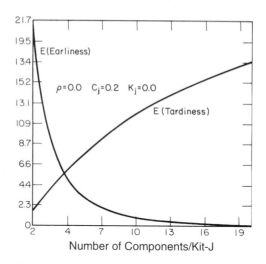

Fig. 8.5   Relationship of expected kit earliness and tardiness to kit size J (Wilhelm and Wang, 1986).

Even though each component is expected to be ready at time 100.0:

$$E[R_j] = H - K_j V^{1/2}[A_j] = 120 - (1)(20) = 100.0$$

the kit cannot be made ready until a much later time, on the average. In addition, the variance of kitting time decreases as J increases, indicating that kitting time will be consistently larger than the ready time of each individual component.

These results explain why kits of larger size are more likely to require long times to kit. This fact is commonly observed by materials managers, but was, apparently, first quantified by Wilhelm and Wang (1986).

The relationship of kit earliness $E[E]$ and expected kit tardiness $E[T]$ to J is given in Fig. 8.5 for the case in which safety lead-time is zero (for $K_j = 0.0$ and $\rho = 0.0$, $C_j = 0.2$). Clearly, performance degrades as the number of components in the kit increases.

The sensitivity of $E[A_j]$ to correlations among component ready times (assumed to be equal for all pairs of components) is depicted in Fig. 8.6, which indicates that expected kitting time decreases as ρ increases. The limiting case is $\rho = 1.0$ for which all components and the kit are ready at the same time 100.0.

If components are ordered from independent vendors, $\rho = 0.0$ and kitting performance is poor. If pairwise correlations ρ are increased, say by producing components in a common facility such as a flexible manufacturing system, performance improves. This result indicates that producing components 'in-house' or ordering them all from the same vendor may improve the consist-

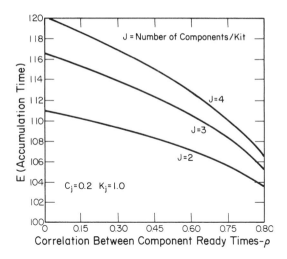

Fig. 8.6 Relationship of expected kitting time to ρ for various kit sizes (Wilhelm and Wang, 1986).

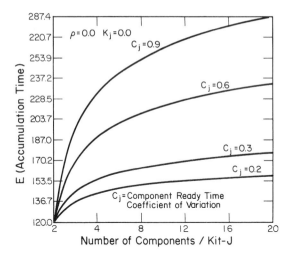

Fig. 8.7 Relationship of expected kitting time to kit size *J* for various coefficients of variation ρ (Wilhelm and Wang, 1986).

ency of kitting operations. It should be noted, however, that the variance of kitting time increases as ρ increases.

In a deterministic environment (i.e. with $C_j = 0.0$ for all *j*), all components, as well as the kit, are ready at time 100.0. For a coefficient of variation greater than zero, there is a probability that some components will be ready before time 100.0. However, the latest component ready time will determine kitting time, and the probability of later delivery causes kitting performance to degrade with larger values of C_j.

Figure 8.7 shows this increase in $E[A_J]$ as C_j increases for each kit size J (with $\rho = 0.0$). Kits consisting of a large number of components with highly variable ready times result in the latest kitting times. This result explains why assembly systems are not able to achieve schedules devised using deterministic procedures.

Section 8.2 deals with one-time production. Models for repetitive operations in small-lot assembly are described in the next section.

8.3 Recursion models for repetitive assembly

This section explores the use of recursion models in estimating the transient performance of repetitive assembly systems. The models can be used to time-manage material flow by using them in a descriptive mode, evaluating policies that influence flow coordination over time, or in a prescriptive mode, setting optimal due times for part deliveries. Furthermore, they provide insights that can be used by the intuitive manager.

These models are based on the observation that the starting time of an operation is limited by the ready times of all resources required by the operation, including, for example, the subassembly from an upstream operation, parts delivered by vendors, the assembly machine and the robot that tends the machine. Repetitive production induces correlations between pairs of operation finishing times because jobs influence each other through sharing resources at upstream operations. In order to estimate these correlations explicitly, the models **assume** that operation times are normal and **approximate** operation start and finishing times by the multivariate normal. This section presents recursion models for the assembly line and the assembly network, demonstrating the capability of recursion models to represent a broad range of features found in actual assembly systems.

8.3.1 *The assembly line*

First, consider a recursion model of the transient operation of a flow-line, which represents an assembly line in which part inventories are managed so that no stockouts occur. Suppose that the assembly line consists of I stations in series and processes a set of J jobs. Job j follows job $(j - 1)$ (for $j = 2, \ldots, J$) down the line and each job is processed in order at stations $i = 1, \ldots, I$.

In general, the influence that a station has on material flow depends upon whether it is a **metering station** (Wilhelm and Ahmadi-Marandi, 1982; Saboo and Wilhelm, 1987; see also p. 214) or not. In stochastic flow-lines, the bottleneck is the station that has the largest expected processing time among all stations in the line and is defined as the first order metering station, since it 'meters' the flow of work through the entire system. The rth order metering station is defined as the one with the largest expected processing time among all stations upstream of the $(r - 1)$st order metering station. The first station is always the highest order metering station.

In deterministic systems, jobs queue only at metering stations; no queuing occurs at non-metering stations. If processing times are random variables, some queuing may also occur at non-metering stations due to this variability. The effects of metering stations are discussed at greater length in subsequent sections.

Operation starting and finishing times can be modeled using:

i = index for stations   = 1, ..., I

j = index for jobs   = 1, ..., J

A_{1j} = the time at which job j is released (or launched) to begin assembly

F_{ij} = finishing time of operation ij

P_{ij} = operation time (i.e. processing time duration) to complete operation ij

S_{ij} = starting time of operation ij

and are:

$$S_{ij} = \max\left(F_{i-1,j}, F_{i,j-1}\right) \tag{8.12}$$

$$F_{ij} = S_{ij} + P_{ij}. \tag{8.13}$$

The time at which operation ij (i.e. for job j on machine i) can start is determined by the times at which the required resources, machine i and job j, are ready. Equation (8.12) relates the ready times of the two resources, job j, which is ready at time $F_{i-1,j}$ when it finishes at the previous station $(i - 1)$, and station i, which is ready at time $F_{i,j-1}$ when it finishes processing the previous job $(j - 1)$. Equations (8.12) and (8.13) describe the transient performance of the line by determining the starting and finishing times of each operation.

Start times at station 1 constitute a special case and are defined by $S_{1j} = \max(A_{1j}, F_{1,j-1})$. Job 1 is also a special case for which:

$$S_{11} = A_{11}$$

and, since no job precedes job 1, equation (8.12) reduces to:

$$S_{i1} = F_{i-1,1} \qquad \text{for } i = 2, \ldots, I.$$

In addition, equation (8.13) specializes to:

$$F_{i1} = S_{i1} + P_{i1} \qquad \text{for } i = 1, \ldots, I.$$

After determining the schedule for job $j = 1$ at all stations, the schedule for job 2 can be determined, considering operations 12, . . . , $I2$ in order. Similarly, the schedules for subsequent jobs can be determined by evaluating operations $1j, . . . , Ij$ in order for $j = 3, . . . , J$.

If the system operates deterministically, calculations are straightforward. However, if operation times P_{ij} are random variables, equations (8.12) and (8.13) are still valid, but they are more difficult to evaluate. In this case, start and finish times will also be random variables with correlations induced by the recursive relationship given in (8.12). This recursive relationship between operation finishing times led to the name, recursion models.

In the stochastic case, mathematical theory does not allow the maximum operator to be evaluated exactly, so recursion models are based on assumptions that allow approximate, but relatively accurate, analysis. If processing times are **assumed** to be independent normal random variables, and if operation starting times are **approximated** by the normal distribution, equation (8.13) indicates that finishing times are (approximately) normally distributed (i.e. since the sum of independent normal random variables is normal). Furthermore, equation (8.12) can be evaluated using equations (8.2)–(8.4).

The ensuing discussion addresses four questions about this model: (1) what is the justification for assuming operation times to be normal; (2) how can the correlation between the finishing times of two operations $r[F_{kl}, F_{ij}]$ be determined so that equations (8.2)–(8.4) can be used; (3) can the model be enhanced to represent more complex assembly operations; and (4) since the distribution of the maximum of normal random variables is not normal, is the approximation sufficiently accurate to be used as a guide for practical decision-making?

Some cases in which operation times can be assumed normal It seems reasonable to assume that operation times are normal in a number of cases. For example, the first case recognizes that operation time may, in fact, be inherently normal. Numerous papers (Payne, Slack and Wild, 1972; McGee and Webster, 1976) study production operations which, apparently, conform to this case; Hira and Pandey (1984) itemize studies that do so. Manual operations are frequently assumed to require normally distributed time durations, since a person completing a number of tasks may introduce variability due to inconsistency (Reeve and Thomas, 1973).

A second case occurs on assembly lines in which a set of elementary tasks is assigned to each station (see Chapter 5). If N elementary tasks are assigned to a station, the Central Limit Theorem indicates that station operation time can be approximated by the normal for large N. It is conjectured that this case occurs frequently in assembly systems.

A normal random variable may take on either positive or negative values so that Monte Carlo sampling procedures could generate negative variates, which have no physical meaning as operation times. The following argument indicates that this concern does not present a problem in modeling many assembly systems.

If the N elementary tasks are i.i.d., each with mean μ and standard deviation σ, the coefficient of variation CV of operation time P is:

$$CV = \sigma/\left(\mu N^{1/2}\right)$$

and decreases as N increases. If P is normal, $\Pr[P \leq 0]$ is given by the unit normal variate:

$$Z = -\left(\mu/\sigma\right)N^{1/2}.$$

Consider the case in which elementary task times are exponentially distributed (according to Muth (1973) this extreme case is not typically found in actual systems and therefore represents a bound) so that:

$$\sigma/\mu = 1, \qquad Z = -N^{1/2} \qquad \text{and} \qquad CV = N^{-1/2}.$$

Table 8.1 shows how $\Pr[P \leq 0]$ and CV decrease as functions of N. Values of $\Pr[P \leq 0]$ associated with $N \geq 12$ are not included in most textbooks since they are negligible. As N increases, the normal becomes a better approximation of the distribution of P. At the same time, $\Pr[P \leq 0]$ becomes negligible and CV decreases to a value less than 0.3. This analysis indicates that the normal is a 'natural' distribution for assembly operations composed of a set of independent elementary tasks. Furthermore, Table 8.1 shows that coefficients of variation less than 0.3 are reasonable for this case.

However, the probability of generating a negative variate for operation time increases with the coefficient of variation. Since a negative operation time has no physical meaning, a typical simulation model would either set it to zero or generate another variate to replace it. Both of these methods of dealing with

Table 8.1 $\Pr[P \leq 0]$ and CV for increasing values of N

N	Z	$\Pr[P \leq 0]$	CV
6	−2.4495	0.0071	0.4082
10	−3.1623	0.00078	0.3162
12	−3.4641		0.2887
18	−4.2426		0.2357
20	−4.4721		0.2236
24	−4.8990		0.2041
25	−5.0000		0.2000

Source: Wilhelm (1987) reproduced with permission from Taylor & Francis.

negative variates change the mean, variance and distribution (actually, two different truncated normal distributions result) of operation time and lead to larger finishing times. In contrast, recursion models allow for the possibility of negative operation times so that comparisons with simulation values may be poor if the simulation program actually models operation time using a truncated normal distribution.

The third case in which operation time is (approximately) normal is when a batch of size Q is processed. Since part processing times are i.i.d., the Central Limit Theorem indicates that the distribution of P_{ij} may be modeled by the normal so long as Q is large.

Automated assembly in which variability is negligible represents a fourth case in which each P_{ij} could validly be assumed to be normal (i.e. with zero variance).

In robotic assembly, variability is introduced if a robot requires a random number of attempts to position a part correctly or a vision system requires a random number of views to identify a part unambiguously. In addition, automated systems may jam or break down due to a variety of reasons. In these cases, the random variable P_{ij} is defined by:

$$P_{ij} = C_i + \sum_k \pi_{ik} T_{jk} \qquad (8.14)$$

in which:

C_i = duration of automated operation i

π_{ik} = Pr$\left[$downtime of type k occurs during operation $i\right]$

T_{jk} = random duration of a type k downtime for job j.

Since recursion models deal with transient rather than steady-state operations, a specific realization of the number of downtimes must be modeled for each operation. Thus if C_i and T_{jk} are independent normals (or constants) in equation (8.14), P_{ij} is normal. If T_{jk} are not normal but a number of i.i.d. downtimes are incurred, the Central Limit Theorem again justifies the use of the normal assumption.

As a more complex example, consider a model of a flexible manufacturing cell in which one robot indexes parts through a series of I assembly stations (see section 9.3). The completed product is then inspected; with probability p it is rejected and must subsequently be disassembled and recycled for another attempt at assembling an acceptable product. Since the robot is busy continuously, only one product can be in-process at a time. Seidmann, Schweizer and Nof (1985) model the total time required to complete one product as well as the time, P, to complete a batch of Q identical products. They give a sketch of the probability density function (pdf) of P (Fig. 4, p. 1177) for batch size Q = 1, which shows that batch completion time is multi-modal with modes that correspond to assembly/rework cycles. Considering all possible outcomes in a steady-state analysis, it is true that completion time is multi-modal (see also Wilhelm (1987), Seidmann, Schweizer and Nof (1987)). The **unconditional** operation time is thus bimodal, but **conditioned** on the particular sequence of events for a particular job, the operation time may still be normal.

In a transient analysis, one must consider the finishing time of the batch (recall $Q = 1$) on the particular cycle at which it is accepted as a good product. This finish time represents the realization of a particular outcome of the stochastic assembly/rework process and is conditioned on making, say, Y assembly/rework cycles. It is thus defined as the convolution of $Y I$ assembly, Y test and $(Y - 1)$ disassembly operation times. Since Seidmann *et al.* (1987) assumed these operation times to be statistically independent, the sum (i.e. the finishing time of the product) is, according to the Central Limit Theorem, approximately normal for large Y. In fact, the time to complete one cycle would be approximately normal if I is large.

Operation (processing) time is a **duration**. Recursion models are based on the **assumption** that operation times are normal. If the assumption does not hold, the model cannot be applied. Operation starting and finishing times are points in time that recursion models **approximate** by the multivariate normal. Even if operation times are normal, finishing times are not exactly normal; hence the models **approximate** the true distributions.

Estimating correlations between finishing times Now, consider the second question: 'How can the correlation between the finishing times of two operations $r[F_{kl}, F_{ij}]$ be determined so that equations (8.2)–(8.4) can be used?' Recursion models were developed in a series of papers that propose various algorithms for estimating correlations, expand the set of practical features that can be modeled and test the accuracy of the approach in a variety of settings. The most efficient algorithm for estimating correlations in complex systems is presented in Wang and Wilhelm (1992). Some basic ideas are described below.

The correlation between any pair of finishing times may be estimated as an operation start time is determined. First, the index number of an operation O_{ij} may be defined as $O_{ij} = (j - 1)I + i$, so that when operation ij is addressed, the moments of earlier operations (i.e. for $O_{kl} < O_{ij}$), as well as the correlations between them, have been determined. In addition, the correlation between some random variable RV and the finishing time of an operation is related to the correlation between RV and the starting time of that operation by:

$$r\left(RV, F_{ij}\right) = \beta_{ij} r\left(RV, S_{ij}\right) \qquad (8.15)$$

in which:

$$\beta_{ij} = V^{1/2}\left(S_{ij}\right)/V^{1/2}\left(F_{ij}\right).$$

Thus, if $RV \equiv F_{kl}$, the correlation $r(F_{kl}, F_{ij})$, may be expressed as:

$$r\left(F_{kl}, F_{ij}\right) = \beta_{ij} r\left(F_{kl}, S_{ij}\right)$$

which may be 'expanded' using equation (8.12) to express S_{ij} in terms of the finishing times of the earlier operations that determine it:

$$r\left(F_{kl}, F_{ij}\right) = \beta_{ij} r\left[F_{kl}, \max\left(F_{i-1,j}, F_{i,j-1}\right)\right]. \tag{8.16}$$

The pairwise correlations between the finishing times in equation (8.16) are known (i.e. $r[F_{kl}, F_{i-1,j}]$, $r[F_{kl}, F_{i,j-1}]$ and $r[F_{i-1,j}, F_{i,j-1}]$), since they were estimated at prior steps in the recursion procedure, so the correlation of interest may be calculated using equation (8.5).

This approach requires the storage of appropriate parameters for operation finishing times of interest. If a pairwise correlation between two finishing times is not known, equations (8.12) and (8.15) must be used to 'expand' one of the finishing times, expressing it in terms of the finishing times that determine it. This expansion process will eventually require the correlation between two 'remote' operation finishing times. Operations are 'remote' if there is a large difference between their operation numbers (e.g. $L_{max} = O_{ij} - O_{kl} \gg 0$). The correlation between two remote finishing times can be assumed to be zero, since the operations are not closely related. Defining remoteness using a **large** value of L_{max} increases runtime, since more steps must be taken in the expansion procedure. However a **large** L_{max} increases accuracy. Thus, an appropriate criterion for remoteness, L_{max}, must be determined to balance the runtime required and the accuracy achieved.

Saboo and Wilhelm (1987) studied the correlation $r[F_{i-1,j}, F_{i,j-1}]$ as a function of job number (i.e. $j = 1, \ldots, 100$) at each station, since that correlation is needed to evaluate equation (8.12). They considered 14 different cases, each consisting of $I = 5$ stations but with $E[P_{ij}]$ and $V[P_{ij}]$ specified differently for each case. In each case, a given station uses the same $E[P_{ij}]$ and $V[P_{ij}]$ for all jobs. In addition, they studied the correlation $[F_{I,100}, F_{I,100-L}]$ for 'lags' $L = 1, \ldots,$ $L_{max} = 99$ to learn more about how the correlation between finishing times at a given station reduces as the 'lag' (or difference in job numbers) increases. Test results led to several observations which are discussed below, since they lend additional insight into the transient flow of materials in assembly lines.

The most fundamental observation is that $r[F_{i-1,j}, F_{i,j-1}]$ monotonically approaches a limiting value as j increases, either increasing at a decreasing rate or decreasing at a decreasing rate. This result leads to the interpretation that the correlations reach a 'steady-state' value, indicating that the system has also reached a steady state. Since the correlations approach their limiting value rather quickly, the runtime of a recursion model can be drastically reduced once 'steady state' has been reached; $r[F_{i-1,j}, F_{i,j-1}]$ need not be calculated beyond the job at which it attains its limiting value within a specified tolerance. In fact, the smooth shapes of the $r[F_{i-1,j}, F_{i,j-1}]$ versus j functions indicate that curve-fitting procedures could be used to extrapolate the correlation values after the first few points have been calculated. This would eliminate most of the computational effort required to implement the recursion model, resulting in a very efficient procedure (Wilhelm and Saboo, 1988).

A second observation is that, as j increases, $r[F_{i-1,j}, F_{i,j-1}]$ decreases if the station is a metering station and increases at all stations downstream of a bottleneck. This result emphasizes the importance of metering stations in material flow. Queues at metering stations buffer against upstream effects, reducing dependence on upstream finishing times.

Thirdly, as lag L increases, $[F_{I,100}, F_{I,100-L}]$ decreases, as expected. The rate of decrease (increasing or decreasing) appears to depend upon the coefficient of variation of processing time. This relationship had, apparently, not been identified previously.

In all test cases, the correlation $r[F_{ij}, F_{i,j-10}]$ increased with j whether station i was metering or not. In fact, $r[F_{ij}, F_{i,j-10}]$ is not negligible even for lags $L = 10$.

Modeling practical features   The third question is now addressed by showing how recursion models can incorporate certain practical features. First, consider modeling finite buffers in the assembly line. Finite buffer capacities dramatically increase system complexity, altering the relationships between operation finishing times. With infinite buffers ahead of all stations, correlations are induced only by upstream effects; the history of upstream operations affects operation ij through $F_{i-1,j}$ in equation (8.12). In contrast, finite buffers also cause downstream operations to affect operation ij and therefore give rise to fundamentally different system characteristics.

If buffer capacities are small, downstream operations exert greater influence at a station. However, no matter what capacities are provided, buffers transmit correlations induced upstream without damping them. As expected, the two cases involving zero and infinite buffer capacities define bounds for the performance of the case with finite capacities.

The influence of finite buffers may be modeled by modifying equation (8.12) to:

$$S_{ij} = \max\left(F_{i-1,j}, T_{i,j-1}\right) \tag{8.17}$$

in which:

$$T_{i,j-1} = \max\left(F_{i,j-1}, T_{i+1,j-B(i+1)-2}\right)$$

$B(i)$ = capacity of buffer ahead of station i.   (8.18)

If $B(i + 1)$ is infinite, then $T_{i,j-1} = F_{i,j-1}$ and equation (8.17) reduces to (8.12). Otherwise, operation ij may be determined by the completion of a downstream operation, the particular one $[j - B(i + 1) - 2]$, which ensures that space will be available for job $(j - 1)$ in the buffer ahead of station $(i + 1)$. Using equation (8.18) to define $T_{i+1,j-B(i+1)-2}$, it is apparent that S_{ij} depends upon the completion of operation $(j - B(i + 1) - B(i + 2) - 3]$ at station $(i + 2)$; in fact, the influence of all downstream stations on S_{ij} is evidenced by the expansion of $T_{i,j-1}$ using equation (8.18) recursively.

Station i is blocked just before starting operation j for duration B_{ij}, which is given by:

$$B_{ij} = \max\left(0,\ T_{i,j-1} - F_{i,j-1}\right).$$

During this time, job $(i - 1)$ resides at station i, since there is no space at station $(i + 1)$ for it.

Computational procedures are described below using two new elements of notation:

$$A_j \quad = \text{time when job } j \text{ will be ready for processing at station 1,}$$

a random variable

$$N(i) = (I - i + 1) + \sum_{k=i+1}^{I} B(k).$$

Assuming that the flowshop starts in an empty and idle state, job 1 is a special case for which operation times may be easily determined as described in section 8.3.1. In order to implement the procedure, the correlation between each pair of operations for job 1 is required. For operations one station 'removed' (i.e. stations i and $(i - 1)$):

$$r\left(F_{i1},\ F_{i-1,1}\right) = \beta_{i1} \qquad \text{for } i \geq 2$$

which can be derived using equation (8.15) to justify:

$$r\left(F_{i1},\ F_{i-1,1}\right) = \beta_{i1} r\left(S_{i1},\ F_{i-1,1}\right)$$

and equation (8.12) to re-express S_{i1}:

$$r\left(F_{i1},\ F_{i-1,1}\right) = \beta_{i1} r\left(F_{i-1,1},\ F_{i-1,1}\right).$$

The result follows, since $r(F_{i-1,1},\ F_{i-1,1}) = 1.0$ by definition. Similar reasoning shows that the correlation between operations that are two stations removed is given by:

$$r\left(F_{i1},\ F_{i-2,1}\right) = \beta_{i1}\beta_{i-1,1} \qquad \text{for } i > 3$$

and ultimately, that the correlation of operations that are m stations removed involves the product of m β factors.

Job 2 is also a special case for which defining equations may be expressed in special form. It is interesting to note that the starting times of job 1 are never affected by the capacity of station buffers. Job $j = B(2) + 2$ is the first one that can be blocked at station 1 by operations at station 2. Similarly, job $j = B(2) + B(3) + 3$ is the first one that can be blocked at station 1 by an operation at station 3. Continuing this evolution, job $j = N(1)$ is the first one for which the first $(I - 1)$ stations in the flow-line can be blocked simultaneously by an operation at station I. For subsequent jobs, there is a finite probability that the first $(I - 1)$ stations will be blocked at each operation. It can be expected, therefore, that station 1 will incur the greatest amount of blocked time in balanced lines, since it has the largest number of downstream stations. Expanding equation (8.17) for station 1, S_{1j} may be expressed as:

$$S_{1j} = \max\left(A_j, F_{1,j-1}, F_{2,j-B(2)-2}, \ldots, F_{In(1)}\right).$$

The maximum operator should be evaluated by taking finishing times in order:

$$S_{ij} = \max\left(F_{I-N(i)}, \ldots, \max\left(F_{i+1,j-B(i+1)-2}, \max\left(F_{i,j-1}, F_{i-1,j}\right)\right)\ldots\right)$$

and performing calculations from right to left (understand that $F_{0,j}$ is A_j by definition) so that equations (8.2)–(8.4) are used to estimate the moments of the maximum of each pair of variables, and equation (8.5) is used to compute correlations in the trivariate case.

Corrections required to evaluate the maximum operators have been estimated using the 'nesting' procedure described in the previous section. The term 'nesting' seems appropriate, since the procedure involves the use of correlations in a nested fashion.

As a second example of incorporating practical features, consider modeling two parallel machines at a station. If two jobs (say j and j') are started simultaneously on the machines at time t, one job will finish at time $t + \min(P_{ij}, P_{ij'})$ and the other will finish at time $t + \max(P_{ij}, P_{ij'})$, but it is not possible to say which machine will finish first. Thus, parallel machines must be modeled on a **conditional** basis. That is, a round robin sequencing rule assigns jobs 1, 3, 5, 7, ... to the first machine, and jobs 2, 4, 6, 8, ..., to the second. A more general approach from Wang and Wilhelm (1991) is described in section 8.3.2.

In addition, other practical features have been modeled in a series of papers that describe the development and computational evaluation of recursion models. The kitting process was studied by Wilhelm and Wang (1986) as described in section 8.2. Wilhelm and Ahmadi-Marandi (1982) first evaluated recursion models in application to the assembly line, and Wilhelm, Saboo and Johnson (1986) later enhanced them to address capacity and material flow considerations. Wilhelm (1986a) developed several fundamental results underlying transient operations of the flow-line, and Saboo and Wilhelm (1987) studied related correlation processes. The flow-line is of fundamental importance because it represents a basic configuration, the assembly line, as indicated in section 8.3.1. Saboo, Wang and Wilhelm (1989) extended capabilities to model the generalized flow-line, a configuration in which each job moves downstream for each successive operation but need not be processed at all machines. They incorporated a variety of practical features, including lot sequencing, parallel machines at a station and test/rework cycles, and described the application of their model to a case of realistic size and scope representing a circuit card assembly system. Perhaps of most interest, assembly networks were addressed by Saboo and Wilhelm (1986) as outlined in section 8.3.2. Recent papers by Wang and Wilhelm (1993) and Wilhelm, Som and Carroll (1992) develop capabilities to time-manage material flow in stochastic assembly networks. An application involving machine tending in robotic cells was described by Wilhelm (1986b), and even more complex operations in a robotic assembly cell were modeled by Wilhelm, Kalkunte and Cash (1987). Wang and Wilhelm (1991) developed the recursion model that is currently the most accurate and most capable of dealing with practical features such as those found in complex cellular configurations.

Most recursion models approximate starting and finishing times using the multivariate normal. However, more general distributions have been evaluated, including the Su (Wilhelm, 1986d) and Coxian-type distributions (Kalkunte and Wilhelm, 1987), the log–normal distribution which Wilhelm (1986c) used in an application to model a flexible manufacturing system, the Erlang (Kalkunte and Wilhelm, 1989), and mixtures of Erlangs which were used by Wang and Wilhelm (1993) and Wilhelm, Som and Carroll (1992) to model assembly networks.

Recursion model accuracy Recursion models offer value by suggesting a new way of viewing assembly operations. For example, research on recursion models has revealed the importance of correlations in understanding system performance and of the maximum operator in describing material flow. Certain inherent properties of assembly systems have been identified and bounds have been devised to develop rough estimates of performance very quickly. In particular, this work has shown that schedules based on the assumption that the system is deterministic are too optimistic and cannot be achieved in a stochastic environment.

However, to be suited for actual implementation, a model must be able to estimate performance with an acceptable level of accuracy. Recursion model inaccuracies are caused primarily by approximating operation start and finishing times by the multivariate normal and by estimating correlations (in particular, by the criterion adopted for 'remoteness'). These potential inaccuracies have been systematically evaluated and results have consistently shown that recursion models give a high level of accuracy in a runtime much less than that required for Monte Carlo simulation, the only other method capable of modeling large-scale, complex assembly systems. Simulation estimates can be made more accurate, but at the expense of lengthy runtimes. This chapter summarizes some computational experience so that the reader can judge the accuracy/runtime trade-off from his own perspective. Of course, the interested reader can also consult the references to learn about other computational tests.

Considering both accuracy and runtime, recursion models provide a viable support for practical decision-making. In addition, they offer a structure to which optimization methods can be applied to prescribe mathematically optimal, time-managed material flow control.

8.3.2 The assembly network

Recent emphasis on methods to ensure just-in-time operations in small-lot production/assembly systems underscores the importance of careful management of material flow to ensure coordination of production and assembly operations so that inventory and customer service objectives can be met. Material flow planning may involve analysis of a one-week to six-month time horizon for which policies that control material flow are prescribed. This activity is distinguished from job sequencing, a short-range problem of combinatorial nature which determines the order in which jobs are produced.

This section describes a generic recursion model of the assembly network along with some fundamental properties that might be used to establish schedules. In addition, an actual industrial case setting is described to demonstrate the application of recursion models to time-manage material flows in assembly networks.

Network structure and models Figure 8.8 depicts an assembly network that relates components required for product assembly in a four-level bill of materials consisting of components (8–15) at level four, subassemblies at level three (4–7) and level two (2–3) and the end product at level one (1). If all components are special-ordered, lots of components (8–15) are required to assemble each lot of the end product. Successive lots may require specializations (e.g. color) to meet customer requirements, so it is assumed that lots are unique and follow one another through the network. Small-lot systems may never reach a steady state because they are subject to numerous disruptions, including engineering change orders, random disturbances and time varying demands.

Figure 8.9 restructures the network to represent assembly operations. Here, components 8 and 9 are assembled to form subassembly 4; in turn, subassemblies 4 and 5 are assembled to form subassembly 2, which is joined with subassembly 3 to yield the end product 1. Thus, stations that assemble 1, 3 and 4 might be viewed as forming a final assembly (flow) line which is fed parts 8 and 9 and subassemblies 5 and 3.

Models that describe the transient behavior of queuing networks could be applied to study small-lot assembly networks, but available models cannot incorporate the range of features found in actual systems. In addition, queuing models typically invoke the assumption that the intervals between successive arrivals are not correlated.

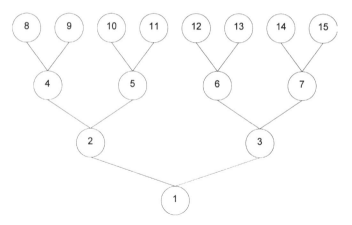

Fig. 8.8   Four-echelon, binary assembly tree (Wilhelm, Som and Carroll, 1992).

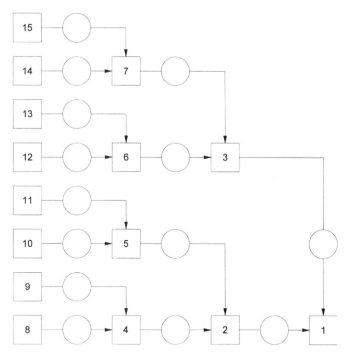

Fig. 8.9   Production/assembly configuration (Saboo and Wilhelm (1986) reproduced with permission from IIE, Georgia, USA).

Markovian models of transient queuing networks are analytically intractable and are typically evaluated numerically using special techniques to speed computations. McGee and Webster (1976) applied a simulation model to study the transient performance of a two-station flow-line in which operation times and interarrival times were assumed to be normally distributed. They observed that interdeparture intervals were correlated and approximately normally distributed. In-plant studies (Carre, 1971) have shown that throughput rates are quite sensitive to correlations.

A recursion model can easily be devised to describe transient operations in an assembly network. Based on the fundamental observation that operation *ij* can start when all resources (the 'kit' or set of components, the subassembly and the station) are ready:

$$S_{ij} = \max\left[A_{ij}, F_{i-1,j}, F_{i,j-1}\right] \tag{8.19}$$

in which:

A_{ij} = ready time of the kit of components needed by operation *ij*

$$= \max\left[A_{ij}^1, A_{ij}^2, \ldots, A_{ij}^{N(ij)}\right]$$

$N(ij)$ = number of components required in the kit for operation *ij*.

Other practical features could easily be incorporated in the model. If components are delivered in lot size Q_i, the variable A_{ij} needs to appear in the model only for operations $j = 1, (Q_i + 1), (2Q_i + 1)\ldots$. A finite buffer of capacity $B(i + 1)$ ahead of station $i + 1$ could be modeled by including $F_{i+1,j-B(i+1)-2}$ within the maximum operator. If a cart (i.e. material handling device) is used to transport a job between stations i and i' and only m carts are available, S_{ij} would also be limited by the availability of a cart, so $F_{i',j-m}$ would be included within the maximum operator to represent this limitation.

Job j follows job $(j - 1)$ (for $j = 2, \ldots, J$) down the assembly line consisting of stations $i = 1, 2, \ldots, I$. As before, equation (8.13) defines operation finishing time, and specializations can be made for station 1 and for job 1.

If kit ready times A_{ij} and/or operation times P_{ij} are random variables, start and finish times will also be random variables with correlations induced by the recursive relationship given by equation (8.19). Mathematical theory does not allow the distribution of the maximum of several dependent random variables to be defined exactly, so equation (8.19) can be evaluated only approximately.

Saboo and Wilhelm (1986) devised a recursion model of the assembly network, assuming that operation times P_{ij} are normally distributed and mutually statistically independent of one another and approximating operation start and finishing times by a multivariate normal. They established five properties of assembly networks which can be used to provide bounds on the expected values of operation starting and finishing times.

Property 1:   In assembly networks, random variables A_{ij}, S_{ij} and F_{ij} (for $i = 1, \ldots, I$ and $j = 1, \ldots, J$) are associated.

Associated random variables have non-negative correlations. This fact implies:

Property 2:   If the correlation between two finish times is assumed to be
               zero, the true correlation will not be overestimated.

Thus bounds can be established for expected start times, either by setting all
variances to zero or by setting all pairwise correlations to zero:

Property 3:   The deterministic estimate which sets all $V[A_{ij}] = V[P_{ij}] = 0.0$
               gives a lower bound for expected start times: $E_D[S_{ij}]$.
Property 4:   Setting the correlations of all pairs of finishing times to 1.0
               gives a lower bound for expected start times: $E_L[S_{ij}]$.
Property 5:   Setting the correlations of all pairs of finishing times to 0.0
               gives an upper bound for expected start times: $E_U[S_{ij}]$.

In general:

$$E_D\left[S_{ij}\right] \le E_L\left[S_{ij}\right] \le E\left[S_{ij}\right] \le E_U\left[S_{ij}\right]$$

and equalities hold only if the system is deterministic (that is, $V[A_{ij}] = V[P_{ij}] = 0.0$ is true for all operations). Corresponding bounds for operation finishing times may be obtained by adding $E[P_{ij}]$ to each of the bounds for starting time, thus:

$$E_D\left[F_{ij}\right] = E_D\left[S_{ij}\right] + E\left[P_{ij}\right]$$
$$E_L\left[F_{ij}\right] = E_L\left[S_{ij}\right] + E\left[P_{ij}\right] \text{ and}$$
$$E_U\left[F_{ij}\right] = E_U\left[S_{ij}\right] + E\left[P_{ij}\right]$$

If a single lot is assembled, the assembly network reduces to a PERT network and Property 5 agrees with bounds determined for expected project completion time.

Saboo and Wilhelm (1986) present a detailed method for implementing the recursion model by estimating the pairwise correlations required to evaluate equation (8.19). The interested reader can also reference their work for a computational evaluation of the recursion model in application to a binary assembly tree (i.e. one in which each node has two predecessors). They compare estimates obtained from their recursion model with those obtained from simulation. Worst-case tests in balanced networks indicate a difference of only 1.0% between the two estimates of expected makespan. In imbalanced cases, the difference is only 0.2%. They also report that simulation requires much more runtime than does the recursion model. In addition, their tests demonstrate that Properties 3–5 establish valid bounds. In particular, these results indicate that deterministic schedules significantly underestimate actual finishing times so that they cannot be achieved in a stochastic environment. Finally, they relate an example which demonstrates the application of recursion models in the following industrial case setting.

Industrial case setting   The purpose of this example is to demonstrate the application of the recursion modeling approach as a management decision aid in a case of realistic size and scope. The case deals with the process of material flow planning and is based on an actual in-plant study; specific numerical values have been modified to observe proprietary interests.

Special-ordered truck cabs, which may be customized by selecting options or specifying unique part designs, are assembled from over 110 parts in a five-level bill of materials. Figure 8.10 depicts the network in which one subassembly is assembled. Seven other subassemblies are assembled in similar networks, and all are transported to final assembly where 10 operations com-

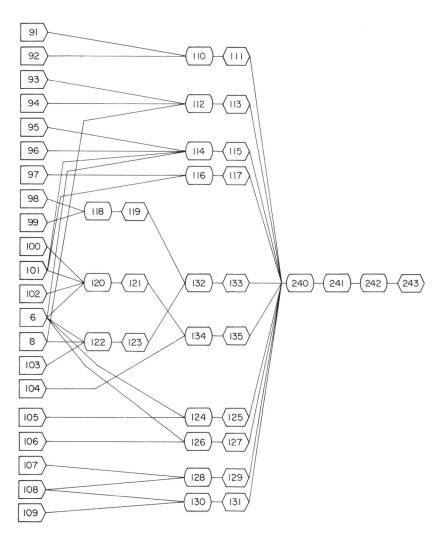

Fig. 8.10   Truck cab production/assembly network 1 (Saboo and Wilhelm, 1986).

plete the cab. In all, some 150 operations are required to process components and assemble each cab.

The case setting objective was to determine the sensitivity of shop perform-ance to management policies that establish safety lead-times for parts ordered from vendors, time intervals for release of parts into assembly and lot size. The sequence in which cabs are assembled is fixed by the sequence in which they are ordered by customers.

A number of features are incorporated to faithfully represent actual operations:

1. moments of operation times are defined specifically for each j to model assembly of the unique products required by each customer;
2. multiple 'copies' of some components are required by a cab;
3. operations that require components are modeled by eliminating A_{ij} in (8.19);
4. lots of size Q are modeled by setting the moments of operation times to be Q times the moments associated with a lot of size 1;
5. delivery of raw materials in lots of size Q are modeled by including part ready time, A_{ij}, in (8.54) only for each Qth part (i.e. for $j = 1, (Q + 1), (2Q + 1), \ldots)$;
6. multiple components are accumulated pairwise for an assembly opera-tion using a binary network in which stations require zero operation times. For example, the network in Fig. 8.9 would simply accumulate a kit of eight components at station 1 if all operation times were 0.
7. bills of materials contain several levels as depicted in Fig. 8.10.

Tests were conducted to demonstrate use of the recursion model in prescribing material flow management policies (release time interval and safety lead-time), in determining the sensitivity of performance to the variability of opera-tion times and in setting lot size. These tests, along with a fourth that evaluated the accuracy of model estimates, are discussed below.

Test 1 demonstrated model application during the process of material flow planning to prescribe two management policies that control material flow: the release time interval and the safety lead-time for part deliveries. To pro-vide a 'worst-case' analysis and an experiment that can readily be duplicated, a 'balanced' network was used so that $E[P_i] = 20$ time units and the coefficient of variation was 0.2 for each operation at station i $(i = 1, \ldots, I)$.

Delivery times of vendor-supplied parts, A_{ij}, were assumed to be independ-ent and normally distributed with mean:

$$E\left[A_{ij}\right] = \left(j - 1\right)RI - KV^{1/2}\left[A_{ij}\right]$$

in which:

    RI = management control variable which determines the time be-tween launching successive lots into the shop (i.e. the release time interval)

K = management control variable, which determines the safety lead-time, $KV^{1/2}[A_{ij}]$, allowed for parts ordered from vendors

$V^{1/2}[A_{ij}]$ = standard deviation of vendor lead-time (arbitrarily) defined to be $0.8E[P_i] = 0.8(20) = 16.0$ to represent the (high) variance in component production/delivery

RI values of 10, 20 and 30 time units were used; $RI = 10$ allows $E[P_i] = 20$ to 'pace' operations, while $RI = 30$ 'paces' operations causing idleness waiting for parts. K values of 0, 1 and 2 were tested; $K = 0$ allows each part to be delivered later than scheduled start time $(j - 1)RI$ with probability 0.5, while larger values reduce the probability of late delivery at the cost of higher levels of in-process (part) inventory.

(RI, K) policies are compared in Table 8.2 using four measures of performance: (1) mean makespan for completing $J = 100$ lots; (2) the standard deviation of makespan; (3) average system time (AST), the average time that all parts spend in the system:

$$AST = \sum_{j=1}^{J}\sum_{i \in I'}\left\{E[F_{Ij}] - E[A_{ij}]\right\}/\{JN\}$$

in which:

I' = the set of initial stations at which parts are delivered
N = the number of initial stations

and (4) average cycle time (ACT), the mean time between completions of lots:

$$ACT = \sum_{j=2}^{J}\left(F_{Ij} - F_{I,j-1}\right)/(J - 1).$$

Results indicate little difference between $K = 1$ or $K = 2$ for $RI = 10$ because queues build up and render makespan relatively insensitive to safety lead-times. $(RI, K) = (20, 2)$ is the best combination tested, although inventory can be reduced using $RI = 30$ to expedite AST. The recursion model required approximately 17 CPU seconds (on an IBM 3081-D) per run, while the simulation model required 212 CPU seconds, a runtime that discouraged the use of more than 50 replications. If all parts are available at time zero (designated as subcase $K = \infty$), makespan and ACT are minimized while AST and, therefore, in-process inventory is maximized. Finally, if a flow planning technique assumed all variances to be zero, makespan, AST and ACT would be dramatically underestimated as indicated. If vendor delivery schedules were based on such an analysis, in-process inventory would build to exceptionally high levels because parts would be delivered long before they could be used.

Table 8.2 Industrial test case 1: $E[P_i] = 20.0$, $V^{1/2}[P_i] = 0.2E[P_i]$, $V^{1/2}[A_{ij}] = 0.8E[P_i]$

	RT = 10.0				RT = 20.0				RT = 30.0			
K	Mean makespan	Std. dev. of makespan	Average system time	Average cycle time	Mean makespan	Std. dev. of makespan	Average system time	Average cycle time	Mean makespan	Std. dev. of makespan	Average system time	Average cycle time
Recursion												
0	2731.71	23.98	1143.08	23.10	2746.82	22.84	661.39	23.25	3456.10	12.92	485.16	30.41
1	2726.58	23.79	1154.04	23.04	2732.47	22.90	664.24	23.10	3440.10	12.92	485.30	30.25
2	2726.48	23.79	1169.51	23.04	2725.32	23.30	674.55	23.03	3424.10	12.92	485.80	30.09
Simulation[1]												
0	2723.74	28.04	1133.56	23.06	2741.43	28.89	652.51	23.24	3455.24	21.98	483.54	30.45
1	2717.95	27.02	1143.59	23.00	2727.24	28.29	655.32	23.09	3439.24	21.76	483.70	30.29
2	2717.95	27.02	1159.15	23.00	2719.10	27.36	664.80	23.01	3423.24	21.75	484.10	30.12

	Mean makespan	Std. dev. of makespan	Avg. system time	Avg. cycle time
For K = 2: Recursion estimates are:	2726.48	23.79	1633.18	23.04
Simulation estimates are:	2717.95	27.02	1622.86	23.00

For the deterministic case with RT = 20.0: Makespan = 2420.00
Average System Time = 440.00
Average Cycle Time = 20.00

[1] 50 replications.
* Outside 95% confidence intervals established by the simulation model.
IBM 3081 – D

Source: Saboo and Wilhelm (1986), reproduced with permission from IIE, Georgia, USA.

Test 2 evaluated the sensitivity of $(RI, K) = (20, K)$ policies to the variance in part delivery times. The rationale behind this test is that 'parts' may be complex subassemblies, so the variance of delivery time can be much larger than that of an individual operation. Use of $V^{1/2}[A_{ij}] = 0.8E[P_i]$ gave the results in Table 8.3 which indicate that makespan is increased correspondingly, but the (RI, K) policy of $(20, 2)$ remains best.

Test 3 evaluated management policies that determine lot size. It was assumed that parts could be ordered, processed and assembled in lots of size 1 (as before), 2 or 4. Results, given in Table 8.4, clearly indicate the superiority of small lots in reducing makespan, AST and in-process inventory.

Test 4 compared simulation and recursion estimates in an example of even larger scale. Data from tests 2 and 3 were used in this experiment, but $J = 200$ cabs (lot size = 1) were assembled. This involved over 30000 operations, and, as shown in Table 8.5, the recursion estimate of expected makespan is within 0.27% of the simulation estimate. The recursion procedure required only 8.5% of the runtime required by the simulation model. In comparison with other tests, this one demonstrates an application in a case of significantly larger scale. Accuracy and runtime compare favorably with simulation capability, providing motivation to apply the model in actual industrial situations.

These tests, based on a case setting of realistic size and scope, demonstrate the use of a recursion model as a decision aid in the process of material flow planning. Management policies related to release time intervals, safety lead-times and lot size were evaluated and the best alternatives were indicated by both recursion and simulation models. Makespan entailed completion of over 15000 operations and estimates of mean makespan from recursion and simu-

Table 8.3   Industrial test case 2: $E[P_i] = 20.0$, $V^{1/2}[P_i] = 0.2E[P_i]$, $V^{1/2}[A_{ij}] = 1.6E[P_i]$, $RT = 20$

K	Mean makespan	Std. dev. of makespan	Average system time	Average cycle time
Recursion				
0	2783.71	23.58	694.86	23.62
1	2750.16	23.34	692.65	23.28
2	2728.69	23.43	707.13	23.07
Simulation[1]				
0	2778.63	32.71	687.02	23.61
1	2743.05	30.48	683.14	23.25
2	2721.59	29.18	697.11	23.04

[1] 50 replications.
IBM 3081–D

Source: Saboo and Wilhelm (1986), reproduced with permission from IIE, Georgia, USA.

Table 8.4 Industrial test case 3: $E[P_i] = 20.0$, $V^{1/2}[P_i] = 0.2E[P_i]$, $V^{1/2}[A_{ij}] = 0.8E[P_i]$, $RT = E[P_i]$ (lot size)

	Lot size = 1				Lot size = 2				Lot size = 3			
K	Mean make-span	Std. dev. of make-span	Average system time	Average cycle time	Mean make-span	Std. dev. of make-span	Average system time	Average cycle time	Mean make-span	Std. Dev. of make-span	Average system time	Average cycle time
Recursion												
0	2746.82	22.84	661.39	23.25	3174.00	24.85	1102.38	46.67	4014.51	29.80	1978.08	93.51
1	2732.47	22.90	664.24	23.10	3154.51	24.74	1107.81	46.27	3988.54	29.22	1988.38	92.42
2	2725.32	23.20	674.55	23.04	3146.14	24.97	1123.92	46.10	3979.57	29.41	2013.37	92.05
x	2726.48	23.79	1633.18	23.04	3146.17	25.13	2059.48	46.10	3978.77	29.48	2911.39	92.02
Simulation[1]												
0	2741.43	28.89	652.51	23.24	3156.31	32.83	1086.88	46.27	3995.25	37.70	1964.32	92.77
1	2727.24	28.29	655.32	23.09	3136.77	31.55	1092.29	45.87	3967.70	36.36	1974.85	91.62
2	2719.10	27.36	664.80	23.01	3127.25	28.91	1107.23	45.68	3956.53	37.93	1998.39	91.16
x	2717.95	27.02	1622.86	23.00	3125.90	28.37	2042.54	45.65	3955.64	38.35	2896.84	91.12

[1] 50 replications.
IBM 3081–D

Source: Saboo and Wilhelm (1986), reproduced with permission from IIE, Georgia, USA.

Table 8.5 Industrial test case 4: $E[P_i] = 20.0$, $V^{1/2}[P_i] = 0.2E[P_i]$, $V^{1/2}[A_{ij}] = 0.8E[P_i]$, $RT = 20$, $J = 200$

K	Mean makespan	Std. dev. of makespan	Average system time	Average cycle time	CPU runtime
Recursion					
0	4865.93	31.27	694.77	22.21	36.68
Simulation[1]					
0	4878.87	31.73	692.91	22.29	431.29

[1]50 replications.
IBM 3081–D

Source: Saboo and Wilhelm (1986), reproduced with permission from IIE, Georgia, USA.

lation models differed by only 0.2%, demonstrating the accuracy of the recursion model.

In all tests, the recursion model required significantly less runtime than the simulation model. This should promote use of the model in optimization routines and on personal computers. Even on a mainframe, runtimes prohibited more extensive simulation analysis.

Transient analysis is important, since requirements may vary over time in a small-lot system which may never reach a steady state. Up to now, simulation has been the only available means of modeling transient operations in assembly networks. The recursion model presents a viable alternative with reasonable accuracy and favorable runtime.

8.4 Optimal time management

Sections 8.2 and 8.3 describe several recursion models. The purpose of this section is to show how the models can be used to optimize time-managed material flow control. In particular, section 8.4.1 describes how managers can use the models in section 8.2 to optimally control the kitting process, and section 8.4.2 relates how managers can use the models in section 8.3.1 to manage part deliveries. Section 8.4.2 also identifies some useful principles that the intuitive manager can use without actually applying recursion models.

The interested reader may also consult Wilhelm, Som and Carroll (1992) who show how time-managed material flow can be optimized in assembly networks using the mixture of Erlangs to provide more flexibility (and perhaps accuracy) in representing operation finishing times. Wilhelm, Som and Carroll (1992) also explore rescheduling methods.

8.4.1   Time-managing the kitting process

Managers can control each parameter that defines the kitting process: ρ may be changed by make versus buy decisions, C_j by operator training or automation, and J by product design. Materials managers implement the final control step by setting the safety lead-time parameter K_j.

Figure 8.11 shows that $E[A_j]$ is a linear function of K_j (for $\rho = 0.0$ and $C_j = 0.2$). However, the expected values of kit tardiness $E[T]$ and total work-in-process time for all components $E[W]$ are non-linear functions of K_j as shown in Fig. 8.12 (for $\rho = 0.0$ and $C_j = 0.2$). Since these functions are convex, some value of K_j optimizes kitting performance.

The cost model presented below may be used to prescribe optimal safety lead-times to balance early and late deliveries of components. The total expected cost associated with a kitting operation, TC, may be defined by:

$$TC = C^T E[T] + C^P E[P] + \sum_{j=1}^{J} C_j^W E[W_j]$$

where:

C_j^W = cost per unit time for carrying component j as in-process inventory
C^T = penalty cost for kit tardiness
C^P = cost per unit time of holding a kit of components

C^T must be determined to reflect the true cost of disrupting an assembly schedule and, perhaps, of making tardy deliveries to end-product customers.

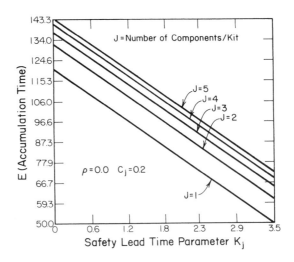

Fig. 8.11 Relationship of expected kitting time to K_j for various kit sizes J (Wilhelm and Wang, 1986).

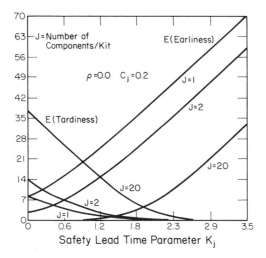

Fig. 8.12 Relationships of expected kit earliness and tardiness to K_j for various kit sizes J (Wilhelm and Wang, 1986).

C^P may include the cost of holding a kit in inventory as well as the cost of specialized containers or storage facilities.

By referring to equations (8.8)–(8.11) for the definitions of $E[T]$, $E[E]$, $E[S_j]$ and $E[W_j]$, it is easy to see that TC is a function of the J management control variable K_j. Total cost TC may thus be minimized by applying any suitable algorithm to optimize this unconstrained non-linear objective function.

8.4.2 *Time-managing part deliveries*

Equation (8.19) gives a model of an assembly network that might represent the low-volume assembly of a series of large expensive products such as jet engines, each of which requires a lengthy lead-time. In such applications, we can think of a 'mainframe' moving from one station to another down the final assembly line. At each station, a kit of components is assembled to the mainframe, and the end product emerges from station 1.

Recursion models can provide numerous measures of schedule performance, including:

$$E[\text{Makespan}] = E[F_{1J}]$$

$$E[\text{End-product earliness}] = E\left\{\max\left[0, \left(\gamma_j - F_{1j}\right)\right]\right\}$$

$$E[\text{End-product tardiness}] = E\left\{\max\left[0, \left(F_{1j} - \gamma_j\right)\right]\right\}$$

$$E[\text{Kit work-in-process time at operation } ij] = E\left[W_{ij}^K\right] = E\left[S_{ij}\right] - E\left[A_{ij}\right]$$

$$E[\text{Mainframe work-in-process time at operation } ij] = E\left[W_{ij}^M\right] = E\left[S_{ij}\right] - E\left[F_{i-1,j}\right]$$

in which:

$$\gamma_j = \text{due time of end product } j.$$

This section describes some principles that might be used by the intuitive manager to time-manage flow in an assembly line. In addition, it presents a method to optimize the time management of material flow in such systems.

Some principles for use by the intuitive manager   Assuming that the system starts empty and idle, each station incurs an unavoidable idleness at the beginning (S_{i1}) and end ($F_{IJ} - F_{Ij}$) of the schedule. This unavoidable idleness may be important in runs of finite size, but it is not considered for long-run, steady-state operations. However, station productivity might depend upon the number and duration of idle periods that result from scheduling policies. In both transient and steady-state operation, scheduling policies determine controllable idleness, I_{ij}, defined as:

$$I_{ij} = S_{ij} - F_{i,j-1}$$

and may lead to numerous idle periods of short duration or few idle periods of long duration. The latter may contribute to productivity by making time available for other useful purposes.

In order to build intuition about some underlying characteristics, assume that the system operates deterministically so that each operation at station i requires time P_i, kit ij is delivered exactly according to schedule (i.e. $A_{ij} = S_{ij}$) and all J mainframes are available ahead of station 1 at time zero so that $W^M_{1j} = S_{1j}$ for $j = 1, \ldots, J$, where:

$$W^M_{ij} = \text{queuing time of the mainframe for operation } ij$$

In this restricted system, only the launch interval RI can be controlled to manage schedule performance. Several possible launch intervals are:

$$RI_1 = P_{m(1)}$$

where $m(1)$ denotes the number of the first order metering station (i.e. the bottleneck):

$$RI_2 = P_{m(2)}$$

where $m(2)$ denotes the number of the second order metering station, and:

$$RI_{r_{max}} = P_{m(r_{max})}$$

where $m(r_{max})$ denotes the number of the highest order metering station, which is always station 1. The launch interval and the arrangement of stations in the line affect starting times S_{ij}, avoidable idle periods I_{ij} and mainframe queuing time at operation ij, W_{ij}^M. Upstream from the bottleneck, the due times for kits change as a function of launch interval, but the bottleneck meters all flow so that due times for kits at stations downstream of the bottleneck should not be changed as a function of launch interval.

Since W_{ij}^M changes with the launch interval, the time required to complete a product from launch to completion also depends upon RI:

$$\text{Time to produce product } j = P_1 + \sum_{i=2}^{I} \left(W_{ij}^M + P_i \right).$$

However, the makespan for the schedule is the same for all launch intervals $RI \leq P_{m(1)}$ and depends only upon the bottleneck:

$$F_{IJ} = \sum_{i=1}^{m(1)-1} P_i + JP_{m(1)} + \sum_{m(1)+1}^{I} P_i \qquad (8.20)$$

since it requires time for the first job to reach the bottleneck, for all J jobs to be processed on the bottleneck, and for the last job to be completed at stations downstream of the bottleneck. This result has several interesting implications. First, the total in-process queuing time of all mainframes is the same regardless of launch interval; changing RI influences only the station(s) at which queuing will occur. The second implication is that station utilization, U_i:

$$U_i = JP_i / F_{IJ}$$

is the same for all launch intervals. Even though station utilization and the total in-process waiting time for mainframes are fixed by the arrangement of stations, launch interval determines system productivity and floor space requirements. RI_1 yields the maximum controllable idleness by requiring:

$$I_{ij} = I_i = P_{m(1)} - P_i \text{ for all } i \text{ and } j$$

$RI_{r_{max}}$ minimizes I_{ij} (ahead of the bottleneck) and may contribute to productivity by allowing the next operation to begin as soon as possible and by making $(F_{IJ} - F_{1j})$ as large as possible so that other, productive work can be done at the end of the schedule.

In this restricted, **deterministic** system, a queue can only develop at metering stations. Thus the arrangement of stations along the line is important, since the mainframe accumulates contributed value as assembly progresses. A fundamental difference between metering and non-metering stations is that $S_{ij} = F_{i,j-1}$ (for $j = 2, \ldots, J$) for the former, and $S_{ij} = F_{i-1,j}$ (for $i = 2, \ldots, I$) for the latter.

Balanced lines with $P_i = P$ (for $i = 1, \ldots, I$) can use only one launch interval, $RI = P$. This configuration is most attractive relative to all performance measures, including productivity and floor space requirements, since no queuing will occur at station $i \geq 2$.

Performance of unbalanced lines may be improved by locating the bottleneck at station 1 so that no in-process queuing will occur at downstream stations. Productivity may be improved, however, by arranging the stations with $P_1 \leq P_2 \leq \ldots \leq P_I$ to minimize the durations of controllable idle times at all stations.

If operation times P_{ij} or kit available times A_{ij} are random variables, operating characteristics change, but the underlying influences of the deterministic system remain. For example, in-process queuing can occur at non-metering stations due to variability, but the largest queuing times can still be expected at metering stations.

The due times of kits must be established to balance the risks of early and late delivery. If a kit is early, it may tie up considerable amounts of capital and floor space. On the other hand, if it is late, it may delay an operation, causing station idleness, the queuing of subassemblies and disruptions that prevent performance to schedule. In either event, random variations inevitably lead to higher levels of work-in-process (for both kits and mainframes). As described in section 8.2.1, the due times of kits may be established by setting:

$$E\left[A_{ij}\right] = H_{ij} - KV^{1/2}\left[PD_{ij}\right]$$

in which:

H_{ij} = scheduled time to start operation ij
K = management-determined parameter that sets the safety lead-time $KV^{1/2}[M_{ij}]$
PD_{ij} = time to produce and deliver kit ij.

If PD_{ij} is a normal random variable, A_{ij} will also be a normal random variable.

Wilhelm and Ahmadi-Marandi (1982) describe a test that evaluated the combined effects of launch interval RI and safety lead-time factor K in a case in which part delivery times A_{ij} were normally distributed random variables and operation ties were deterministic (i.e. $V[P_{ij}] = 0$). This case represents the typical situation in which part delivery times are more variable than are individual assembly operations. Launch interval RI_1 degraded performance because random variations allowed the average time between departures at the bottleneck to exceed $P_{m(1)}$. Other launch intervals in their example reduced expected controllable idleness ahead of the bottleneck and resulted in the same average makespan. Using $K = 0$ represents the situation in which a deterministic approach is taken to manage a probabilistic system. Setting $K = 0$ resulted in the largest expected makespan, since tardy parts adversely affected schedule performance. Expected makespan was reduced by setting $K = 1$, and reduced further by $K = 2$. As expected, larger safety lead-times compensate for random deliveries at the expense of increasing the average in-process queuing of parts. The expected in-process queuing of mainframes was largest at metering stations due to the imbalance among operation times. Part tardiness resulted in some mainframe queuing at all stations, even at non-metering stations downstream of the bottleneck.

Optimal time management   Sarin and Das (1987) develop a means of prescribing safety lead-time control variables, K_{ij}, to optimize time-managed flow in an assembly line. They use dynamic programming first to optimize the single job problem, then to apply that result to the multiple job case.

The objective of the single job problem is to minimize the sum of expected in-process queuing times for mainframes and kits. The dynamic programming formulation defines each stage to represent an operation and uses the single-stage return function:

$$SR_i = \begin{cases} C_i^M(A_i - F_{i-1}) & \text{for Case 1 in which } A_i > F_{i-1} \\ C_i^K(F_{i-1} - A_i) & \text{for Case 2 in which } A_i < F_{i-1} \end{cases}$$

in which the index j is omitted because only one job is considered and:

SR_i = single-stage return at stage i
C_i^M = cost/unit time for a mainframe to wait ahead of stage i
C_i^K = cost/unit time for a kit to wait ahead of stage i.

Operation i is limited by the availability of the kit in Case 1 and by the mainframe in Case 2. Sarin and Das define kit delivery time as a normally distributed random variable:

$$A_i = E[D_i] + V^{1/2}[D_i]X_i$$

in which:

$E[D_i]$ = the decision variable setting the expected kit delivery time
$V^{1/2}[D_i]$ = (known) standard deviation of the kit delivery time distribution
X_i = a standard normal random variable

Since A_i is normal by definition and F_{i-1} is approximated as a normal, the random variable $(F_{i-1} - A_i)$ can be approximated as a normal with mean $(E[F_{i-1}] - E[D_i])$ and variance $(V[F_{i-1}] + V[D_i])$. The probability that Case 1 will occur can be estimated, assuming that operation times P_i are deterministic and kit delivery times A_i are normal. The expected value of the single-stage return $E[SR_i]$ can be expressed as the expected value of a truncated normal:

$$E[SR_i] = (C_i^M + C_i^K)(a_i/\sqrt{2\pi})\exp(-\alpha_i^2/2) + C_i^K(E[F_{i-1}] - E[D_i])$$
$$- (E[F_{i-1}] - E[D_i])(C_i^M + C_i^K)\Phi(-\alpha_i)$$

in which:

$\Phi(-\alpha_i) =$ distribution function of the standard normal evaluated at $-\alpha_i$

$\alpha_i \quad = \left(E[F_{i-1}] - E[D_i]\right)/a_i$

$a_i^2 \quad = V[F_{i-1}] + V[D_i].$

Figure 8.13 depicts the dynamic programming formulation for the assembly of a single job. The input state variable to stage i is F_{i-1}, a random variable, and the input to the first stage is F_0, the ready time of the mainframe to begin assembly. The return function at stage i, $E[SR_i]$, gives the expected cost of waiting for both the mainframe and the kit at stage i. The decision at stage i is denoted $E[D_i]$ and sets the due time of the kit for that operation.

Since the $E[SR_i]$ function is convex, the optimal decision for an independent (i.e. single) stage $E^*[D_i]$ is obtained by setting the derivative of $E[SR_i]$ with respect to $E[D_i]$ to zero, and solving for $E^*[D_i]$:

$$E^*[D_i] = E[F_{i-1}] - N_1 \left(V[F_{i-1}] - V[D_i] \right)^{1/2}$$

in which:

$$N_1 = \Phi^{-1}\left[C_i^M / C_i^M + C_i^K\right].$$

The decision $E^*[D_i]$ yields the optimal single-stage return:

$$E^*[SR_i] = N_2 \left(V[F_{i-1}] + V[D_i] \right)^{1/2}$$

in which:

$$N_2 = \left(C_i^M + C_i^K\right)\left\{N_1\Phi(N_1) + \left[\exp\left(-N_1^2/2\right)\sqrt{(2\pi)}\right]\right\} - C_i^M N_1.$$

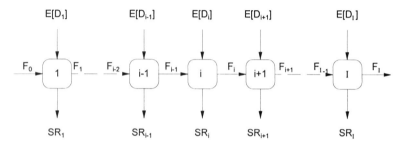

Fig. 8.13 Dynamic programming formulation for the single job case (Sarin and Das, 1987).

A two-stage return, say from the combination of stages i and $i + 1$, is described by f_i:

$$f_i = E[SR_i] + E*[SR_{i+1}].$$

Using backward recursion, stage $(i + 1)$ is optimized first. Both $E[SR_i]$ and $E*[SR_{i+1}]$ are functions of the decision at stage i, $E[D_i]$, so the partial derivative $\dfrac{\partial f_i}{\partial E[D_i]}$ is:

$$\frac{\partial f_i}{\partial E[D_i]} = -\left(C_i^M + C_i^K\right)\phi\left(\alpha_i\right) + C_i^M + \frac{\partial E*[SR_{i+1}]}{\partial E[D_i]}.$$

Since the computation of the partial derivative $\dfrac{\partial E*[SR_{i+1}]}{\partial E[D_i]}$ is 'cumbersome', Sarin and Das (1987) develop a regression model to determine it. They use four sets of 1024 data records to establish a regression model of the partial derivative PD_i:

$$PD_i = \theta_1 V^{1/2}\left[D_i\right]^{\theta_2} N_2\Phi\left(\alpha_i\right) + \theta_3 V^{1/2}\left[D_i\right]^{\theta_4} N_2 + \theta_5 N_2 + \theta_6 N_2 N_3^{\theta_7}$$

in which:

$$N_3 = V[D_i]$$

$\theta_1 = 0.04577093$, $\theta_2 = 0.47805193$, $\theta_3 = 2.39312763$, $\theta_4 = 0.03059004$, $\theta_5 = 2.3500$, $\theta_6 = 0.02354503$ and $\theta_7 = 0.58349074$. The residual error is 0.0542.

Stage i requires the following calculations (Sarin and Das, 1987):

$$G_{i,k,1} = \theta_1 V^{1/2}\left[D_i\right]^{\theta_2} N_{i-1,k+1,1}$$

$$G_{i,k,2} = N_{i-1,k+1,1}\left[\theta_3 V^{1/2}\left[D_i\right]^{\theta_4} + \theta_5 + \theta_6\left[N_{i-1,k+1,2}\right]^{\theta_7}\right]$$

$$SG1_i = \sum_{k=2}^{I-i+1} G_{i,I-k,1}$$

$$SG2_i = \sum_{k=2}^{I-i+1} G_{i,I-k,2}$$

$$T_i = \Phi^{-1}\left[\frac{C^{M_i} + SG2_i}{C^{M_i} + C^{K_i} - SG1_i}\right]$$

$$Z_{i,1} = T_i^2\left[\Phi(T_i) - \Phi^2(T_i)\right] + \Phi(T_i) - \Psi^2(T_i) + T_i\Psi(T_i)\left[1 - 2\Phi(T_i)\right]$$

$$Z_{i,2} = Z_{i,1}V[D_i] + V[D_i]\left[1 - 2\Phi(T_i)\right]$$

$$N_{i,1,1} = \left(C_i^M + C_i^K\right)\left[T_i\Phi\left(T_i\right) + \frac{\exp\left(-T_i^2/2\right)}{\sqrt{(2\pi)}}\right] - C_i^M T_i$$

$$N_{i,1,2} = V\left[D_i\right]$$

$$N_{i,k,1} = N_{i-1,k+1,1} Z_{i,1}^{1/2}$$

$$N_{i,k,2} = \left[Z_{i,2} + \frac{N_{i-1,k+1,2}^{1/2}}{Z_{i,1}}\right]^{1/2}.$$

The optimal decision is then:

$$E^*\left[D_i\right] = E\left[F_i\right] - T_i\left\{V\left[F_{i-1}\right] + V\left[D_i\right]\right\}^{1/2}$$

and the optimal return is:

$$f^* = \sum_{k=1}^{I-i} N_{i,k,1}\left[V\left[F_{i-1}^2\right] + N_{i,k,2}\right]^{1/2}.$$

After completing the backward recursion, the optimal decision for each stage is computed in the forward direction (i.e. from stage 1 to I).

Sarin and Das (1987) describe some limited test results on a 10-stage assembly line, which indicate that their dynamic programming procedure is 12% better than the deterministic scheduling rule (i.e. $K = 0$), 3% better than the $K = 1$ rule, 9% better than the $K = 1.5$ rule and 26% better than the $K = 2$ rule. Their single-stage rule, which neglects stage interactions, was within 2% of their dynamic programming solution and better than all of the K rules. In reporting runtimes, they note that their dynamic programming procedure ran in 0.03 seconds while a single simulation run required 2 seconds; thus, their procedure can be expected to be more practical in large-scale cases than would be the use of simulation to prescribe K_{ij} values.

They also describe extensions, including a means of prescribing customer due time and a means of charging a premium for early kit delivery. For example, the former is effected by defining an additional station, $I + 1$, with $V[D_{I+1}] = 0$, C_{I+1}^K = tardiness penalty per unit time, and C_{I+1}^M = cost per unit time to hold end-product inventory. The optimal decision $E^*[D_{I+1}]$ gives the customer due time. If the customer due time is, in fact, specified, $E^*[D_{I+1}]$ is set to that value and other $E^*[D_i]$ are translated appropriately.

Subsequently, Das and Sarin (1988) described a means of embedding their dynamic programming procedure in a method to resolve the J job problem. A sketch representing the relationships between operations (i.e. stages) in an $I = 3, J = 3$ schedule is given in Fig. 8.14. Each row represents the processing of one job at each of the three stations. Directed arcs into each stage represent the finishing and kit delivery times that determine starting time, and vertical lines represent machine availability. The return for each stage is indicated by a downward-pointing arrow.

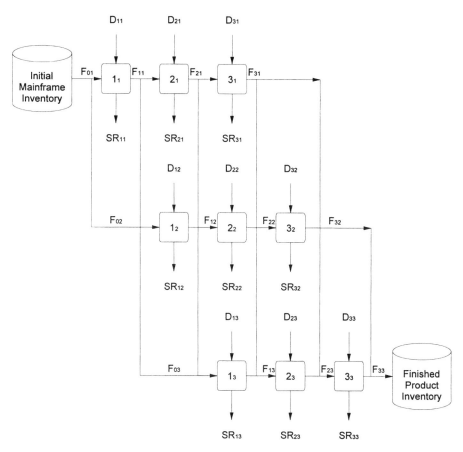

Fig. 8.14 Relationships between operations in an $I = 3, J = 3$ schedule (Das and Sarin (1988) reproduced with permission from IIE, Georgia, USA).

The longest path through this network from mainframe inventory to finished goods inventory determines the makespan of the schedule and is called the critical path. If operation times P_{ij} and kit ready times A_{ij} are known with certainty (i.e. deterministic), a simple critical path analysis would identify schedule makespan. In fact, this path would trace the first job to the bottleneck, process all jobs at the bottleneck, then trace the last job from the bottleneck to the end of the line as reflected in equation (8.20). We are interested in the case in which kit ready times A_{ij} are random variables, and thus must contend with the large number of combinations of arcs that might comprise the critical path. The ultimate objective is to prescribe kit due times to minimize the total cost of the schedule, but Das and Sarin conjecture that the path that determines the makespan is positively correlated with the one that minimizes cost and propose a heuristic that applies their dynamic programming procedure to determine due times for operations in selected chains of operations. Das and Sarin include stage $(I + 1)$ to prescribe customer due time.

Table 8.6   The Das and Sarin heuristic

---

1. Identify the critical path (Das and Sarin assume that $V[D_i] = 0$ for $i = 1, \ldots, I$ and apply a critical path analysis) and apply the dynamic programming procedure to prescribe due times $E[D_{ij}]$ for operations on the critical path.

2. Identify the second order metering station. If none exists or if $P_{m(2)}/P_{m(1)} < 0.90$, GOTO Step 4. Otherwise, a 'secondary critical path' exists (i.e. one that has a significant probability of defining the critical path on a realization of the assembly process) and is composed of the chain of operations denoted by:

   $$\{(m(2), j) \mid j = 2, \ldots, J\} \cup \{(k, J) \mid k = m(2) + 1, \ldots, m(1) - 1\}$$

3. Consider the 'island' of operations 'trapped' between the critical and secondary critical paths. For each job $j = 2, \ldots, J - 1$, apply the dynamic programming procedure to prescribe $E^*[D_{ij}]$ for operations in the chain:

   $$\{(k, j) \mid k = m(2) + 1, \ldots, m(1), - 1\}$$

   and expected starting time of stage $(m(1), j)$ as the customer due time for this chain.

4. Consider the remaining operations **upstream** of the bottleneck. For each job $j = 2, \ldots, J$ apply the dynamic programming procedure to prescribe $E^*[D_{ij}]$ for operations in the chain $\{(i, j) \mid i = 1, \ldots, m\}$, where m is $(m(2) - 1)$ if a secondary critical path exists and $(m(1) - 1)$ otherwise. Use the starting time of metering station $(m + 1)$ as the customer due time for this chain.

5. Consider the operations **downstream** of the bottleneck. Because in-process queuing cost increases as the mainframe progresses down the line, this step works to delay flow between the bottleneck and $\underline{m}(2)$, the station with the longest processing time downstream of $\underline{m}(1)$. The delay critical path consists of the operations in the chain:

   $$\{(k, 1) \mid k = \underline{m}(1) + 1, \ldots, \underline{m}(2)\} \cup \{(\underline{m}(2), j) \mid j = 2, \ldots, J - 1\}$$

   The dynamic programming procedure is applied to operations in this delay critical path using the starting time of operation $(\underline{m}(2), j)$ as the customer due time for the chain.

   In the second part of this step, the island of operations between the critical and delay critical paths are scheduled as in Step 3. For each job $j = 2, \ldots, J - 1$ the chain:

   $$\{(k, j) \mid k = \underline{m}(1) + 1, \ldots, \underline{m}(2) - 1\}$$

   is scheduled using the dynamic programming procedure.

6. This step repeats Step 4 iteratively until $\underline{m}(2) = I$. The delay critical path is taken to be the critical path at each of these iterations.

Source: Das and Sarin (1988), reproduced with permission from IIE, Georgia, USA.

The Das and Sarin heuristic is itemized in Table 8.6. It consists of six steps that identify various types of chains, each of which is optimized independently by the dynamic programming procedure.

Total cost is a random variable and can be defined as the sum of the in-process holding costs for the kit and mainframe at each operation and of the cost of station (controllable) idleness preceding each operation:

$$TC = \sum_{i=1}^{I+1} \sum_{j=1}^{J} \left[C_i^K \left(S_{ij} - A_{ij} \right) + C_i^M \left(S_{ij} - F_{i-1,j} \right) + C_i^I \left(S_{ij} - F_{i,j-1} \right) \right]$$

in which:

C_i^I = the cost per unit time for station i being idle.

At station $(I + 1)$, these costs may be interpreted as a tardiness penalty for missing the customer due time and an earliness cost associated with holding a completed job in finished goods inventory until the customer due time. Because the schedule is determined by aggregating results for each chain, the expected total cost can be estimated most accurately using a single simulation run, averaging cost over a set of replications.

Das and Sarin (1988) tested their approach in a set of 30 problems that involved from 5 to 14 jobs and from 5 to 24 stations. They devised a ratio to measure the effect of randomness on total cost and found that randomness increased total cost by an average of 13% in these cases. Thus, randomness exerts a significant influence over system performance and these results show again that deterministic scheduling methods should not be used in stochastic environments, since they cannot compensate for randomness. The total cost incurred by the heuristic averaged about 35% less than a pure $K = 1$ strategy over all 30 cases.

These results show that time-managed material flow control can be achieved in large-scale assembly systems, improving schedule performance and reducing cost in comparison with alternative methods. In particular, it is most important to emphasize that results demonstrate clearly that scheduling methods based on the assumption that the system operates deterministically cannot deal effectively with actual stochastic environments.

8.5 Summary

Section 8.2 presents a conceptual model of the kitting process which is used to approximate expected kitting time, kit earliness and tardiness, and the time that components spend as in-process inventory. Empirical tests indicate that the accuracy of the approximation is acceptable for practical purposes. The models are used to describe fundamental characteristics of the kitting process such as the functional relationship between expected kitting time and the number of required components.

The approach offers two advantages in comparison with simulation. The first is runtime; a comprehensive sensitivity analysis required 30 milliseconds while an equivalent simulation model took over three minutes to develop estimates of the same set of measures. The second is that the conceptual model led to derivation of closed-form models to compute performance measures such as the expected values of kit earliness and tardiness.

These advantages appear to make the approach well suited for application to the large-scale problems encountered in industry; no other quantitative methods are available for use in such settings. The approach complements traditional 'push' and 'pull' strategies for managing material flow by providing a way to time-manage material inputs and outputs.

Conceptual models may be extended for application in other settings. For example, if a component is inventoried rather than special-ordered, its ready time R_j must be used in the maximum operator for each Q_jth item (where Q_j is the replenishment lot size for component j), since it is known that on-hand stock will be available to fill kits at other times.

Section 8.3 describes models for estimating the transient performance of some repetitive assembly systems. The models assume that operation times are normally distributed and approximate the joint distributions of starting and finishing times by the multivariate normal. A number of cases in which operation times can validly be assumed normal are described; they appear to represent typical manual and automated operations in assembly systems. A recursion model of the assembly line is presented along with a method to estimate the correlations between certain pairs of operation finishing times, and the basic model is extended to model features such as finite buffers and parallel machines at a station. Tests show that recursion model estimates compare favorably with those of simulation and require much less runtime.

A recursion model of the assembly network is also presented in section 8.3 along with some fundamental properties that might be used in establishing schedules. The model is evaluated in application to an industrial case setting involving the assembly of truck cabs. Modeling capabilities are extended to incorporate such features as release time intervals, setting safety lead-times for part deliveries in lots of specified size and multi-level bills of materials. Tests show that the recursion model gives accurate estimates of transient performance in large-scale networks and is able to describe the sensitivity of operations to release time intervals, safety lead-times, delivery time variance and lot size.

Section 8.4 demonstrates the use of recursion models to optimize the time management of material flow in the kitting process and in repetitive assembly lines. Fundamental relationships are used to propose a means of managing the kitting process by determining safety lead-time parameters that optimize total cost, providing an appropriate balance between the risks of early and late delivery of components.

Methods for time-managing material flow in the assembly network are then described. Model outputs are used to establish several principles that might be used by the intuitive manager, then a method to optimize time-managed material flow is presented. Perhaps the most important guiding principle is that applying deterministic scheduling procedures in a stochastic assembly environment leads to overly optimistic schedules that cannot be achieved in the stochastic system.

These results indicate the practical importance of recursion models. Because they can represent a broad range of features, they can be used in a variety of applications. For example, the model might be used to evaluate a number of design and operating alternatives relatively quickly or to prescribe a short-term schedule in the 'real-time' control of a cell. Most importantly, these models of transient performance provide a means of 'looking ahead', allowing time-managed material flow to achieve the business goals of assembly systems.

The interested reader may wish to refer to two other papers that complement this chapter. Wilhelm, Som and Carroll (1992) present a time-managed material flow model that seems well-suited for an assembly system that is subject to random disruptions. The model focuses on the underlying structure of an assembly network in which production operations are independent. Certain fundamental operating characteristics of assembly systems are identified in this environment, including those related to kitting and to the principle that schedules are most economically controlled by coordinating vendor deliveries rather than by deliberately delaying materials in-process.

Numerical tests indicate that the approximating model is highly accurate and suggest that greater accuracy may be achieved only with runtimes that are prohibitive for practical applications. In particular, examples demonstrate how material flows could be time-managed to minimize expected cost, an important consideration not often used in prescribing material management policies. The time management of material flow appears to be better suited than other paradigms for the job shop and small-lot operations.

Wang and Wilhelm (1991) model the transient performance of a cell, gaining insights into system operation and evaluating a new recursion model of small-lot assembly systems. They present a recursion model capable of representing a broad variety of features found in cellular production/assembly systems, including assembly lines, flow-lines with finite buffers, generalized flow-lines, parallel machines, test/rework cycles, job shop routing, lot sequencing and material handling. They evaluate the model in a numerical example; in modeling a number of design and operating features including parallel machines, finite input/output buffers, the degree of balance among processing times, lot routing and lot sequencing; and in a hypothetical, yet representative, production setting involving circuit card assembly. Furthermore, the model is compared with a simulation model that is 'equivalent' in the sense that the two models share a common processing logic and a number of subroutines. In all cases, the recursion model gave estimates of performance that compare closely with those derived from simulation with significantly less runtime. In fact, the accuracy of recursion estimates of transient performance is better than typical approximations of steady-state performance (see Chapter 6).

Quality issues are essential in assembly and have been addressed indirectly up to this point. The next chapter focuses on inspection and test technologies, design of rework and repair functions and new methods to assure quality by integration and communication networks, error diagnostics and recovery.

8.6 Review questions

1. Describe the crucial role of kitting in material flow management.
2. Why do stochastic kitting operations incur both in-process holding and job tardiness costs, each with some finite probability?
3. Why is it important to consider variability in system operations rather than to suppress it using deterministic models in stochastic environments?
4. What relationship is there between the expected values of part and kit ready times?
5. Is it appropriate to model distributions using the normal in assembly systems?
6. How well does the normal approximate the distribution of the maximum?
7. How can correlations between pairs of operation finishing times be determined?
8. Describe pitfalls in estimating lead-times as currently done in the MRP environment.
9. Why are part ready times often correlated in actual assembly systems?
10. Why is kit ready time a random variable that is different from the random variables describing part ready times?
11. List parameters that affect kitting and describe the influence of each.
12. Why are kits that require more parts likely to require longer kitting times?
13. Describe how safety lead-times can be controlled to manage the kitting process.
14. How can models of transient operations be used to time-manage material flow?
15. What fundamental observation describes when an assembly operation can begin?
16. What causes the finishing times of operations to be correlated?
17. In what cases is it appropriate to model the distribution of operation time as normal?
18. How can the correlation between two operation finishing times be estimated?
19. Describe how the basic model in equations (8.12) and (8.13) may be enhanced to incorporate finite buffers, parallel machines at a station and job sequencing?

20. What effects do metering stations have on the correlations between finishing times?
21. Do recursion models approximate performance with a level of accuracy sufficient for practical decision-making?
22. Why do deterministic scheduling methods give overly optimistic schedules that cannot be attained in stochastic environments?
23. List some principles established by recursion models that can be used by the intuitive manager in time-managing material flow.
24. Derive the distribution function of $A = \max[R_1, R_2]$ if R_1 and R_2 are independent and identically distributed as uniform random variables on $[a, b]$.
25. Approximate the distribution of the random variable $Z = \max[\eta_1, \eta_2]$, where (η_1, η_2) is a bivariate normal with parameters $\mu_1 = 1000$, $\mu_2 = 1000$, $\sigma_1 = 150$, $\sigma_2 = 150$ and coefficient of correlation $\rho = r[\eta_1, \eta_2] = 0.15$. Specifically, calculate a^2, α, $\phi(\alpha)$ and $\Phi(\alpha)$. Using these values calculate $E[Z]$ and $V[Z]$. Discuss what distributions might be used to approximate the distribution of Z in this case.
26. Model $A = \max[\eta_1, \eta_2, \eta_3]$, where (η_1, η_2, η_3) is a trivariate normal with parameters $\mu_1 = 1000$, $\mu_2 = 1000$, $\mu_3 = 1000$, $\sigma_1 = 150$, $\sigma_2 = 150$, $\sigma_3 = 150$ and coefficients of correlation $\rho = r[\eta_1, \eta_2] = 0.15$, $\rho_1 = r[\eta_3, \eta_1] = 0.15$ and $\rho_2 = r[\eta_3, \eta_2] = 0.15$.
27. Determine all metering stations in a 10-station assembly line with expected operation times: 4, 3, 6, 3, 7, 2, 10, 5, 4 and 9.

References

Carre, A.S. (1971) 'Correlation theory applied to work flow measurement', *International Journal of Production Research*, **9** (2), 209–18.
Clark, E.C. (1961) 'The greatest of a finite set of random variables', *Operations Research*, March–April, 145–62.
Das, S.K. and Sarin, S.C. (1988) 'Selection of a set of part delivery dates in a multi-job stochastic assembly system', *IIE Transactions*, **20** (1), 4–11.
Greer, W.R. and La Cava, G.J. (1979) 'Normal approximations for the greater of two normal random variables', *Omega*, **7** (4), 361–3.
Hira, D.S. and Pandey, P.C. (1984) 'A computer simulation study of manual flow lines', *Journal of Manufacturing Systems*, **2** (2), 117–25.
Johnson, N.L. and Kotz, S. (1972) *Distributions in Statistics, Continuous Multivariate Distributions*, John Wiley, New York.
Kalkunte, M.V. and Wilhelm, W.E. (1987) *Use of Coxian-Type Distributions in a General Model of the Maximum of Bivariate Random Variables*, Working paper, Department of Industrial and Systems Engineering, Ohio State University.
Kalkunte, M.V. and Wilhelm, W.E. (1989) *Kit Completion Time in Two-Stage Assembly Systems*, Working Paper, Department of Industrial Engineering, Texas A&M University.
Lambrecht, M.R., Muckstadt, J.A. and Luyten, R. (1982) *Protective Stocks in Multi-Stage Production Systems*, Technical Report 526, School of Operations Research and Industrial Engineering, Cornell University.

Maxwell, W.L. and Muckstadt, J.A. (1983) *Establishing Consistent and Realistic Reorder Intervals in Production-Distribution Systems*, Technical Report 561, School of Operations Research and Industrial Engineering, Cornell University.

McGee, G.R. and Webster, D.B. (1976) 'An investigation of a two-stage production line with normally distributed interarrival and service time distributions', *International Journal of Production Research*, **14** (2), 251–61.

McGinnis, L.F. and Bozer, Y.A. (1984) *Kitting: A Generic Descriptive Model*, Working Paper, Material Handling Research Center, Georgia Institute of Technology.

Muth, E. (1973) 'The production rate of a series of work stations with variable service times', *International Journal of Production Research*, **11** (2), 155–69.

Orlicky, J.A. (1975) *Material Requirements Planning*, McGraw-Hill, New York.

Payne, S., Slack, N. and Wild, R. (1972) 'A note on the operating characteristics of "balanced" and "unbalanced" production flow lines', *International Journal of Production Research*, **10** (1), 3.

Reeve, N.R. and Thomas, W.H. (1973) 'Balancing stochastic assembly lines', *AIIE Transactions*, **5** (3), 223–9.

Saboo, S. and Wilhelm, W.E. (1986) 'An approach for modeling small-lot assembly networks', *IIE Transactions*, **19** (4), 322–34.

Saboo, S. and Wilhelm, W.E. (1987) 'Description of certain correlation processes in flowlines', *International Journal of Production Research*, **25** (9), 1355–92.

Saboo, S., Wang, L. and Wilhelm, W.E. (1989) 'Recursion models for describing and managing transient flow of materials in generalized flowlines', *Management Science*, **35** (6), 722–42.

Sarin, S.C. and Das, S.K. (1987) 'Determination of optimal part delivery dates in a stochastic assembly line', *International Journal of Production Research*, **25** (7), 1013–28.

Seidmann, A., Schweitzer, P.J. and Nof, S.Y. (1985) 'Performance evaluation of a flexible manufacturing cell with random multiproduct feedback flow', *International Journal of Production Research*, **23** (6), 1171–84.

Seidmann, A., Schweitzer, P.J. and Nof, S.Y. (1987) 'Observations on the normality of batch production times in flexible manufacturing cells', *International Journal of Production Research*, **25** (1), 151–4.

Sellers, C.J. and Nof, S.Y. (1986) 'Part kitting in robotic facilities', *Material Flow*, **3**, 163–74.

Sellers, C.J. and Nof, S.Y. (1989) 'Performance analysis of robotic kitting systems', *Robotics and Computer-Integrated Manufacturing*, **6** (1), 15–23.

Tamaki, K. and Nof, S.Y. (1991) 'Design method of robot kitting system for flexible assembly', *Robotics and Autonomous Systems*, **8** (4), 255–73.

Wang, L. and Wilhelm, W.E. (1991) 'An algorithm for applying recursion models to cellular production/assembly systems', *International Journal of Flexible Manufacturing*, **4** (2), 129–58.

Wang, L. and Wilhelm, W.E. (1992) 'A recursion model for cellular production/assembly systems', *International Journal of Flexible Manufacturing*, **4**, 129–58.

Wang, L. and Wilhelm, W.E. (1993) 'A PERT-based paradigm for modeling assembly operations', *IIE Transactions*, **25** (4), 88–103.

Wilhelm, W.E. (1986a) 'A model to approximate transient performance of the flowshop', *International Journal of Production Research*, **24** (1), 33–50.

Wilhelm, W.E. (1986b) 'An approach for modeling transient material flows in robotized manufacturing cells', *Material Flow*, **3** (1–3), 55–68.

Wilhelm, W.E. (1986c) 'Lognormal models of transient operations in the flexible manufacturing environment', *Journal of Manufacturing Systems*, **5** (4), 253–66.

Wilhelm, W.E. (1986d) 'Su models of the maximum of several dependent random variables', *Computers in Industrial Engineering*, **10** (4), 335–46.

Wilhelm, W.E. (1987) 'On the normality of operation times in small-lot assembly systems', *International Journal of Production Research*, **25** (1), 145–9.

Wilhelm, W.E. and Ahmadi-Marandi, S. (1982) 'A methodology to describe operating characteristics of assembly systems', *IIE Transactions*, **14** (3), 204–13.

Wilhelm, W.E. and Saboo, S. (1988) *A Fast, Approximate Means of Estimating Transient Performance of the Flowline*, Working Paper, Department of Industrial and Systems Engineering, Ohio State University.

Wilhelm, W.E. and Wang, L. (1986) 'Management of component accumulation in small-lot assembly systems', *Journal of Manufacturing Systems*, **5** (1), 27–39.

Wilhelm, W.E., Kalkunte, M.V. and Cash, C. (1987) 'A modeling approach to aid in designing robotized manufacturing cells', *Journal of Robotic Systems*, **4** (1), 25–48.

Wilhelm, W.E., Saboo, S. and Johnson, R.A. (1986) 'An approach for designing capacity and managing material flow in small-lot assembly lines', *Journal of Manufacturing Systems*, **5** (3), 147–60.

Wilhelm, W.E., Som, P. and Carroll, B. (1992) 'A model for implementing a paradigm of time-managed, material flow control in certain assembly systems', *International Journal of Production Research*, **30** (9), 2063–86.

9

Quality and inspection in assembly

9.1 Introduction

Assembly is often described as the 'moment of truth' for products, when finally complete subassemblies, and then the full assembly, can be tested. But it is also recognized that assembly is where previous defects may be 'buried and hidden' not to be discovered until much later. By emphasizing 'quality at the source' – essentially at every step of the production process – the potential for hidden defects decreases, and the need for the correction of errors by costly repair and rework at later stages diminishes. But inspection **during** assembly is becoming more necessary because most tests for functionality are still possible only **during** and **after** an assembly is built.

The modern and still evolving concept of quality has emphasized two directions of continuous improvement:

1. deeper, throughout the production system, seeking to build quality into the product at the source and prevent errors before they impact on subsequent steps;
2. wider, with the total quality management (TQM) approach that integrates technical and human factors in the entire organization, from supplier to customer, to yield cost-effective products at a quality level which satisfies customers.

These trends are healthy from a competitive point of view. The objective of this chapter is to describe how quality assurance is integrated into assembly.

The chapter is organized as follows. Quality issues related to assembly are described in section 9.2. Integrated inspection, test and assembly are described in section 9.3 and models for comparing alternative inspection technologies are presented. Design criteria and analysis of rework and repair in assembly are described in section 9.4. The effectiveness of error diagnosis and recovery in assembly is discussed in section 9.5. Finally, the role of communication networks and information integration for assembly quality assurance are the subject of section 9.6.

9.2   Quality issues related to assembly

Three important topics that are fundamental to assembly quality are calibration, tolerance and standards. They are discussed in this section.

9.2.1   Calibration

Calibration is any procedure the objective of which is to increase the accuracy and precision of a system. Accuracy and precision, as general terms, refer to the quality, state or degree of conformance to a recognized standard or specification. A formal definition is given by ANSI B5.54-1992 as follows:

> **Accuracy** is a quantitative measure of the degree of conformance to recognized national or international standards of measurement.
> **Repeatability** is a measure of the ability of a machine to position a tool sequentially with respect to a workpiece under similar conditions.

Note that in the definition the term **repeatability** has replaced the undefined term precision. Repeatability is a statistical measure of repeated actions and motions, while accuracy is associated with each individual action or motion. For instance, assembly robots have a repeatability that is better than 1 mm, but their accuracy in reaching points specified as position and orientation relative to a work cell is orders of magnitude worse.

Causes for inaccuracy and lack of repeatability include:

- environmental factors such as temperature and humidity;
- machine factors such as structure, friction, backlash;
- measurement factors such as round-off, computation errors;
- operation factors such as tool wear, workpiece clamping errors.

Calibration can remedy these problems to a certain extent. For example, the Grumman Aircraft company describes calibration as a key technology in achieving flexible robotic assembly and its effective integration to the factory systems (Barone, 1989). Calibration means two main types of activities in relation to assembly (Warnecke, Schraft and Wanner, 1985; Whitney, Lozinski and Rourke, 1986):

1.   initializing the configuration of an assembly workstation, machine or robot before and during each start-up to ensure the accuracy and quality of assembly;
2.   increasing the absolute positioning accuracy of assembly workheads and robots for accurate motions that cannot be taught directly, as in the case of off-line programming.

During the Industrial Revolution, the invention of precise mechanical gauges enabled calibration and the assembly of interchangeable components into relatively complex mechanisms. Over the years, optical measurements improved the effectiveness of calibration. Electro-optical calibration systems and the use of lasers have further improved the degree of achievable accuracy. Gradually, calibration science has integrated precision measurement techniques with the machine or robot control system, such that the initialization involves accurate setting and resetting of mechanical, control and software subsystems (Hayati, 1988; Zupancic, 1994).

Most calibration techniques address the geometry of positioning and insertion motions. For instance, in electronic component placement or insertion, accurate positions are needed for the multiple pins of each component. The quality of soldering and electrical contact is determined by positioning accuracy. Any deviations can be corrected by measuring the current actual position of a workhead or robot end-effector (hand or tool) and comparing it with the absolute required position (usually a reference position which is considered 'true'). For multi-link robot arms, a robot kinematic model is used to analyze how to correct the position of each individual joint.

Several techniques have been developed for calibration, for example calibration of robot motions based on force sensors (Duelen *et al.*, 1990), on electro-optical distance sensor (Partaatmadja, Benhabib and Goldenberg, 1990), on a theodolite system for spatial coordinate determination (Duelen, Muench and Surdilovic, 1991), and on vision-guided grasping for truss assembly in space (Nicewarner and Kelley, 1993). A joint ESPRIT project CAR, involving the KUKA robot company, Leica instrument company and the Fraunhofer-IPK in Berlin, among others, has addressed calibration applied to quality control and maintenance (Schroer *et al.*, 1995).

Calibration is often concerned with geometric dimensions of components or their placement, as exemplified by positioning and path-following accuracy. However, the calibration of tools is also concerned with the results of the assembly tool function, for instance the opening of a glue deposition tool, the temperature calibration of heating and drying elements and of test and inspection instruments. Figure 9.1 shows a Nordson proportional dispenser of adhesive or sealant material that is used for robotic assembly of various components, for instance in automobile assembly. The dispenser controller can adjust the flow at the nozzle for changes in robot velocity to provide precise, uniform material deposition. Adjustments are based on signals from the robot controller. With a pressure transducer at the nozzle, the dispenser controller automatically compensates also for pressure fluctuations in material delivery. It can be calibrated to learn the necessary material viscosity for a wide variety of materials, and adjust material temperature for optimum viscosity. In addition, the controller provides system status and diagnostic data, and records information for statistical process control.

9.2.2 Tolerance

A key measure of quality is tolerance. When pioneers of the Industrial Revolution created the notion of part interchangeability, they invented the concept

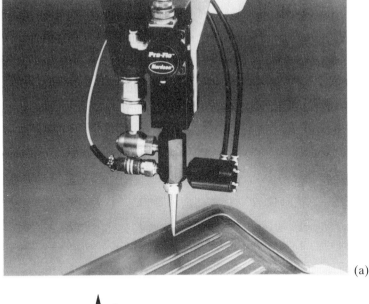

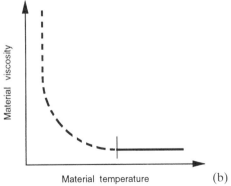

Fig. 9.1 Adhesive or sealant dispenser for robotic assembly: (a) proportional flow dispenser; (b) typical temperature–viscosity curve (with the optimum at the flat range) (courtesy of Nordson Corp.).

of tolerance, a compromise with the realization that no two parts can be manufactured absolutely identical. Tolerance is defined as maximum upper and lower deviations from a nominal specification of a component that are still tolerated as acceptable for the component's purpose. For example, a peg can be inserted in a hole if the peg's diameter remains below the maximum tolerance deviation and the hole's diameter remains above its lowest tolerated diameter. In printed circuit board through-hole assembly, for instance, the laminate board material is closely controlled for thickness and warping tolerances. The diameter of holes drilled in boards for component insertion is also controlled, and so is the width of inserted pins.

Tolerance is considered a design specification for components, for instance to ensure a critical fit or clearance in an assembly, and for processes, for instance to ensure a temperature range. Methods have been developed to allocate tolerance to assembled components, e.g. equal allocation, equal repeatability or least-cost methods (Spotts, 1973; Chase *et al.*, 1990). Optimal design of tolerance is important to assembled product quality and cost. It is a topic under active research, especially with the influence of concurrent engineering.

On-line monitoring and 100% inspection of the quality of crimping in the assembly of wire terminals is shown in Fig. 9.2. Crimp quality is monitored in terms of crimp height to the fourth decimal place, at a rate of 5000 crimps per hour. Control data for up to 200 different parts can be stored in the monitor unit. Inspection information is provided to the monitor by force and position

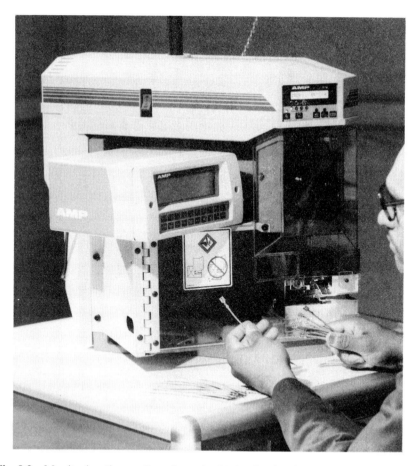

Fig. 9.2 Monitoring the quality of terminal crimping in wire assembly. The crimp quality monitor is mounted on a wire termination station (courtesy of AMP Inc.).

sensors which are built into the assembly applicators. The device can also monitor quality aspects in terms of missing wire strands, short wire brush and crimped insulation. An audible alarm is sounded and a signal light turned on to alert operators to any out-of-tolerance quality problem that is detected. Display screens provide a variety of statistical process control information. The unit also has a built-in serial network interface to download inspection information for subsequent analysis.

9.2.3   Standards

Standards are designed for quality acceptance of products and services and are therefore essential for industrial assembly. There are several motivations to develop standards for quality:

- standards provide general and specific guidelines to determine if quality requirements are satisfied;
- standards serve as a uniform, formal basis for comparison between alternative suppliers and, when needed, for their certification;
- standards provide a formal basis for contractual agreements between suppliers and customers;
- standards help guide the design of new products or product improvements.

Standards are becoming especially important in the global market economy (Wadsworth, 1992). Three types of standard are common: (1) internal company standards which are based on the company's interests and best industry-wide practice; (2) industry-specific standards, for instance for wire terminals and connectors, or the automotive industry standard ESO-9000; (3) international standards, for instance those developed by ISO, the International Organization for Standardization.

The ISO 9000 series of international quality standards includes: ISO 9000 containing the terminology and format of the standard; ISO 9001/2/3 to be used for contractual evaluation and qualification; ISO 9004 containing general guidelines meant for internal audits only. The main elements of ISO 9001 are shown in Table 9.1.

9.3   Inspection and test during assembly

New approaches to the function of quality control are emerging that are drastically altering the way inspection and test are performed. These new approaches are based on advanced sensor technologies often combined with computer-based systems to interpret the sensor signals. For example, machine-vision systems can identify defective components at the feeder and 'weed them out'. Optical inspection machines scan assembled circuit boards, diagnose all faults and these data are immediately used for repairing the

Table 9.1 Structure of ISO 9001 quality system standard

1. Management responsibility	11. Quality system
2. Contract review	12. Design control
3. Document control	13. Purchasing
4. Purchaser–supplier relationship	14. Product identification and traceability
5. Process control	15. Inspection and testing
6. Test equipment	16. Test status
7. Control of nonconformance	17. Corrective actions
8. Handling and storage	18. Quality records
9. Internal quality audits	19. Training
10. Servicing	20. Statistical techniques

board. In addition, new software tools are being developed to automate the operation of complex sensor systems involved in assembly workstations and to statistically analyze the sensor measurements.

This section describes a variety of techniques for inspection and test during assembly. The focus is on decision models for the selection of an appropriate technique when a number of alternatives are available.

9.3.1 Inspection and test technologies

Examples of inspection requirements for printed circuit board assembly are described by Kear (1987), including: component lead forming; wave soldering; quality standards; printed circuit cleaning standards; material handling and storage standards. The instructions for visual inspection of a PCB, either manual, semi-automatic or automatic, typically have a drawing of the PCB with notes such as 'Insulation on transformer leads to be within 0.030″ of solder joint and undamaged', or 'Resistor R1 to clear PCB by 0.200″ and be parallel with board surface'. Testing printed circuit board assemblies is usually preceded by 100% testing of components before assembly. A recommended sequence is:

1. Test components and printed circuit boards.
2. Assemble.
3. Solder.
4. Trim leads.
5. Inspect/test.
6. Repair.
7. Inspect/test

and so on.

While some of the inspection and testing tasks can be performed by non-contact techniques, certain testing usually requires contact. Arabian (1989) compares the advantages of both and lists various testing techniques of electronic assemblies. **Contact testing** involves the use of conductive pins, probes or connector elements. The main disadvantages of contact testing are potential mechanical damage, e.g. abrasion at contact points; solder degradation; and potential electrical damage, such as causing a short circuit by contacting tightly adjacent components, or wave-form distortion in high-speed circuits. **Non-contact testing** is more costly to set up but has advantages where it can be applied, mainly by avoiding the above damages and saving manufacturing time. Examples of non-contact testing are:

- electron beam methods used to detect excess or lack of electrons in memory arrays;
- laser probing with a weak beam (tens of microwatts) to detect photo-currents of sub-micro amps;
- X-ray techniques to inspect multi-layer PWBs for flaws;
- optical techniques to detect electrical signals, e.g. on/off indications, or fiber-optic techniques used to monitor high frequency devices.

Automatic optical inspection (AOI) systems have been developed with computerized checks of design rules. Automatic test equipment (ATE) has been developed for four main types of testing:

1. bare board tester (BBT);
2. pre-screen circuit tester (PSC);
3. in-circuit tester (ICT) for verifying the electrical connection and correct operation of components already assembled;
4. functional tester, which tests the assembly for its intended functions.

The objective of **in-circuit testing** is to diagnose manufacturing faults, e.g. wrong components, bent or broken pins, wrong orientations and component static problems. The objective of **functional testing** is to diagnose dynamic, high-speed electrical faults such as timing, delay and feedback problems. These tests are usually followed by a **unit test** and a **system test** to diagnose any operational faults.

An important goal in modern production is the pretesting of all incoming boards and components by the supplier, which is assumed to be completed at the supplier's location prior to just-in-time delivery. Another important goal is to monitor quality or conditions as assembly progresses and minimize the cost of rework or scrap by identifying faulty items before adding more cost to already defective material. Therefore, computer-aided inspection, test and repair which are integrated with assembly stations have the following objectives:

1. reduce assembly time by reducing test and repair time, mainly by eliminating extra moves and set ups;
2. automate to reduce errors and the skill level required of operators;
3. collect real-time communication of test and repair information for cost-effective repair decisions and for other assembly operations that may be affected;

4.   provide real-time communication and repair data to implement design changes as soon as they are authorized;
5.   maintain cumulative statistics of test and repair data for long-term quality and reliability analyses, and as per inter-company standards such as ISO 9000.

In addition to the test and inspection techniques for electronic assembly described above, typical tests integrated with assembly in general include (Hoffmann, 1993):

1.   **part presence**, tested by either a probe or a machine-vision system to ensure that each part is present and is positioned correctly;
2.   gauging to verify **dimensional requirements**, typically with linear variable differential transformer (LVDT) probes that will check the dimension in question, most often after a crimping or staking operation;
3.   **leak testing** of a part or an assembly of parts – leak rates can be checked by a variety of methods including mass flow pressure, decay/increase in differential pressure and helium leakage tests;
4.   **functional testing**, to ensure that a part or an assembly meets performance operating conditions – typical tests verify electrical, pneumatic and mechanical characteristics;
5.   **electrical testing** of parameters such as continuity, voltage, current and contact.

Modern computerized instrumentation not only provides the benefits of data acquisition and analysis technologies to such production-line inspection procedures as gauging and leak testing, but also enables users to meet just-in-time requirements. Based on sensor information, errors due to machine, part or procedure failure can be discovered at the assembly workstations. Once an error occurs, actions can be taken immediately for repair, correction or error recovery. The next section describes models to assess and select alternative sensor technologies for inspection and test.

9.3.2   Selection of inspection technologies

The ability to monitor a process, measure its variables and express the results quantitatively is the only way to fully understand and control a process. This idea is basic to statistical process control (SPC) and on-line inspection; however, some measurement techniques are relatively costly.

Different off-the-shelf sensors and inspection systems can measure a part's dimensions while it is in process. Furthermore, advances in information technology and machine tools have made integration and implementation of on-line inspection relatively simple. However, there is a need for a systematic assessment of alternative inspection technologies, particularly when 100% inspection is desirable.

Two assessment models to select on-line inspection systems have been developed by the PRISM Program at Purdue University (Remski and Nof, 1993): a **productivity measurement model** (PMM) and a **performance assessment model** (PAM). The models compare laser-based and machine-vision based systems, two prominent non-contact inspection techniques. Similar models can be developed to assess and select other sensor-based inspection techniques.

The selection emphasis is on the technological inspection abilities such as accuracy, consistency and response time, and on economics. Data for assessment are obtained through field and laboratory experiments. The goal is to implement such assessment models in virtual manufacturing to inform users which inspection technology, if any, would be a better choice in a given situation.

On-line quality inspection Sensors are a key element of an on-line feedback system on a machine tool or assembly system. Without sensors, an **in-process measurement and control system** (IPMCS) or an **on-line inspection system** (OLIS) cannot be implemented (Murphy, 1990).

Many sensor types are available for inspection, and there may be a problem in choosing the most appropriate sensor for a given process. Sensor types can be contact or non-contact, direct or indirect. These categories are determined by how measurement of a workpiece is obtained. **Contact sensors** have a measurement head that touches the workpiece. Contact disadvantages are wear, error due to sensor deflection and possible surface marking. **Non-contacting sensors** do not touch the workpiece but apply some flow energy (e.g. electrical, air, light or sound waves) to measure a surface. These sensors are preferred since they have none of the contact disadvantages.

Indirect measurements are derived by encoding a signal received from a sensor. **Direct** measurements are made by sensors whose output signal is directly proportional to the measurement, e.g. the distance the head moves. Therefore a direct measurement can be made independently of encoders. Sensors in this category are more versatile and applicable to both on-line as well as in-process measurement systems.

Relevant sensor characteristics are:

- sensor size;
- stand-off, the distance from a measured item;
- resolution, the smallest increment of a measurement device;
- sensor accuracy, which is its measurement error;
- effective spot-size that can be measured.

These characteristics are used in the performance assessment model described below.

Productivity measurement model The productivity measurement model (PMM) can be used to evaluate the potential productivity of an assembly and inspection system, following a multiple inputs approach (Stewart, 1983). For assessment (Remski and Nof, 1993), either spreadsheet or dynamic programming can be applied to compare outputs to inputs (Fig. 9.3), removing the effects of inflation and accounting for a product mix. The result is a productivity estimate based on the number of assemblies inspected by an on-line inspection system (OLIS). The input can include any controllable resources associated with inspection: direct labor, manufacturing support, production equipment, energy, materials or inventory.

The actual productivity (AP) is calculated as the general productivity (GP)

Variable	Definition
P_k	manufacturing or assembly process k
I_k	inspection process k
G_k	proportion of acceptable parts from process k
B_k	proportion of defective parts from process k
M_{Ik}	Type I Error probability from process k
M_{IIk}	Type II Error probability from process k
Q^*_{ijk}	output quantity of good product i over period j from process k
S^*_{ijk}	scrap quantity of product i over period j from process k

Productivity Measurement Model Quantities

$$Q^*_{ijk} = B_k Q^*_{ij(k-1)} M_{IIk} + G_1 Q^*_{ij(k-1)} (1 - M_{Ik})$$

$$S^*_{ijk} = B_k Q^*_{ij(k-1)} (1 - M_{IIk}) + G_1 Q^*_{ij(k-1)} M_{Ik}$$

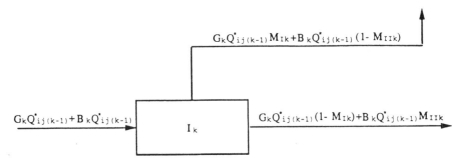

Fig. 9.3 PMM definitions and model.

minus the lost productivity (*LP*) due to bad parts and assemblies. In general, any process has four categories of output: good products designated correctly as such by inspection (Pass | Good), good products designated by mistake as defective (Fail | Good), bad products designated by mistake as good (Pass | Bad), and bad products designated correctly as defective (Fail | Bad). As a result, the output can be categorized as follows (see variable definitions in Fig. 9.3):

	Accepted production	**Lost production**
'Scrap' output, $S_{ijk}^{*} = B_k Q_{ij(k-1)}^{*}\left(1 - M_{IIk}\right)$		$G_k Q_{ij(k-1)}^{*} M_{Ik}$
'Good' output, $Q_{ijk}^{*} = G_k Q_{ij(k-1)}^{*}\left(1 - M_{Ik}\right)$		$B_k Q_{ij(k-1)}^{*} M_{IIk}$

In a network of assembly stations, the accepted output of each inspection process will become input to subsequent stations. The ratio of the total accepted portion of Q_{ijk}^{*} for all periods considered over the total resource inputs over the same periods yields the productivity measure. Alternatively, a composite measure of productivity can be applied.

PMM results for a case study comparing seven alternative inspection technologies for a particular assembly example (Remski, 1993) are shown in Table 9.2. From the results, the best choice is LDR laser inspection, followed by LV22 vision inspection.

The advantage of PMM is the consideration of how a particular inspection technology affects the overall productivity of a production system. A more detailed, technical evaluation of inspection performance is described in the next section.

Performance assessment model   The performance assessment model (PAM) is based on technical and cost criteria. Before the model can be described further, several non-contact inspection technologies are briefly reviewed.

Laser-based inspection   Light sources offer non-contact measurement techniques. Lasers have made these techniques highly accurate, and with decreas-

Table 9.2   Comparison of alternative inspection technologies

Technology	Type	Productivity measurement
LDR	Laser	803.54
LVBG	Laser	743.96
NCLPG	Laser	748.79
MCV56	Machine-vision	775.52
MCV28	Machine-vision	764.61
LV22	Machine-vision	799.35
LV02	Machine-vision	746.89

ing cost they become more affordable. The two main measuring approaches are the **shadow technique** and **laser triangulation**.

- **Shadow technique.** The blocking of a beam of scanned laser light is sensed as follows. A laser is shone onto a rotating mirror which reflects the light in a divergent manner. A lens is used to refract this light into a plane of parallel rays which intersect the inspected workpiece. This intersection causes the formation of a shadow. The edges of the shadow are detected by a linear array of photo-diodes on the opposite side of the light source. The size of the workpiece is determined by the amount of time the photo-diodes 'see' the shadow. Therefore, the mirror's rotation needs to be tightly controlled in order to guarantee accuracy. The shadow technique's accuracy and resolution are about 1×10^{-5} inch. Unfortunately, they are limited to convex workpieces, since other features may be shadowed by a different part geometry. The stand-off distance can be from several inches to a few feet. At larger distances, variations in the air tend to affect the accuracy.

- **Laser triangulation.** This technique uses a laser light source and the principle of triangulation to determine measurements. For inspection, it is capable of relatively high data rates (Shade, 1992). A single point of laser light is projected onto a workpiece, and either the diffusely or specularly reflected light is imaged by a detector. The light source and a photo-diode detector array are set away from the workpiece at fixed settings. The detector is usually a CCD (charge couple device) array, or a lateral-effect photo-diode. The position of a spot on the detector corresponds to the distance from the object. Typically, a laser triangulation system would be used to inspect parts immediately after they have been formed or machined. Inspection can be accomplished by installing a laser sensor in a machine tool-holder. Once a process is completed, the sensor can be loaded into the machine tool-holder, e.g. a spindle, and an inspection program can begin. The resolution of laser triangulation, achieved by magnifying the image onto the photo-diodes, is about 5×10^{-5} inch. Typical accuracy obtainable by laser triangulation ranges from 5×10^{-5} to 5×10^{-4} inch. Stand-off distances span from 1 to 10 inches.

Machine-vision inspection Machine-vision is the newest, relatively most flexible non-contact sensor with potential applications for IPMCS/OLIS. This technology is based on a video machine-vision coupled with a computer-based image analyzer. Even though this technology dates back to the first generations of digital computers and televisions, it has become cost-effective for industrial applications only recently. The main elements are a camera, interface and computer. The basic configuration consists of a machine-vision unit and its magnifying optics, frame grabber (high-speed memory that collects image-frames from the machine-vision unit) and image analyzer. In order to use machine-vision as an IMPCS/OLIS system, features in the field of view must be located and defined. Resolution depends on the number and spacing of individual diodes, or pixels, that the machine has in its array. By using magnifying optics, the resolution can be increased to a range of 1×10^{-4} to 1×10^{-3} inch, compared with a range of 1×10^{-3} to 5×10^{-3} inch without magnification. The accuracy of machine-vision is within the range of 1×10^{-4} inch, which is one order of magnitude larger than the accuracy obtained in a laser or fiber-optic system. The stand-off distances may range from 0.05 inch to several inches, or up to several feet if macro-focusing telescopic lenses are used. Contrary to other optical methods, however, the spot-size does not have a significant effect on the performance of machine-vision, since it can capture the view of an entire scene at one time.

The performance assessment model (PAM) combines the economics and technical abilities of compared technologies in four parts: technological ratios, detection time, inspection time and economic measures.

Technological ratios   Every sensor has five common attributes: size, stand-off, resolution, accuracy and effective spot-size. They can be combined as ratios that characterize each technology.

- **Size to stand-off ratio** (Sz/St): relates how close a sensor can be placed to the inspected scene, e.g. a crimping point in a machine operation. If the ratio is too high, the sensor would be too large to permit effective measurement immediately behind the crimped edge.
- **Resolution to accuracy ratio** (Re/Ac): is relevant to the inspected quality requirements. Resolution is the smallest dimensional increment that a sensor can measure. Accuracy indicates how effectively the sensor can be integrated into the measurement and control system. Together, they indicate the versatility of a sensor system in accuracy terms.
- **Resolution to effective spot-size ratio** (Re/Ef): aids in determining the accuracy of measurements, especially when unlike geometries are involved. For a sensor whose purpose is to measure surface roughness, this ratio should be less than 1.

The vector $[(Sz/St);(Re/Ac);(Re/Ef)]$ can be used to directly compare candidate technologies and select those appropriate. For selected candidates, a combined technological ratio, r, is calculated as:

$$r = \left(Sz/St\right)\left(Re/Ac\right)\left(Re/Ef\right)$$

Detection time   This is the amount of time spent by an inspection system on clearing designated faults after an inspection.

PAM model variables are defined as follows:

F = number of sensors/inspected features per assembly
x = expected proportion of defective parts

t = sensor cycle time (seconds)
T_c = time to clear a defect from the system (seconds)
m_1 = type I error probabiliy (good designation when bad)
m_2 = type II error probability (bad designation when good)
m_3 = bad–bad designation probability
m_4 = good–good designation probability
N = number of assemblies to be inspected

An inspection can result in a rejected part because of true defects, but also because of random errors in the inspection procedure and/or in the sensor. Ideally, of course, m_1 and m_2 are negligible. With the defined variables, $m_1 + m_3 = x$ and $m_2 + m_4 = 1 - x$. It is assumed that a clearing time T_c is required for all the faults designated by the inspection system before the next inspection can start. The ratio of this clearing time to (FNt), the total inspection time, is defined as d.

Inspection time   The next portion of the model addresses the average inspection time of acceptable parts. The total time to inspect all N assemblies, including unacceptable parts, is $[(FNt) + (m_2 + m_3) FNT_c]$. (Note that only m_4N parts would be truly acceptable.) The average time for inspection, designated t_i, is this total time to inspect N assemblies, divided by the number of acceptable parts.

Economic measures   The fourth criterion is a dimensionless cost of inspection performance. Following Boothroyd's (1992) definition of dimensionless cost per assembly, the following variables are applied:

V = annual production volume
Q = cost of the inspection equipment equivalent to the work
     of one manual inspector during one shift
W = annual cost of employing one manual inspector
S = number of shifts
U = total cost of equipment, engineering, setup
G = total wage rate for inspectors
t_{mi} = average manual inspection time (seconds)
f = plant efficiency
t = technology's average inspection time per assembly (seconds)
$R = G/(W F)$

The total cost of inspection per assembly, C, is determined by:

$$C = t\left(G + \left(UW/SQ\right)\right) \tag{9.1}$$

If the cost of inspection is determined not per assembly but on an annual basis, dividing it by V, the annual production volume, provides the inspection cost per assembly. However, dimensions may be different for different technologies. To ensure uniform comparison, C is normalized to a dimensionless quantity. Thus the inspection cost under a given technology is divided by the cost of one manual inspector W, by the number of features to be inspected F, and by the average equivalent manual inspection time per part t_{mi}. As a result, the dimensionless cost of inspection per assembly, C', after substitutions is:

$$C' = (t/t_{mi})(R + (U/F)/(SQ)) \qquad (9.2)$$

For uniform comparison, one manual inspector working alone, without any equipment, is considered as the basic unit.

Composite performance measure A composite measure applies a weighted factorial combination of the technological ratios, detection time proportion, average inspection time per assembly and dimensionless inspection cost per assembly, or:

$$PAM = (r^{w1})(d^{w2})(t_i^{w3})(C'^{w4}) \qquad (9.3)$$

Where Wi, $i = 1, 2, 3, 4$, indicates the relative weight or significance of each of the four measures for a particular company. The objective is to select the technology that will minimize the value of PAM. For the example that follows, all weights are assumed equal.

PAM illustration Three inspection technologies applicable for assembly are examined: non-contact laser profile gage (NCLPG); tri-beam gage (VBG); machine-vision system (MCV). The three are described by Shade (1992), Harding (1991) and Murphy (1990), respectively. Common assumptions for all three technologies are:

$$
\begin{aligned}
x &= 0.02 &&\text{– proportion of defective parts} \\
Q &= \$10\,000 &&\text{– equivalent cost of inspection equipment} \\
W &= \$12\,000 &&\text{– annual cost of employing one inspector} \\
S &= 2 &&\text{– number of shifts} \\
G &= \$36\,000 &&\text{– total wage rate for inspectors} \\
t_{mi} &= 25\,\text{sec} &&\text{– average manual inspection time}
\end{aligned}
$$

F is assumed as one. Cycle times and clearing times of all three technologies, in seconds, are varied from 1 to 50 for the cycle time t, and 1 to 5 for the clearing time T_c. Figure 9.4(a) shows the effects observed. It can be seen that crossover points between technologies occur around $t/T_c \approx 1.25$ for VBG and NCLPG, and between 0.6 and 0.8 for MCV and VBG. As the ratio of t/T_c increases, MCV dominates in terms of the performance measure, followed by NCLPG and then VBG.

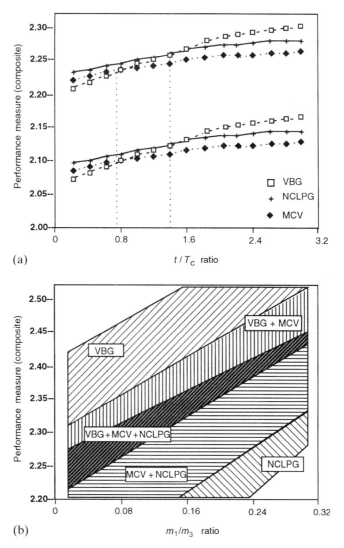

Fig. 9.4 Three inspection technologies compared: (a) PAM vs. t/T_c ratio; (b) PAM vs. m_1/m_3 ratio (Remski and Nof, 1993).

A second part of the analysis is to vary the Type I error probability m_1, and the bad–bad (when a truly bad part is designated as bad) designation probability m_3. For the former, the range is varied from 1% to 30%, and the latter ranges from 71% to 99%. Figure 9.4(b) shows that under the assumptions of this analysis the NCLPG displays an advantage over the other two technologies. The machine-vision system is also advantageous up to a ratio of about 0.15.

The purpose of assessment models is to help in determining which on-line inspection technology should be implemented for a particular assembly system. The two models presented, PMM and PAM, take into account not only the economics of each assessed technology, they also consider manual inspectors or inspection stations that can be improved, their specific technological abilities and error detection probabilities. Such models, implemented in a decision-support system, can speed up the design and implementation of on-line inspection systems in assembly.

9.4 Rework and repair in assembly

Parts and assemblies found to be faulty by test or inspection can be either discarded or brought to a satisfactory state. The latter is often called **rework**. Rework is defined as any repeated or additional work needed to improve the quality of a product to an acceptable level. Rework includes repairing faulty components or a faulty assembly. It may also include adding a missing component or rearranging the assembly. Theoretically, these activities may be considered non-value-added, but without them products are usually rejected. In this section, the functions of rework and repair are described, and models are presented for the design of effective assembly systems that include rework.

9.4.1 Rework and repair functions

Three observations can be made about rework.

- Ideally, there should be no need for rework if components and products are produced with zero defects. In practice, test and rework are needed in order to weed out inevitable defects.
- Practically, rework is an economic issue. The true cost and value of a reworked product should be compared with the true scrap or recycling value of the defective item (whether it is a finished or partial product). 'True' means an overall economic consideration; for instance, a costly repair can be justified if it is the only way to complete an order for timely shipping. Clearly, as rework cost and time are lower, it becomes more desirable.
- Rework cost is usually lower the earlier it is performed along the production and assembly process.

Three key questions to consider in an effort to rationalize rework are:

1. How can rework be designed so that it is most cost-effective?
2. What is the trade-off between more expensive components which have less flaws versus more rework during assembly?
3. What is the trade-off between rework versus scrapping of parts and subassemblies?

Questions 2 and 3 have to be answered for various different stages throughout the assembly operation. All three questions have been studied by spreadsheet and simulation analyses (Dooley, 1983; Wilhelm, Nof and Seidmann, 1986; Fu, Gonzalez and Lee, 1987). The RPFF model described next offers a simple tool to rationalize these decisions without extensive computer simulations. (A summary of the RPFF model was introduced in section 6.4.2.)

9.4.2   The rework product feedback flow model (RPFF)

RPFF analysis focuses on products in any assembly station with integrated rework. The model assumes that in order to satisfy required quality standards a certain proportion of the items must iterate (recycle), once or more, through repair and rework tasks. Probability theory is employed (Seidmann and Nof, 1985, 1989) to formulate the product flow. General equations provide estimates of total assembly time, number of expected rework cycles and sensitivity to the process and component quality, i.e. the probability of test failure.

The advantages of this model are that estimates can be derived directly from equations without any detailed computer simulation, and no specific distributions must be assumed for assembly or rework processes. The model provides estimates to answer the three questions listed in section 9.4.1. Although the RPFF model is based on simplifying assumptions, it addresses practical issues such as variable test failure probabilities and reasonable but limited number of allowable rework cycles. The estimates are also useful for evaluating capacity, in-process buffer space and dynamic control policies. Later in this section, application to flexible assembly design includes interacting effects of robot speed, rework rate, station size and attainment of production schedules.

Assumptions and modeling examples   The RPFF model assumptions are as follows (see Fig. 9.5).

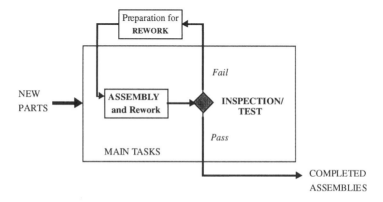

Fig. 9.5   General structure of a station with assembly, inspection/test and rework.

1. Assembly, inspection and test tasks are all termed **main tasks**.
2. After each assembly cycle a product is tested. The probability of passing the test is q $(0 \leq q \leq 1)$, whereas the probability that rework is required is p, $p = 1 - q$.
3. When an assembled product has to recycle for rework, there are certain preparation tasks such as straightening faulty components, obtaining replacement components or partial disassembly, followed by a repeat of certain assembly tasks in the same workstation. Assuming that rework is performed immediately in the same station is a simplifying yet reasonable assumption, especially for robotic and automatic assembly. Note that **preparation for rework** is not the rework itself, which is considered part of the main tasks.
4. For simplicity, only one work order (for either one or a batch of assemblies) is assumed to be in process at the workstation at a time. This 'unitary assumption' means that station resources are set up and devoted to complete an acceptable product order before it is allowed to leave the station and move on.

A typical cell for assembly and rework is depicted in Fig. 9.6.

The RPFF model is presented in Seidmann and Nof (1985) and Seidmann, Schweitzer and Nof (1985, 1987) and comparisons of model results to simulation analysis are described in Wilhelm, Nof and Seidmann (1986). RPFF equations are shown in Table 9.3, which is divided into cases where rework probability is constant and unlimited rework iterations are permitted (top), and cases where rework probabilities are variable and only a finite number of iterations are permitted (bottom). Because of the simplifying unitary assumption, the equations are correct for any bounded time distribution of assembly and rework operations; only mean and variance time estimates are needed.

Example 1 An assembly robot has to insert a component which is picked up from one of three feeders. The robot can continue operating as long as at least one feeder is functioning. If insertion fails, the component is discarded, a new one is picked up from a feeder. Insertion would then be attempted again. Travel time to different feeders is different. Insertion failures may occur if either a component or a workpiece are faulty or misaligned. All these uncertainties result in variable task time. In this case, 'main tasks' are pick-up and insert, and 'rework preparation' includes discarding old components and travel to a feeder to pick up another component. A 'test' is successful if insertion is completed, as can be sensed by the robot controller. There is a certain probability of insertion failure which implies the need for rework.

Example 2 Electronic components are inserted or mounted on a board. Then the board is picked up by a manipulator, moved to a vision-based tester, then to a functional tester. All these tasks are considered as main tasks by the model. The total failure probability in any test is the rework probability. The

(a)

1. Place rear endbell into fixture
2. Set first bearing into endbell
3. Set rotor into bearing-endbell
4. Set stator around armature
5. Set second bearing on top of rotor shaft
6. Set front endbell over bearing-rotor-stator
7. Insert first screw
8. Insert second screw
9. Drive both screws
10. Insert first brush holder
11. Insert second brush holder
12. Press both brush holders
13. Test motor for function and balancing, identify rework needs, and execute/rework the necessary steps
14. Repeat step #13 up to three times
15. Off-load completed motor

(b)

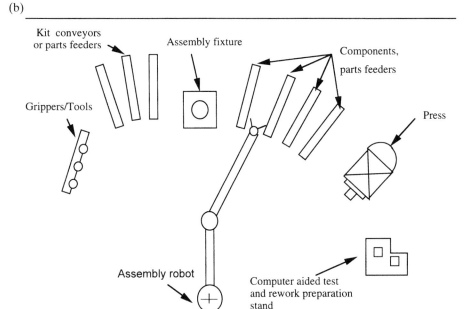

Fig. 9.6 An integrated robotic assembly-and-test station with rework capabilities: (a) simplified sequence of steps to assemble electric motors by a robotic workstation; (b) the structure of the assembly workstation.

interpretation of diagnosed faults, either by computer or operator, and the removal of faulty components are considered rework preparation tasks. In the next iteration, the manipulator picks up and moves the board, now without any faulty components, back to the main tasks.

Suppose the data shown in Table 9.4 are relevant for both examples. RPFF equations were used to calculate the estimates shown in the table. For simplicity, assume an assembly order or batch size of one. Based on calculated estimates and given cost parameters, one can decide which assembly strategy would be preferred.

Measure	Mean	Variance
For constant p, unlimited reworks		
N = Number of cycles to obtain a good assembly	Q/q	Qp/q^2
T_A = Total time in assembly and rework	$QE[A]/q$	$Q(pE^2[A]/q + V[A])/q$
T_T = Total time in preparation for rework	$QE[T]p/q$	$Qp(E^2[T]/q + V[T])/q$
θ = Total time to complete Q accepted assemblies	$Q(E[A]+pE[T])/q$	$Qp(E[A]+E[T])^2/q^2 + Q(V[A]+pV[T])/q$
Y = Yield (good assemblies per time unit)		$Q/\theta = q/(E[A]+pE[T])$
Covariance of θ and N		$\mathrm{COV}(\theta,N) = Qp(E[A]+E[T])/q^2$
Coefficient of correlation between θ and N		$\rho(\theta,N) = \mathrm{COV}(\theta,N)/(V[\theta]V[N])^{1/2}$
For variable p, up to J cycles (assume $Q = 1$ for brevity)		
N = Number of cycles to obtain a good assembly	$1 + \displaystyle\sum_{i=1}^{J-1}\left(\prod_{j=1}^{i} p_j\right)$	$\displaystyle\sum_{k=1}^{J} k^2 g(k) - E^2[N]$
θ = Total time to complete one accepted assembly	$E[A]E[N]+E[T](E[N]-1)$	$(E[N]-1)(V[A]+V[T])+(E[A]+E[T])^2 V[N]+V[A]$
Y = Yield (good assemblies) per time unit	$Y(J,p)=1-\left(\displaystyle\prod_{i=1}^{J} p_i\right)$	$Y(J,p)=1-\left(\displaystyle\prod_{i=1}^{J} p_i\right)\Big/E[\theta]$

Variables: p = rework probability $(q = 1-p)$; $E[A]$, $V[A]$ = mean, variance of single passage time through assembly or rework: $E[T]$, $V[T]$ = mean, variance of single passage time in preparation for rework; Q = lot size.
For calculating N when p is variable, $g(n) = (1 - p_1)$ for $n = 1$; $g(n) = p_1 p_2 \ldots p_{n-1}(1 - p_n)$ for $2 \leq n \leq J - 1$; $g(J) = p_1 p_2 \ldots p_{J-1}$, where J is the limit on the total number of cycles.

Table 9.4 Parameters and estimates for RPFF analysis of examples 1 and 2

Case	Parameters					Performance estimates								
	$E[A]$ (min)	$V[A]$ (min²)	$E[T]$ (min)	$V[T]$ (min²)	p	$E[N]$	$V[N]$	$E[T_A]$ (min)	$V[T_A]$ (min²)	$E[T_T]$ (min)	$V[T_T]$ (min²)	$E[\theta]$ (min)	$V[\theta]$ (min²)	Y (1/min)
1	0.20	0.04	0.20	0.04	0.20	1.25	0.31	0.25	0.06	0.05	0.02	0.30	0.11	3.33
2	0.80	0.64	0.80	0.64	0.20	1.25	0.31	1.00	1.00	0.20	0.36	1.20	1.76	0.83
3	1.00	0.20	1.00	0.20	0.20	1.25	0.31	1.25	0.56	0.25	0.36	1.50	1.55	0.67
4	1.00	0.80	1.00	0.80	0.20	1.25	0.31	1.25	1.31	0.25	0.51	1.50	2.45	0.67
5	0.60	0.12	0.60	0.12	0.20	1.25	0.31	0.75	0.26	0.15	0.14	0.90	0.63	1.11
6	2.40	1.92	2.40	1.92	0.05	1.05	0.06	3.00	4.20	0.60	2.28	3.60	10.08	0.28
7	0.20	0.04	0.20	0.04	0.05	1.05	0.06	0.21	0.04	0.01	0.00	0.22	0.05	4.55
8	0.80	0.64	0.80	0.64	0.05	1.05	0.06	0.84	0.71	0.04	0.07	0.88	0.85	1.14
9	1.00	0.20	1.00	0.20	0.05	1.05	0.06	1.05	0.27	0.05	0.07	1.11	0.44	0.90
10	2.40	1.92	2.40	1.92	0.05	1.05	0.06	2.53	2.34	0.13	0.42	2.65	3.40	0.38

Certain observations from Table 9.4 are as follows. The first six cases assume a rework probability of 0.20, representing lower quality of incoming components and less precise assembly tasks. The next four cases assume a lower, 0.05 probability of rework, representing higher quality, more reliable and expensive components. Obviously, with less expected rework in cases 7–10, the expected throughput time $E(\theta)$ for acceptable assemblies is significantly shorter, e.g. 0.22 vs 0.30; 2.65 vs 3.60 min, about −26%. However, the calculated estimates can now be used for evaluating the economic trade-offs discussed above. For instance, cases with a relatively higher p will be preferred as long as the value of 26% time savings is **not** sufficient to justify investment in better quality components and processes.

A failure probability depends on the dimensional or functional tolerance designed for parts and components. It is often determined by the type of industry and customer expectations. Traditionally, a 'three sigma' distribution (meaning a tolerance of three standard deviations above and below the mean dimensional or functional value) could be acceptable. Assuming a normal distribution of parts' attributes, the total of six sigma range (2×3) implies that 0.3%, or 3000 ppm (parts per million), of the tested parts fail. In certain industries, particularly with electronic components and assemblies, companies such as IBM and Motorola have aimed at a 'six sigma' quality. The total of twelve sigma range (2×6) means that 99.9999998% of the parts are of acceptable quality, and only 2 ppb (parts per billion) may fail.

Limited number of reworks with unequal probabilities In practice, rework iterations are always limited and the rework probability is not constant. To illustrate the equations shown in the lower part of Table 9.3, suppose the number of iterations is limited to $J = 3$ and the rework probabilities are p_1, p_2, p_3, respectively, for the first, second and third iterations. In this case, $g(n)$ is as follows: $g(1) = 1 - p_1$; $g(2) = p_1(1 - p_2)$; $g(3) = p_1 p_2$; and:

$$Y(3, \, p) = \left(1 - p_1 p_2 p_3\right) \Big/ E[\theta] \tag{9.4}$$

Note that the expected 'yield' of rejected assemblies in this case is:

$$X(3, \, p) = \left(\prod_{i=1}^{J} p_j\right) \Big/ E[\theta] = \left(p_1 p_2 p_3\right) \Big/ E[\theta] \tag{9.5}$$

In studies (Seidmann and Nof, 1989) it was found that the simpler models assuming unlimited rework can serve as good approximations when there are relatively low values of p and $E[T]$.

The rework probabilities need not be identical for all the rework attempts. These probabilities can be decreasing from cycle to cycle as the assembled product improves, or increasing from cycle to cycle as a result of deteriorating performance.

9.4.3 Assembly station performance evaluation

Evaluating the performance of an assembly station by RPFF can be under-taken at three levels: task, product and system. At **task level**, the mean and variance of task time indicate the required total process time and the range of total assembly operation times. At task level, the probability of successful assembly completion indicates the efficiency of a station at performing given tasks. At **product level**, the product throughput time indicates the production capacity requirement for given products. The **yield**, a ratio of good product output to product input, indicates a station's efficiency at the product level. At **assembly system level**, measures of utilization and cost can be used to evaluate plans of automatic assembly, layouts and operating strategies for assembly system design.

If rework is **unlimited** then all output is considered good quality, and **operational efficiency**, η, can be defined as follows:

$$\eta = E'[A]/E[\theta] = E'[A]Y \qquad (9.6)$$

This is the ratio of the mean **ideal** cycle time $E'[A]$ when no rework is needed (i.e. $q = 1$) to the **actual** mean product assembly time including rework $E[\theta]$. The operational efficiency when rework is **limited** can be computed similarly:

$$\eta(J, p) = E'[A]Y(J, p). \qquad (9.7)$$

Cost considerations   The activity-based cost per assembled product (or order) is taken as the total of setup, tooling, material and operating cost. The **setup cost** C_S per assembly of a batch of Q identical products is a function of time spent on facility changeover, feeders and fixtures reconfiguration and tester changes required during part-type changeover. Therefore:

$$\text{Setup cost per assembled product} = C_S/Q.$$

The **tooling cost** for each cycle is C_t. Tooling costs include replacement of worn or broken tools and fixtures, disposable kitting trays and time required for ongoing adjustments of equipment. Csakvary (1985) indicates the major cost factors in robotic assembly are part jamming and related tool or feeder breakdowns. Therefore, tooling cost per completed unit depends on the number of repeated trials. On average, tooling cost per assembled product = $C_t \cdot E[N]$. The **material cost** also depends on the number of iterations, since during each cycle some parts and materials are replaced at cost C_m. Thus, on average, material cost per assembled product = $C_m \cdot E[N]$. Finally, given **operating cost** per time unit C_o, the total operating cost is relative to the amount of time each assembly occupies the station. On average, operating cost per assembled product = $C_o \cdot E[\theta]$.

The **total average cost** per assembled product C is the sum of the above four cost components, as follows:

$$C = \left(C_S/Q + C_t \cdot E[N] + C_m \cdot E[N] + C_o \cdot E[\theta]\right). \tag{9.8}$$

For illustration, case 2 in Table 9.4 is compared to case 8. Suppose $Q = 12$, $C_S = \$60$, $C_o = \$6$; for case 2 lower grade components and tooling are applied so $C_t = \$4$ and $C_m = \$3$, while in case 8, $C_t = \$5$ and $C_m = \$4$. Applying equation 9.8:

$$C(\text{case 2}) = [60/12 + 4 \times 1.25 + 3 \times 1.25 + 6 \times 1.20] = \$20.95 \text{ per assembly}$$
$$C(\text{case 8}) = [60/12 + 5 \times 1.05 + 4 \times 1.05 + 6 \times 0.88] = \$19.73 \text{ per assembly}$$

The slight advantage of case 8, about 6% lower cost per assembly despite the higher material and tooling costs, is influenced by the 27% shorter throughput time. However, yet another alternative must be considered, a strategy of no rework. It will be preferred if the scrap value per assembly is at least \$19.73. (Actually, the comparison should be made with the least costly, feasible configuration from Table 9.4.)

In the above discussion only variable costs are considered, and it is assumed that revenue for rejected items and faulty parts is negligible. This is not always the case. It is necessary to consider adding specific variables and modify some of the above assumptions so that these models are made more relevant for a particular firm. For example, the model can be expanded by adding costs of special mechanical feeding devices and part magazines, or inventory lot sizing considerations.

Due date delivery The probability of meeting a prescribed due date T_D for a given batch of size Q is an important measure at the system level. Suppose the probability distribution of total batch time θ is approximately normal with mean $E[\theta]$ and variance $V[\theta]$ computed from Table 9.3. The probability that a batch requires less than T_D time units is:

$$\Pr[\theta < T_D] = \Pr\left(Z < \left(T_D - E[\theta]\right)/V^{1/2}[\theta]\right) \tag{9.9}$$

where Z is the standard normal variable. It is also possible to compute an approximate confidence interval for the batch completion time. Since θ is the actual batch completion time, then:

$$\Pr\left(E[\theta] - Z_{\alpha/2}V^{1/2}[\theta] \leq \theta \leq E[\theta] + Z_{\alpha/2}V^{1/2}[\theta]\right) = 1 - \alpha \tag{9.10}$$

where $Z_{\alpha/2}$ denotes a percentage point of the standard normal distribution such that $\Pr[Z > Z_{\alpha/2}] = \alpha/2$. Thus the approximate $100(1 - \alpha)$ percent confidence interval is:

$$\left(E[\theta] - Z_{\alpha/2} V^{1/2}[\theta] \right) \le \theta \le \left(E[\theta] + Z_{\alpha/2} V^{1/2}[\theta] \right). \tag{9.11}$$

For example, suppose a batch with $Q = 41$, $E[\theta] = 242\,\text{h}$, $V[\theta] = 119\,\text{h}^2$ and $T_D = 256\,\text{h}$. The probability of meeting this due date is:

$$\Pr[\theta < 256] = \Pr\left[Z < \frac{256 - 242}{\sqrt{(119)}} \right] = 0.899.$$

The confidence interval on the batch completion time is $\{242 - (1.96)\sqrt{(119)} \le \theta \le 242 + (1.96)\sqrt{(119)}\}$, or $\{220.6 \le \theta \le 263.4\}$.

9.4.4 Design of a flexible assembly robotic station with inspection and rework

To illustrate application of RPFF in station design, two examples are considered in this section: a printed circuit board assembly station, and a combined machining and assembly station. Two design issues are discussed: selection of robot speed, and station size in terms of the best number of tasks planned for a station.

Assembly station A printed circuit board assembly station is common in industry (Warnecke and Schraft, 1982; Riley, 1983) and operates as follows (similar to Fig. 9.6). A robot or workhead picks up components out of a feeder, orients each in turn and inserts it. Suppose the number of main tasks (e.g. moving to a feeder, picking up components, moving and orienting them, and finally inserting them) is L and total robot travel time during assembly is τ, assumed deterministic. For simplicity of discussion, assume $Q = 1$ throughout the following section. Several practical problems may occur, including feeder jamming, wrong components, etc. Such problems are usually cleared during initial station start-up (see also discussion on error recovery in section 9.5) or strict quality control of incoming components.

A typical problem which is difficult to eliminate is the alignment of component pins that have to be positioned precisely. The robot control program, when realizing that an insertion attempt has failed, will automatically activate a recovery procedure, bring the component to a realignment device and have the component realigned. Suppose a recovery procedure requires a total time β (assumed exponentially distributed), where β is the expected value of rework or recovery time. Assume also constant rework probability p, and suppose the operation time for each of the L assembly tasks in the station has, for simplicity, a log-normal distribution with mean $E[A_i]$ and variance $V[A_i]$.

Assume that $E[A_i]$ for each task i can be related to the recovery time β, as follows:

$$E[A_i] = j\beta$$

where $j > 0$, $i = 1, \ldots, L$, β is the expected time for rework preparation functions and j is a known constant. It is reasonable to assume that τ, the total travel time between main tasks, is relative to L, and can be expressed as $\tau = zL\beta$, where $z > 0$ and is a known technical factor relating τ to the rework preparation time. Thus for the L tasks in the workstation $E[A] = Lj\beta + zL\beta$ and the equation for $E[\theta]$ in Table 9.3 can now be rewritten, after substitutions:

$$E\left[\theta\right] = q^{-1}\beta\left(zL + Lj + p\right) \tag{9.12}$$

and the yield:

$$Y = q\left[\beta\left(zL + Lj + p\right)\right]^{-1}. \tag{9.13}$$

Hence the operational efficiency can be calculated:

$$\eta = \frac{\beta L\left(z + j\right)q}{\beta L\left(z + j + p/L\right)} = q\left[1 + p/\left(L\left(z + j\right)\right)\right]^{-1}. \tag{9.14}$$

Robot speed and rework rate Fast arm acceleration is important in assembly because paths are relatively short, but how fast should the motion speed be? Relatively quick assembly operations may result in more need for rework and recovery. The relationship between speed and rework will now be explored.

System yield for alternative designs A fast robot will require a relatively short τ. To compare two alternative designs, suppose they apply different robots but the same L, β, j and z. One can arbitrarily set β_1, in the first design, as $\beta_1 = 1\,\text{min}$. Note that with these assumptions, the value of β determines the assumed robot speed. With **unlimited** rework (p_1 and p_2 are the rework rates at the first and second station designs), for the two stations to have the same production yield Y, $Y(p_1) = Y(p_2)$, or from equation 9.13:

$$\left(1 - p_1\right)\left[\beta_1\left(zL + Lj + p_1\right)\right]^{-1} = \left(1 - p_2\right)\left[\beta_2\left(zL + Lj + p_2\right)\right]^{-1}. \tag{9.15}$$

From here the mean recovery time in the second design (for equal yield) should then be:

$$\beta_2 = \left[1\left(zL + Lj + p_1\right)\cdot\left(1 - p_2\right)\right]/\left[\left(zL + Lj + p_2\right)\left(1 - p_1\right)\right]. \tag{9.16}$$

For example, for $L = 3$, $j = 5$, $z = 0.92$ and $p_1 = 0.3$, then β_2 is given by $\beta_2 = 25.8(1 - p_2)/(17.76 + p_2)$. The first design has $Y(p_1 = 0.3) = (0.7)\,[0.92(3) + 3(5) + 0.3]^{-1} = 0.039$ assemblies per time unit. Thus, in the same cell size ($L = 3$) and operation ($j = 5$, $z = 0.92$) a different robot speed is associated with a different value of p. Since p_2 is positive, it can be shown in this case that $\beta_2 < 1.45\,\text{min}$ is required in the second design to result in equal yield. In other words, in order to obtain similar yield (approximately 0.04) with a slower robot a smaller rework rate p_2 must be allowed. For instance, with a 40% slower robot, i.e. $\beta_2 = 1.40\,\text{min}$, a rework rate of only $p_2 = 3.4\%$ can be tolerated. Achieving such a low rate of rework will require a major effort to reduce all factors of variability, and may be unjustifiable compared to the savings generated by installing a slower robot.

This analysis implies that when robot speed is decreased by 40% (in this case) the rework probability must be changed by one order of magnitude. It is also interesting to note that the operational efficiency of the two designs above does not directly depend on the value of β. $E_1 = 0.69$ for the station with the faster robot and higher rework probability, and $E_2 = 0.96$ for the second design. Although the two designs result in the same yield, the station with the smaller rework rate and slower robot will have to operate at significantly higher operational efficiency.

Up to this point, station designs are considered for providing equivalent yield of good assemblies but applying different robots. Next, designs are compared for their task and cost measures.

Task measures for alternative designs   At **task level**, the two designs can be compared as follows (again, assume $Q = 1$).

- **Design 1:** ($\beta = 1.0\,\text{min}$, $p_1 = 0.3$), $E[\theta] = [(0.92)(3) + (3)(5) + 0.3]/0.7 = 25.800\,\text{min}$ and since $V[A] = LV[A_i] = 3V[A_i]$ and $V[T] = \beta^2$ then $V[\theta] = 215.900 + 4.285V[A_i]\,\text{min}^2$.
- **Design 2:** ($\beta = 1.4\,\text{min}$, $p_2 = 0.034$), $E[\theta] = [((0.92)(3) + (3)(5))1.4 + (0.034)1.4]/0.966 = 25.788\,\text{min}$ and $V[\theta] = 25.201 + 3.105V[A_i]\,\text{min}^2$.

Note that the variance $V[\theta]$ in the second design is relatively smaller because of the smaller rework rate. This attribute is desirable in facilities where production flow must be tightly controlled.

Operating costs for alternative designs   Operating unit cost C_o and material cost C_m can be assumed equal for both designs, but the tooling cost C_t can be higher for lower p, which corresponds with tighter tolerances. Suppose $C_t(\text{Design 2}) = K \cdot C_t(\text{Design 1})$ where $K > 1$. Then, following equation (9.8): $C_1 = 25.800C_o + 1.429(C_m + C_t)$ for Design 1, and $C_2 = 25.788C_o + 1.035(C_m + KC_t)$ for Design 2. For illustration: $C_m = D$, $C_o = 5D$, $C_t = 10D$ and $D > 0$, where D is an illustration constant. Comparing unit costs, the second design will be less costly when $C_2 < C_1$ or when $K < 1.425$. It means that in this case a 42.5% increase can be allowed for tooling costs in the second plan with smaller rework probabilities.

Limited rework   A more realistic situation of limited rework is considered next. Suppose the number of rework attempts is limited to $J = 3$, and rework probabilities reduce from cycle to cycle, with $p_1 = [0.3, 0.15, 0.075]$ and $p_2 = [0.034, 0.017, 0.0085]$ for Designs 1 and 2, respectively. Assembly yields will remain the same, as long as $Y_1(3, p_1) = Y_2(3, p_2)$. From Table 9.3 for Design 1, $E[N] = 1.345$; $E[\theta] = 24.232 \min$; $Y_1(3, p_1) = (1 - (0.3)(0.15)(0.075))/24.232 = 0.0411 \, l/min$. Similarly, for Design 2 (assuming general $\beta_2 > 0$), $E[N] = 1.033$; $E[\theta] = 18.379\beta_2$. The following equation defines the β_2 for Design 2 to result in similar yield to Design 1: $Y_2(3, p_2) = (1 - (0.034)(0.017) \cdot (0.0085))/(18.379\beta_2) = 0.0411 \, l/min$. From here, $\beta_2 = 1.36 \min$.

It is interesting to note that the performance results are close to those obtained above for unlimited rework with constant rework probabilities. (This observation implies that the hypothetical unlimited rework model can provide useful robust results that are also easier to calculate.) In terms of operational efficiency, the results are slightly higher than in the unlimited rework case: $\eta_1 = E[T_{A1}]Y(3, p_1) = (5 \cdot 3 + 0.92 \cdot 3)(1)(0.0411) = 0.729$, and $\eta_2 = E[T_{A2}]Y(3, p_2) = (5 \cdot 3 + 0.92 \cdot 3)(1.36)(0.0411) = 0.992$.

Station size and rework rate   Consider a station where components are machined and assembled. A design issue is the **number** of tasks to include in the station, hence its size, as a function of rework rates. It can be observed from the earlier analysis that in general, assembly yield increases when: (1) p decreases; (2) robot speed, represented by $\tau = zL\beta$, increases; (3) the expected time spent on each assembly task, represented by $E[A_i] = j\beta$, is relatively smaller. On the other hand, different station designs can yield similar yields. Consider the following alternatives, all having $z = 1$:

- **Design A:** For yield of approximately 3.3 assemblies per hour, a station can include three long main tasks ($L = 3$), with $p = 0.1, j = 10, \tau = 1.5 \min$ and $\beta = 0.5 \min$.
- **Design B:** Alternatively, a station with the same L but faster tasks, $j = 2$, a slower robot, $\tau = 3 \min$, rework preparation time of $\beta = 1 \min$ and rework rate of $p = 0.5$ can yield a comparable 3.2 assemblies per hour.
- **Design C:** A larger cell with 10 short tasks, $j = 0.65$, a very slow robot, $\tau = 10 \min$, $\beta = 1 \min$ but higher production quality ($p = 0.1$) will also result in an average yield of 3.3 assemblies per hour.

Several alternative designs, all technologically feasible, can be compared by economic analysis as discussed above. Suppose for simplicity identical α, β, C_S, C_m, C_t and C_o. After analysis, it can be shown that Design 1 will be more economical than Design 2 when:

$$\left(C_m + C_t\right)\left(1/q_1 - 1/q_2\right) < C_o\beta\left(\left(L_2\left(z_2 + j_2\right) + p_2\right)/q_2 - \left(L_1\left(z_1 + j_1\right) + p_1\right)/q_1\right)$$

(9.17)

where $q_1 = 1 - p_1$ and $q_2 = 1 - p_2$. Consider the case where a station is comprised of $L = 3$ tasks and has $\beta = 1 \min$. Design 1 calls for relatively long tasks, with $j = 2$ but relatively low $p = 0.1$; Design 2 calls for fast tasks, $j = 0.5$ and higher

$p = 0.5$. In both designs $z = 1$ and the yield is about 6 assemblies per hour. According to equation (9.17), Design 1 is preferred to Design 2 only if $0.11C_o < 0.89(C_m + C_t)$. Sometimes a design is preferred for its lower variability; however, the variance of total unit stay in the station is almost identical here: 102.46 $(min)^2$ in Design 1 and 102.00 $(min)^2$ in Design 2.

In summary, the RPFF model can help designers in quantitative evaluation of alternative designs, considering relations between assembly quality and operations. By applying RPFF equations, one can obtain comparative design estimates promptly without the need for detailed simulation efforts.

9.5 Error diagnosis and recovery in assembly

Quality assurance includes the prevention of quality errors. Five general approaches to error prevention are (Hancock, Sathe and Edosomwan, 1992):

1. **attitude:** orientation and training to correct poor attitudes that cause errors;
2. **control:** process control by self-verification and monitoring;
3. **rewards:** recognition and rewards to those who creatively eliminate error sources;
4. **verification:** verification of output by inspection, review and audit methods;
5. **team review:** correct operating procedures, methods and group meetings to review quality errors.

Clearly, prevention of errors **during design** and **at the source** is the goal. In computerized and robotic systems, errors due to machine, part or procedure failure can be detected during assembly. Once an error is detected, actions must be taken immediately for correction or recovery based on diagnostic information. A significant benefit is that with robotic recovery abilities, less operator intervention may be needed. This subject is termed error detection (or diagnosis) and recovery (EDR). Effectiveness of error recovery has been studied by Nof, Maimon and Wilhelm (1987), Momot and Shoureshi (1990) and others. The objective of this section is to explain how EDR methods work, and how to design them for effective assembly.

9.5.1 Objectives of error diagnosis and recovery

A recent trend in assembly is to develop programs, including EDR, according to the **principle of increased intelligence with decreased accuracy**. Artificial intelligence techniques can compensate for uncertainties and errors that are considered too costly to eliminate. For instance, Hutchinson and Kak (1990) and Borchelt and Alptekin (1994) describe various reasoning and knowledge-based systems for EDR in assembly. Computer programming and software aspects of error recovery have also been studied by Williams, Rogers and

Upton (1986) and DiCesare *et al.* (1993). It was realized that a framework is needed for:

1. analysis of recovery requirements;
2. selection of recovery strategies and solutions;
3. development of effective computer logic of recovery routines.

Studies of planning error recovery programs in robotic work (Nof, Maimon and Wilhelm, 1987) address the above issues. A five-phase framework for planning effective error recovery programs is described through laboratory experiments with mechanical assembly of carburetors and electronic assembly of printed circuit boards. The new concept of a **recovery tree** has been suggested as a means of analyzing alternative recovery strategies. The logic of a selected recovery tree can be the basis for effectiveness analysis and program implementation.

Some error types can be eliminated by improved design or changes in the facility. Other error types may require recovery programs. Typical errors encountered are classified as follows: $Ej(t)$ = all errors of type j encountered during assembly at time t are divided into $DEj(t)$ = design errors; $LEj(t)$ = logistic errors; $SEj(t)$ = spatial errors. The three error classes are illustrated in Table 9.5.

Table 9.5 Some errors encountered in robotic work

Error event	Typical causes	Class	Possible remedy	Note
Part missing	Part has not arrived	$LE_j(t)$	• Increase local stock of parts • Use different part • Call operator	(1) (2)
Part slipping from gripper	Part too small	$DE_j(t)$	• Change part dimensions • Change gripper • Try another part • Pick up part and tighten gripper • Call operator	(1) (1) (2) (2)
	Gripper not tight	$DE_j(t)$	• Repair gripper • Correct control program	(1) (1)
		$SE_j(t)$	• Adapt gripping force • Call operator	(2)
Insertion failure	Bad component position	$SE_j(t)$	• Correct the control program • Try another position/orientation (wiggle, or search)	(1) (2)
	Deformed component	$DE_j(t)$	• Reshape, straighten component • Try another component • Call operator	(2) (2)

Notes: (1) requires modifications in design; (2) may necessitate recovery routines.

Error recovery is defined as 'an ability in intelligent systems to detect errors and, through programming, take corrective action to resolve the problem and complete a desired process' (Nof, 1985). Obviously, recovery may not always be economic, and not all potential errors can be predefined (Maimon and Nof, 1986). Logistic errors $LEj(t)$ stem from logistic problems, e.g. part shortages or delayed parts. Design errors $DEj(t)$ are associated with component design, e.g. design of parts, tools, grippers, etc. Spatial errors $SEj(t)$ are errors of position and orientation, e.g. of a robot, or in the relative position of components. Examples of errors in pin insertion are illustrated in Fig. 9.7.

9.5.2   Planning error recovery programs

The planning framework includes five phases and the following terms are defined: RNj = recovery needs for error type j; RTj = recovery tree representing recovery strategy j; $PER(RTj)$ = performance/effectiveness measure of RTj; $PROG(RTj)$ = a solution program implementing the selected RTj. Planning phases are as follows.

1.   Preliminary task analysis, resulting in identification of recovery needs (RNj).
2.   Development of alternative recovery strategies, expressed as recovery trees (RTj).
3.   Selection of the best recovery strategy ($RTj*$).
4.   Programming recovery routines.
5.   Update of recovery programs (ongoing) to revise recovery capabilities.

The set $Ej(t)$ of potential errors divides into two exclusive subsets: errors caused because of faulty design during cell development which must be corrected prior to regular cell operation, and errors that will continue in normal

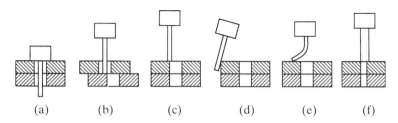

(a)   (b)   (c)   (d)   (e)   (f)

Fig. 9.7   Possible states in a pin insertion task: (a) successful insertion; (b) spatial error ($SEj(t)$) – bad relative position of components; (c) and (d) spatial errors – wrong pin position and orientation caused either by robot or by bad relative positions; could also be caused by control program error; (e) design error ($DEj(t)$) – deformed component; (f) design error ($DEj(t)$) – component too big or hole too small (Nof, Maimon and Wilhelm, 1987).

operation. For example, if a gripper is too large for small components, another gripper should be designed. On the other hand, if during operation a particular component is smaller than specified, then it should be discarded and another one fetched. Often, sensory data must be used because it is impractical to predefine positions to account for all part variations. Furthermore, because a flexible system is designed to handle a variety of different tasks, dedicated fixtures for precise positioning are not economic. In summary, the first phase in recovery planning is to determine the errors that are of the developmental type, and those that belong to RNj and require ongoing recovery.

In the next phase, actions to recover from the most frequent errors can be expressed in a recovery tree (RTj) as potential errors and corresponding steps for recovery (Fig. 9.8). For each error, several recovery levels can be programmed. Performance and economic evaluation of alternative RTjs, each representing an alternative strategy, indicate which strategy is preferred.

A recovery strategy is selected (phase 3) by considering cell design and PER(RTj). PER(RTj) indicates the desirability of each alternative strategy, e.g. cost or time required to achieve an acceptable recovery level. The selected RTj can now serve as a basis for implementing real-time, sensor-based recovery programs. With new needs and accumulated experience, a recovery method can be reviewed periodically and revised. This phase is ongoing, since design changes occur from time to time, changing RNj.

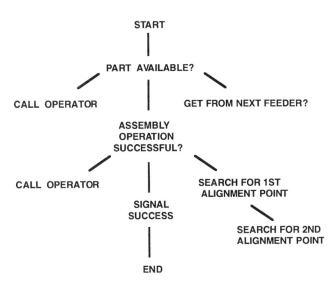

Fig. 9.8 A recovery tree for component insertion. Note: a branch may only be entered once; on success branch downward; on failure branch to right if possible, otherwise branch left; when the end of a branch is reached, unless otherwise specified return to last sensing position; ? signifies sensing position where sensors or variables are evaluated.

9.5.3 Error recovery experiments

A robot was used for the assembly of an automobile carburetor and of electronic components to a printed circuit board. The robot gripper had touch sensors and a light (LED) sensor, which could be monitored during motion. Experiments followed the first four phases of the framework specified above.

Mechanical assembly experiments The carburetor assembly included insertion of a float, a top cover and an intake linkage into a housing. The orientation of the float and linkage in the housing was found to be highly variable; therefore, they would have to be redesigned or precise fixtures must be considered. Activities include positioning a top cover and inserting four bolts into a carburetor housing (Fig. 9.9(a)). Each housing is brought into the cell positioned on a standard tray. Bolts and tops are fed by gravity metal tracks. Thus,

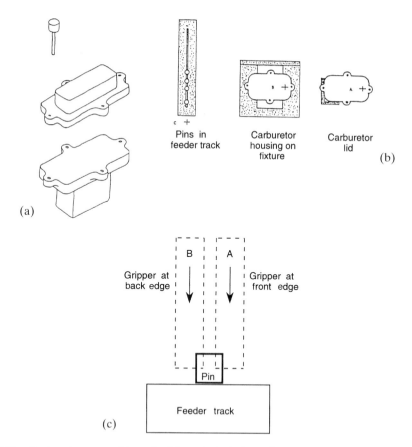

(a)

Pins in
feeder track

Carburetor
housing on
fixture

Carburetor
lid

(b)

Gripper at
back edge

B | A

Gripper at
front edge

Pin

Feeder track

(c)

Fig. 9.9 Carburetor assembly task: (a) exploded view of assembly components; (b) workspace and component feeding; (c) LED sensing to locate bolt position.

a consistent initial state is established for the task, preventing most of the subsequent location and orientation uncertainties (Fig. 9.9(b)).

Two potential error types are a missing component or a bolt insertion failure. A component could be missing because of delayed arrival or because it has slipped from the gripper during motion. An insertion could fail because a top cover is positioned incorrectly over a housing, or arm position and orientation are inaccurate. Since such situations cannot be practically eliminated, recovery needs must be established. Several recovery trees can be developed in which LED sensory data are used to guide the robot when bolts are picked up. In Fig. 9.9(c), the robot program stores the robot position (A) at the front edge of the bolt. The robot then moves until the LED is unobstructed at position (B) which defines the bolt's back edge. The bolt center position relative to the robot is then calculated and used as an approach target when the robot picks up the bolt. This type of procedure is necessary because the bolt dimensions and location are never precisely known. Table 9.6 displays for each assembly task the corresponding sensing procedures and error diagnoses.

A single-level recovery strategy for the top cover sub-task, and a two-level recovery strategy for the insertion sub-task are selected as most appropriate, as follows.

Top cover sub-task with recovery routine

- Step 1: Pick up top cover from feeder.
  (a) If not found, call operator.
- Step 2: Position top cover on housing.
  (a) If not successful, try once more.
  (b) If not successful, call operator.
- Step 3: Continue (insert bolts when ready).

Table 9.6 Sensing procedures and error diagnosis for the mechanical assembly task

Assembly sub-task	Sensing procedure	Error diagnosis
• Pick up top	Grip to specified force	Top present?
• Put top on housing	Motion stopped on force	Housing present?
• Pick up bolt	Find bolt position Grip to specified force	Bolt present?
• Insert bolt	Motion stopped on force	Top present? Proper insertion position? Bolt in hole?

Four-bolt insertion sub-task with recovery routine

- Step 1: Pick up bolt *i* from feeder (*i* = 1, ... *L*).
  (a) If bolt not found, call operator.
- Step 2: Insert bolt *i* in hole *i*.
  First level of recovery: search.
  (a) If insertion unsuccessful, execute search procedure.
  (b) If search procedure not successful, discard bolt, pick up a second bolt (same *i*) and proceed with Step 2.

  Second level of recovery: realign top cover.
- Step 3: Tap top cover to realign with housing.
  Proceed with Step 1 (same *i*).
- Step 4: If all four bolts are inserted, continue (wait for next carburetor assembly to start), otherwise proceed with Step 1.

The search procedure is a repeated poking attempt (to insert a bolt) with fine lateral increments for repositioning and reorienting the bolt. It can employ a variety of search algorithms, for instance a rectangular search map around the initial position.

Electronic assembly experiments   These experiments involve rheostat insertion into a printed circuit board (PCB). The cell consists of two feeders, one for rheostats and one for boards, and a fixture. The feeders and fixture are trays, each with two diagonal alignment pins. The assembly sequence begins with board positioning in the fixture, followed by inserting 10 rheostats in predefined board locations. The program includes a calibration routine relative to reference pins on the feeders and fixture.

In a preliminary task evaluation, a number of potential problems can be identified. When a PC board is picked up at the feeder: (a) a board may be misaligned, causing it to be misaligned in a gripper; (b) a feeder may be misaligned which results in a misaligned board; (c) a board may be missing. When a PC board is inserted into the assembly fixture: (a) a board in the gripper may be misaligned due to the above reasons, and in addition the board could have been wiggled when picked up; (b) a fixture may be misaligned.

Similarly, there are potential problems when a rheostat is picked up at a feeder, e.g. a rheostat may be misaligned, etc. When a rheostat is inserted into a board, the rheostat pins may be misoriented for insertion due to turnaround of the rheostat when picked up. Other problems may be caused by inaccuracies, for instance dimensional variations of alignment pins or variable orientation of fixture and feeders. After evaluation in terms of recovery needs, sequence of error occurrences and logic of recovery, recovery trees can be developed (Fig. 9.10).

The development of a recovery program involves a hierarchy of routines to handle the errors common to a task according to the recovery trees. A selected

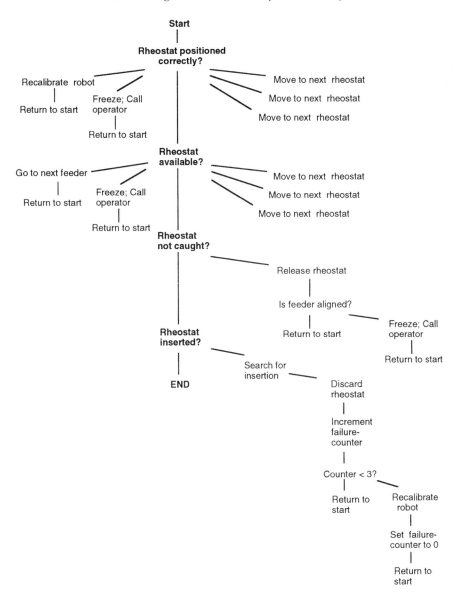

Fig. 9.10 Recovery tree for rheostat pick-up and insertion recovery.

recovery strategy for some of the problems encountered during assembly is shown in Table 9.7. Upon failure of a PC board positioning and insertion attempt, a wiggle routine is applied. It changes, in fine increments, both the orientation and position of a board relative to the fixture. As a result, in most cases a board would be positioned successfully. A search routine that is ap-

Table 9.7 Recovery strategy for the electronic assembly task

Problem/error	Recovery procedure
Misalignment of board	• Sense and calculate the misalignment • Recalibrate arm measurements
Positioning of board in fixture	• Apply wiggle routine
Board missing	• Try next board feeder
Rheostat stuck in feeder	• Try next rheostat feeder • Apply search routine
Rheostat insertion failure	• Recalibrate

plied to recover from rheostat insertion failures is similar to the one applied in the bolt insertion described above.

Experimental observations To compare the effectiveness of recovery programs, experiments were run with and without error recovery (Nof, Maimon and Wilhelm, 1987). Results show that when error recovery is employed, times to execute assembly tasks vary considerably but less operator involvement is required. When no recovery is used, task times vary less but an operator must intervene more often. From a chronology of insertion tasks with and without recovery, it was established that successful insertion on a first try occurs only about 50% of the time, probably because of the imprecise fixtures used. However, with recovery the success rate is significantly higher. The results also show that insertion errors occur often enough to warrant recovery in this experiment. The economic trade-off between a more precise – and costly – fixture compared with the use of error recovery to compensate for less precise – and less costly – resources is similar to the situation described in section 9.5 concerning rework trade-off decisions.

Standard recovery routines, considered natural for well-trained manual workers, can be integrated in robotic devices. Routines for particular needs can be developed as required for given situations. Knowledge-based recovery (Delchamber and Coupez, 1988) and fuzzy logic (Cao and Sanderson, 1992) can further increase recovery effectiveness. With error recovery, better quality assembly can be achieved with less costly fixtures and feeders. The planning framework which has been described above can help in implementing effective error recovery programs.

9.6 Communication and integration of assembly and inspection

Complex information exchange by computer networking within or between companies is becoming a reality of the global market. The objectives of networking are: to integrate distributed resources, to interact with customers, to coordinate suppliers' activities and to integrate distributed design and data to optimize production. In assembly, networking can significantly influence quality and yield. An important goal is to streamline information exchanges and eliminate errors while providing all necessary information when and where it is needed for the best decisions. This section describes several new communication and networking models that have been developed recently to support competitive industrial assembly.

9.6.1 Characteristics of distributed assembly operations

Two major issues in the design of integrated, distributed networks are (Nicol, Wilkes and Manola, 1993): (1) **inter-connectivity**, the ability to exchange information and utilize services provided to distributed applications; (2) **inter-operability**, the ability to collaborate and share resources. Integration is required both at the hardware and logic levels. Examples of integration in manufacturing are described by Weston *et al.* (1989).

The ability to communicate effectively is not sufficient; what exactly is communicated and when, who is responsible and who has priority for using limited resources are major issues that determine the overall effectiveness of communicating parties. The goal of integration is to lead to effective operations while maintaining flexibility. An important part of integration is computer-supported collaboration because integration **with** collaboration enables (Nof, Eberts and Papastavrou, 1993; Nof, 1994a):

- effective concurrent engineering and lifecycle production engineering;
- agile cooperation among participating companies.

Integration and collaboration are designed with three main dimensions which are relevant for assembly applications: (1) vertically – between higher-level planning and lower-level control; (2) laterally – among different planning functions at the same level; (3) time-based – along progression in time. Four recent projects of computer-supported integration with collaboration in distributed assembly and inspection are described in this section. They include:

- assembly communication networks;
- client–server integration protocols for assembly-and-test operations;
- best matching integration of distributed assembly components;
- the TestLAN approach to design assembly-and-test networks.

9.6.2 Assembly communication networks for resource sharing

Distributed processing in assembly-and-testing is increasingly performed with the support of computer networks in response to challenging problems. The main challenge is that poor inspection and testing procedures can cause longer production lead times and increased work-in-progress and capital investment in excessive test equipment. On the other hand, a distributed testing environment can rely on a network of shared testing resources and exploit computer communication via the network to optimize testing performance. Esfarjani and Nof (1994, 1995) describe a distributed assembly-and-test network designed for a large manufacturer of electronic subassemblies. The objective is to overcome typical problems that have been observed as follows.

1.	Lack of a systematic method to access a common expensive tester results in long idle time for operators. The bottleneck tester problem is often solved by purchasing multiple dedicated testers for individual workstations, which is a waste when actual testing time is negligible relative to assembly and repair time.
2.	Inefficiency of information flow between designers and testers causes long lead times to implement changes in test programs or in testing requirements.
3.	Any tester breakdown stops assembly operations until the tester is repaired or another one can be borrowed.
4.	Repeated physical modifications of a tester when moving it frequently between workstations cause electrical connectors to fatigue and result in wrong unreliable test readings.

A communication network allows sharing the common expensive tester among multiple operators. A logical arrangement of the central test work cell is presented in Fig. 9.11. Each remote workstation is connected to the shared tester and can communicate with it. Reliability can be increased by an additional tester on the same network.

Several computerized coordination and collaboration protocols are included in the network solution: (1) protocols for communication between operators and tester control for requesting access to the tester and obtaining diagnosis information from the tester; (2) work flow protocols for organizing operators' priority and access time allocation; (3) communication interface between the tester and tested assembly for on-line signal transfer and diagnosis. For example, wire harnesses are assembled on form boards at individual stations and by connecting their leads to a local test interface harnesses can be tested in place without moving them physically to the tester or bringing the tester to them.

Implementation of the protocols allows the horizontal integration of various assembly cells at the test work cell level, and enhances efficiency and reliability of the testing process while eliminating unnecessary additional testers. A study by Esfarjani and Nof (1995) of wire harness assembly applies simulation and stochastic models to evaluate alternative information exchange protocols. The organization of the network is shown in Fig. 9.12.

### 9.6.3	*Client–server design changes in distributed assembly-and-test*

Integration of distributed functions and cells in modern manufacturing is essential because of the complex non-deterministic relationships among func-

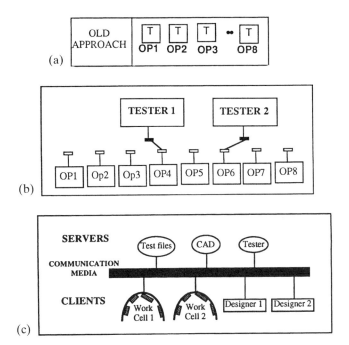

Fig. 9.11 Networking distributed assembly workstations (OP_i) with a shared tester (T) instead of dedicating individual testers or moving testers around saves significant expense: (a) dedicated local testers; (b) moving shared testers to local stations; (c) preferred network approach (Esfarjani and Nof, 1994).

tions and processes (Lenart and Nof, 1993). In distributed assembly-and-test, such integration is important because of the interdependency between design changes and the resulting changes in testing requirements. Beyond advantages of networking in general, it is necessary to establish procedures to achieve cost-effective flexibility of testing by horizontal and vertical information integration. In addition to the problems in testing during assembly discussed in the previous section, design changes in test requirements take a relatively long time to implement. Delays are worse when customers require just-in-time delivery, including a prompt response to ongoing design changes.

Client–server work flow protocols are an effective means of gaining networking advantages. The client–server methodology developed by Esfarjani and Nof (1994, 1995) is based on collaboration protocols with two purposes: (1) inter-operability at test work cell level (**horizontal integration**) for resource sharing, concurrence control and synchronization; (2) work and information flow between design and testing (**vertical integration**) for handling the transfer and propagation of ongoing design changes and test instruction modifications. Such changes and modifications can be generated either by customers' engineers or by local company engineers.

A client–server model of computing is defined (Linthicum, 1994) as a network of computer hardware, communication media and software which

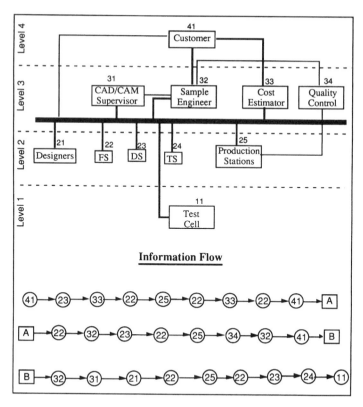

Fig. 9.12 Organization of information network and illustration of information flow with a 'running assembly change' protocol (Esfarjani and Nof, 1995).

is shared among multiple users. Unlike computing, the objective here is to develop client–server collaboration between design, manufacturing, assembly and test cells. The 'clients' are the users interfaced for testing, and the 'server' is a shared design and test file-server (Fig. 9.13). This environment allows the planning of effective integration protocols through quantitative evaluation of alternative protocol logic.

Three general work flow protocol types have been developed:

- **time-out**, defining the maximum length of time a tester is dedicated to a client workstation for testing and repair of diagnosed problems, after which another client can be served if any is waiting;
- **frequent request**, providing different access levels for frequent clients;
- **weighted priority collaboration**, where clients receive priority access based on a combination of characteristics.

In addition, two particular protocols developed for the design-to-testing integration are **send–wait–acknowledge** which treats changes individually, and a **union** protocol which accumulates design changes for periodic implementa-

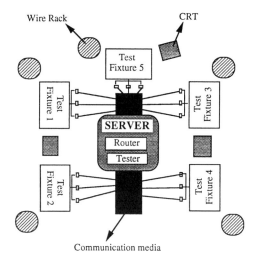

Fig. 9.13 A client–server integrated assembly-and-test approach (Esfarjani and Nof, 1995).

tion. A case study of an electronic component manufacturing facility in Indiana has been used as a test-bed for the new client–server approach. The resulting advantages are summarized in Table 9.8. In comparing networked operations with and without the protocols it was found that the protocols have an immediate benefit by reducing delays (requiring only 40% of the design changes processing time) and providing greater capacity (being able to handle effectively up to 120 changes at a time compared with only about 15 without the protocols, in this study). Another outcome is an approach to planning and evaluation of alternative integration protocols through quantitative analysis.

9.6.4 Inspection-based best matching of distributed assembly components

Another motivation to develop information exchange protocols is the ability to match parts which have 'optimal' joint functionality **prior** to final assembly, thus eliminating any further rework. With this approach, subassembly parts from distributed suppliers are manufactured, tested and prepared with a consideration of the specific characteristics and dimensions of counterparts.

As part of the approach called 'best matching' (Nof and Kang, 1996) three layers of inspection information architecture are identified: the information station module, the information integrator module and the assembly simulator module. A prototype has been developed to illustrate and examine the feasibility of the new inspection and assembly methodology. Parallel computer-based simulation was performed using TIE 1.1 parallel simulator (Nof, 1994b) to compare the logic alternatives for information flow handling.

Two distributed component source locations are shown in Fig. 9.14. The

Table 9.8 Summary of client–server assembly-and-test integration advantages study

Results	Advantage
1. Higher throughput	Operators can work in parallel on multiple product assembly types.
2. Better learning curve	Operators have to know less and have a standard procedure to operate the system.
3. Reduced setup time	Operators have to load the test program without any physical modification in the system.
4. Higher quality control	The operator can more easily identify the status of the test.
5. Statistical analysis	The operational history can be retrieved and verified to check for each given assembly type.
6. Increased system flexibility	Systematic and prompt handling of design changes and fluctuations to unexpected changes in demand.
7. Client and server expansion flexibility	Up to eight operators can use one server. The server can be expanded to 32 interconnected units.
8. Increased worker satisfaction	Less effort for operators to set up the machine. Less communication and conflict with other operators as to who gets to use the shared resource.
9. Increased system reliability	The physical connections do not have to be changed and the input wires are expected to always stay in contact.

Source: Esfarjani and Nof (1995).

results of their inspection information are brought together and integrated in an assembly testing graphic simulation. Simulation results lead to best matching of individual components for assembly and decisions about necessary rework **before** shipping components to final assembly.

Two advantages of the best matching approach have been found: improving the quality of an assembled product, and reducing significantly the premature shipment of components from remote supplier sites. Implementation of the information exchange protocols has provided better information exchange between inspection and other manufacturing functions (local and remote); the ability to predict the functionality of assembled parts; a manufacturing decision-making tool for part dimensional and functional control leading to a lower rejection rate during assembly; and elimination of the costs to transport incompatible assembly parts back to their remote suppliers.

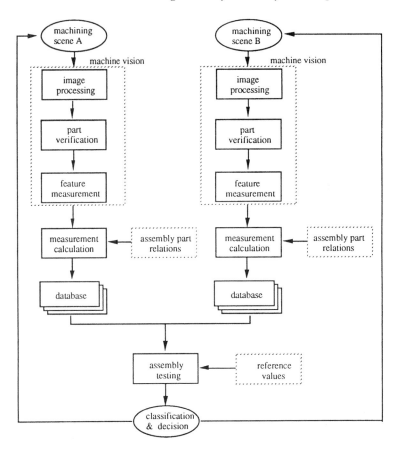

Fig. 9.14   The best matching assembly-and-test approach (Nof and Kang, 1996).

9.6.5   *The TestLAN approach to design assembly-and-test networks*

An electronic assembly case study (Williams, 1994; Williams and Nof, 1995) revealed that failure to create an integrated test system resulted in excessive delays and unnecessary idle times. The focus of this work was the design of a distributed testing network system that provides effective management of resources and is shared by distributed users (Fig. 9.15). The TestLAN approach is based on communication between distributed clients and central shared test resources. It is a generalization of the client–server model discussed above in that the TestLAN can communicate among a multiple number of test and design resources and allow clients to also become servers when necessary.

In a statistical study and queuing simulation (Fig. 9.16) it has been established that assembly performance **with** TestLAN can reduce waiting time by an average of 82% and reduce total flow time by 74% (both significant at the

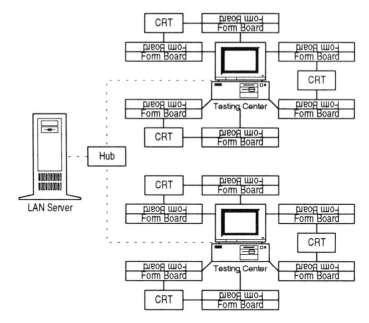

Fig. 9.15   An example LAN configuration for the assembly of wire harnesses (Williams, 1994).

95% confidence level) as well as reduce queue length by over 73%. The different protocols represent different logic of access assignment to shared testers. For comparison, the result without TestLAN is 4.446 workstations (customers) waiting in queue. Note that even without protocols a network approach is advantageous.

Additionally, protocols to coordinate the TestLAN system operation have been investigated. The protocol types analysed are priority assignment, time-out and resource allocation based on negotiated priorities. Analysis of these protocols in different combinations through queuing simulation has indicated that the time-out protocol is the most beneficial (Fig. 9.16). Further analysis and cost–benefit evaluations must be performed to ascertain the usefulness of TestLAN in different types of distributed assembly-and-test facilities.

9.7   Summary

Quality is defined by the American National Standards Institute (ANSI) and the American Society for Quality Control (ASQC) as 'the totality of features and characteristics of a product or service that bears on its ability to satisfy given needs'. Given needs are relative to customers' expectations and may change over time. Satisfying these needs begins with product design and continues through every step of the production process. From the point of

Comparison of integrated and non-integrated testing configurations

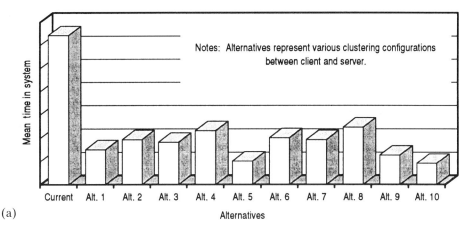

(a)

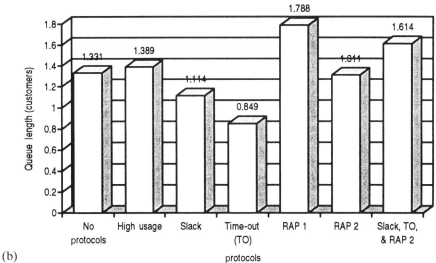

(b)

Fig. 9.16 Performance improvements with TestLAN: (a) relative mean time in the system without ('current') and with TestLAN integration; (b) M/M/C queuing analysis results for queue length in TestLAN (Williams and Nof, 1995).

view of assembly, quality is a function of each individual component's quality, and of the operations that assemble them into a whole product.

This chapter begins with three basic elements of assembly quality: calibration, tolerance and standards. These elements combine the needs and expectations of customers on the one hand, and guidelines for actions that assemblers can follow to satisfy them on the other. Key actions to ensure quality are test and inspection, but test and inspection must be rationalized so that the efforts are effective **and** economic.

Models to evaluate the relative cost-effectiveness of test and inspection techniques are described in this chapter. Models and methods are also included for the design of integrated test and rework assembly stations, and for the analysis and development of effective in-process recovery to eliminate defects. The common theme is the cost minimization of inspection and test operations while, at the same time, maximizing the probability of satisfying a designed level of quality.

The final section in the chapter describes further cost reduction by the recent evolvement of communication-based assembly networks. A major benefit is that distributed assembly operations can now be provided with timely and effective test services by the better utilization of inspection and test resources, while integrating relevant information from its distributed sources. It is expected that in the future, computer-supported assembly networks will continue to evolve. The next chapter is devoted to emerging trends in assembly.

9.8   Review questions

1. What is the purpose of calibration in assembly? Illustrate several different calibration types in:
    (a) SMT component positioning;
    (b) sealant dispensing;
    (c) car windshield assembly.

2. In adhesive dispensing, identify types of quality defects that are caused by problems in:
    (a) accuracy;
    (b) repeatability.

3. Repeat question 2 for the task of crimping a terminal on a wire.

4. Explain the different purposes of tests during various stages in the assembly of:
    (a) electronic components on a PCB;
    (b) a conveyor;
    (c) a car headlight lamp.

5. Explain the characteristics of non-contact inspection sensors and the factors that determine which of them would be most suitable for particular assembly applications.

6. In selecting alternative inspection technologies, compare the different objectives of using the productivity measurement model (PMM) versus the performance assessment model (PAM).

7. Write the PMM equations for the input, good product output quantity and scrap output for inspection process I_k, assuming the following cases:

(a) Type II error is equal to zero;
(b) Type I error is equal to zero;
(c) both error types are equal to zero.

Explain the economic and the operational meaning of each case.

8. Develop the PMM equations for two consecutive inspection stations I_K and $I_{(K+1)}$.
9. Describe the different focus of the four main parts of the performance assessment model.
10. Explain the three inspection technology ratios and the four parts of the performance assessment model.
11. Calculate performance estimates for the example in Table 9.4, cases 1 and 9 for limited rework ($J = 3$), and find the maximum percentage difference compared with infinite rework.
12. Calculate the operational efficiency (η) for cases 1 and 4 in Table 9.4 with limited rework ($J = 3$.)
13. With which of the following production modes is the RPFF model compatible? Explain your answer.

(a) just-in-time
(b) continuous improvement
(c) zero defects

14. Define possible errors in the operation of Fig. 3.7 and classify them as design, logistic or spatial errors.
15. Propose when and how the errors you defined in question 14 can be detected and diagnosed, and consider the economic and effectiveness aspects of your proposals. How would you assess them?
16. Suggest recovery actions for the errors you defined in question 14, and describe them using a recovery tree.
17. Distinguish between a recovery graph (Chapter 3) and a recovery tree.
18. Describe several types of computational logic that are useful for error diagnosis and recovery.
19. What are assembly communication networks and how can they contribute to product quality?
20. Describe several assembly work flow protocols and explain their role.

References

Arabian, J. (1989) *Computer Integrated Electronics Manufacturing and Testing*, Marcel Dekker, New York.
Barone, P.A. (1989) 'Techniques for developing a multi-robot cell for production processing of complex airframe assemblies', *SME Technical MS Series Paper*.
Boothroyd, G. (1992) *Assembly Automation and Product Design*, Marcel Dekker, New York.
Borchelt, R.D. and Alptekin, S. (1994) 'Error recovery in intelligent robotic workcells',

456 *Quality and inspection in assembly*

International Journal of Production Research, Vol. 32, No. 1, pp. 65–73.

Cao, T. and Sanderson, A.C. (1992) 'Sensor-based error recovery for robotic task sequences using fuzzy Petri nets', *Proceedings of the IEEE International Conference on Robotics and Automation*, Nice, France, Vol. 2, pp. 1063–9.

Chase, K.W., Greenwood, W.H., Loosli, B.G. and Hauglund, H.F. (1990) 'Least cost tolerance allocation for mechanical assemblies with automated process selection', *Manufacturing Review*, Vol. 3, No. 1, pp. 49–59.

Csakvary, T. (1985) 'Planning robot applications in assembly', *Handbook of Industrial Robotics* (ed. S.Y. Nof), Wiley, New York, pp. 1054–83.

Delchamber, A. and Coupez, D. (1988) 'Knowledge based error recovery in robotized assembly', *Proceedings of the 9th International Conference on Assembly Automation (ICAA)*, London, UK, pp. 137–54.

DiCesare, F., Goldbogen, G., Feicht, D. and Lee, D.Y. (1993) 'Extending error recovery capability in manufacturing by machine reasoning', *IEEE Transactions on Systems, Man and Cybernetics*, Vol. 23, No. 1, pp. 221–8.

Dooley, B.J. (1983) 'A model for the prediction of assembly, rework, and test yields', *IBM Journal of Research and Development*, Vol. 27, No. 1, pp. 59–67.

Duelen, G., Muench, H. and Surdilovic, D. (1991) 'Advanced robot control system for manufacturing processes', *Annals of CIRP*, Vol. 40, No. 1, pp. 387–90.

Duelen, G., Muench, H., Schmidt, W. and Schroer, K. (1990) 'Control systems for robots – wider functional capabilities' (in English), *Schweissen Schneiden*, Vol. 42, No. 9, pp. 147–9.

Esfarjani, K. and Nof, S.Y. (1994) 'Application of distributed processing in electronic assembly testing', in *PRISM: An Overview*, IE Research Memo 94-22, Purdue University, West Lafayette, IN, pp. 5–6.

Esfarjani, K. and Nof, S.Y. (1995) *Client-Server Integration Models and Protocols for Assembly*, IE Research Memo 95-4, Purdue University, West Lafayette, IN.

Fu, K.S, Gonzalez, R.C. and Lee, C.S.G. (1987) *Robotics: Control, Sensing, Vision, and Intelligence*, McGraw-Hill, New York.

Hancock, W.M., Sathe, P. and Edosomwan, J.A. (1992) 'Quality assurance', *Handbook of Industrial Engineering* (ed. G. Salvendy), Wiley, New York, pp. 2221–34.

Harding, K.G. (1991) 'Application of an on-machine gage for diameter measurement', Progress Report submitted to USAF MANTECH, Contract F33615-91-C-5704.

Hayati, S. (1988) 'Calibration', *International Encyclopedia of Robotics, Applications and Automation* (eds R.C. Dorf and S.Y. Nof), Wiley, New York, pp. 165–7.

Hoffmann, J.E. (1993) 'Assembly/test combo cuts rejects, aids SPC', *Tooling & Production*, Vol. 59, No. 9, pp. 30–3.

Hutchinson, S.A. and Kak, A.C. (1990) 'SPAR, a planner that satisfies operational and geometric goals in uncertain environments', *AI Magazine*, Vol. 11, No. 1, pp. 30–61.

Kear, F.W. (1987) *Printed Circuit Assembly Manufacturing*, Marcel Dekker, New York.

Lenart, G. and Nof, S.Y. (1993) *Object-Oriented Integration of Concurrent Engineering and Laser Processing Cell*, Research Memo 93-1, School of Industrial Engineering, Purdue University, West Lafayette, IN.

Linthicum, D.S. (1994) 'Client-server strategy', *DBMS*, April, pp. 46–55.

Maimon, O.Z. and Nof, S.Y. (1986) 'Analysis of multi-robot systems', *IIE Transactions*, Vol. 18, No. 3, pp. 226–34.

Momot, M. and Shoureshi, R. (1990) 'Dynamic error recovery for an automated assembly workcell', *Proceedings of ASME Winter Annual Meeting on Intelligent Control Systems*, Vol. 23, pp. 43–7.

Murphy, S. (1990) *In-Process Measurement and Control*, Marcel Dekker, New York.

Nicewarner, K.E. and Kelley, R.B. (1993) 'Reliable vision-guided grasping', *Proceedings of the SPIE 1992 Conference on Cooperative Intelligent Robotics in Space*, Vol. 1829, pp. 274–83.

Nicol, J., Wilkes, T.C. and Manola, F.A. (1993) 'Object orientation in heterogeneous distributed computing systems', *Computer*, Vol. 26, No. 6, pp. 57–67.

Nof, S.Y. (ed.) (1985) *Handbook of Industrial Robotics*, Wiley, New York.

Nof, S.Y. (1994a) *Information and Collaboration Models of Integration*, Kluwer Academic, Dordrecht, Netherlands.

Nof, S.Y. (1994b) 'Recent developments in simulation of integrated engineering environments', *Proceedings of the SCS Symposium on AI & Computer Simulation*, Mexico City.

Nof, S.Y. and Kang, H. (1996) *Inspection-Based Integration Protocols for Best Matching of Components from Distributed Suppliers*, IE Research Memo 96-K, Purdue University, West Lafayette, IN.

Nof, S.Y., Eberts, R.E. and Papastavrou, J.D. (1993) 'Computer-based collaborative integration of distributed manufacturing engineering', *Proceedings of NSF Conference on Design and Manufacturing Systems*, January 1993, Charlotte, NC.

Nof, S.Y., Maimon, O.Z. and Wilhelm, R.G. (1987) 'Experiments for planning error-recovery programs in robotic work', *Proceedings of ASME International Conference on Computers in Engineering*, Vol. 2, New York, pp. 253–64.

Partaatmadja, O., Benhabib, B. and Goldenberg, A.A. (1990) 'Development of a distance sensor', *Proceedings of the ASME 2nd Conference in Flexible Assembly Systems*, Chicago, IL, Vol. DE 28, pp. 171–6.

Remski, R.M. (1993) *Assessment Models of On-Line Inspection Technologies*, MSc thesis, Purdue University, West Lafayette, IN.

Remski, R.M. and Nof, S.Y. (1993) 'Analytic and empirical assessment models of on-line inspection technologies', *Computers & IE*, Vol. 25, Nos 1–4, pp. 439–44.

Riley, F.J. (1983) *Assembly Automation: A Management Handbook*, Industrial Press, New York.

Schroer, K., Bernhardt, R., Albright, S., Woern, H., Kyle, S., van Albada, D., Smyth, J. and Meyer, R. (1995) 'Calibration applied to quality control in robot production', *Control Engineering Practice*, Vol. 3, No. 4, pp. 575–80.

Seidmann, A. and Nof, S.Y. (1985) 'Unitary manufacturing cell design with random product feedback flow', *IIE Transactions*, Vol. 17, No. 2, pp. 188–93.

Seidmann, A. and Nof, S.Y. (1989) 'Operational analysis of an autonomous assembly robotic station', *IEEE Transactions on Robotics and Automation*, Vol. 5, No. 1, pp. 4–15.

Seidmann, A., Schweitzer, P.J. and Nof, S.Y. (1985) 'Performance evaluation of a flexible manufacturing cell with random multiproduct feedback flow', *International Journal of Production Research*, Vol. 23, No. 6, pp. 1171–84.

Seidmann, A., Schweitzer, P.J. and Nof, S.Y. (1987) 'Observations on the normality of batch production lines in flexible manufacturing cells', *International Journal of Production Research*, Vol. 25, No. 1, pp. 151–4.

Shade, W.L. (1992) *Machine Tool 3D Laser Measurement System*, Technical Report MS92-180, Society of Manufacturing Engineers, Dearborn, MI.

Spotts, M.F. (1973) 'Allocation of tolerances to minimize cost of assembly', *ASME Journal of Engineering for Industry*, Vol. 95, pp. 762–4.

Stewart, W.T. (1983) 'Multiple input productivity measurement of production systems', *International Journal of Production Research*, Vol. 24, No. 5, pp. 745–53.

Wadsworth, H.M. (1992) 'Standardization and certification of quality assurance', *Handbook of Industrial Engineering* (ed. G. Salvendy), Wiley, New York, pp. 2382–96.

Warnecke, H.-J. and Schraft, R.D. (1982) *Industrial Robotics Applications Experience*, IFS Publications, London.

Warnecke, H.-J., Schraft, R.D. and Wanner, M.C. (1985) 'Performance Testing', *Handbook of Industrial Robotics* (ed. S.Y. Nof), Wiley, New York, pp. 158–66.

Weston, R.H. *et al.* (1989) 'Configuration methods and tools for manufacturing systems integration', *International Journal of CIM*, Vol. 2, No. 2, pp. 77–85.

Whitney, D.E., Lozinski, C.A. and Rourke, J.M. (1986) Industrial robot forward calibration method and results, *ASME Trans. Dynamic Systems, Measurement and Control*, Vol. 108, pp. 1–8.

Wilhelm, R.G., Nof, S.Y. and Seidmann, A. (1986) 'Analysis of recirculation in robotic systems using feedback models', *Robotics and Material Flow* (ed. S.Y. Nof), Elsevier Science, Amsterdam, pp. 69–82.

Williams, D.J., Rogers, P. and Upton, D.M. (1986) 'Programming and recovery in cells for factory automation', *International Journal of Advanced Manufacturing Technique*, Vol. 1, No. 2, pp. 37–47.

Williams, N.P. (1994) *The TestLAN Approach to the Design of Testing Systems*, MSIE thesis, Purdue University, West Lafayette, IN.

Williams, N.P. and Nof, S.Y. (1995) *The TestLAN Approach to Integrated Assembly-and-test Networks*, IE Research Memo, 95-7, Purdue University, West Lafayette, IN.

Zupancic, J. (1994) 'Calibration of an SMT robot assembly cell', *Journal of Robotic Systems*, Vol. 11, No. 4, pp. 301–10.

10

Emerging trends in assembly

10.1 Introduction

In early studies of assembly by Walker and Guest (1952) after World War II, the focus was on manual assembly work in the context of mass production. An emerging technological trend at that time was the assembly line, and the investigation revealed an increasing worker dissatisfaction with machine-paced work. It was predicted that the approach of mass production would prevail and even be 'exported' from the US to the rest of the industrial world. The major conclusion was that it would be necessary in the future to develop human-oriented systems and work methods.

What can be said now about the future of industrial assembly? Emerging trends are examined in this chapter from the logistic and the technological aspects, telling us some of what can be expected. While interest in well-designed human-oriented systems continues, mass production has shifted to flexible production, and flexible assembly has been growing in importance. From a logistic point of view, there are structural changes in the global market of customers, suppliers and distributors along the supply chain. From the technological aspect, there are new techniques, materials and information technologies that are already transforming the way products are assembled (Riley, 1988; Schwartz, 1988).

This chapter concludes the book with a review of key emerging trends in three areas. In the next section, trends in technological elements of assembly operations and systems are reviewed. The following two sections describe trends in the design of assembled products and assembly systems, followed by the emerging influence of information technologies. The chapter ends with an overview of relevant research issues and directions.

10.2 Emerging trends in assembly technology

Technological developments that benefit assembly emerge in response to new approaches to rationalize and improve assembly. For instance, emphasis on design for assembly with less insertion tasks and simpler fastening operations has motivated the development of new adhesives and materials used in surface mount technology. Manual workstations are increasingly designed for teamwork. Quality, reliability and variability demands justify the development of more flexible and accurate assembly machines. Several important technology developments are reviewed in this section.

10.2.1 Manual assembly workstations

While automation can be applied by companies whenever it is possible and justifiable to reach their goals, there are still many jobs that companies prefer to assign to humans. These jobs are often repetitive, involving visual inspection by a single worker at a single station. Manual workstations tend to involve bulk or limp materials and relatively more unstructured requirements. Facing increasing competition, attention to the efficiency of manual assembly work is growing.

An important trend in manual workstations is the focus on ergonomics. Ergonomic design considers the interaction between workers and their workplace to increase safety and improve work effectiveness. This trend is also influenced by the 1990 Americans with Disabilities Act which requires the design of work environments that are accessible and usable also by disabled workers. New workstations and accessories are developed in the US with adherence to ergonomic guidelines issued by OSHA, the US Occupational Safety and Health Administration, and NIOSH, the US National Institute of Occupational Safety and Health (Benson and Iverson, 1994).

The trend towards teamwork and flexible work groups further emphasizes the need for adjustability to individual workers (Fig. 10.1). For instance, such adjustability may imply: (1) the work surface height should be adaptable to individual height; (2) the work surface should be adjustable to improve variable workpiece positioning for increased comfort and efficiency; (3) an operator should be able to reposition the work frequently to accommodate changes in posture.

Workstation modularity and flexibility are also becoming important features. In the increasingly flexible assembly environment, a team should be able to quickly customize the station to the specific needs of variable tasks and alternating assemblers.

10.2.2 Assembly machines

The emerging focus on quality, along with an improving economy, are factors of growth in the demand for assembly machines and systems and their justification. There is a growing awareness of the need to design more flexible assembly machine systems that can be adapted more readily and more quickly to product changes. The replacement of manual labor by assembly automation can reduce production error rates significantly, while lowering direct labor costs (and associated costs of labor insurance and benefits) for assembly,

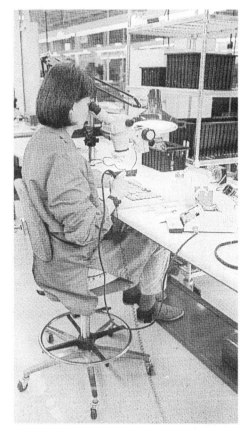

Fig. 10.1 A manual assembly-and-test workstation with ergonomic design considerations (courtesy of Altron Inc.).

inspection, testing and rework. US annual revenues from assembly machine systems and components totaled around $1 billion in the 1990s, and is expected to grow.

Another factor is the trend of miniaturization and increasing complexity of certain products, which means they can no longer be assembled manually (for instance, a tiny disk drive made for portable computers). This trend also increases the demand for handling and transfer systems that are more accurate to provide correct component placement. Another trend is the integration of sensors and machine vision in applications of part location, placement and operation verification.

10.2.3 Fasteners

Mechanical fasteners – screws, bolts, rivets and others, made of both metal and plastic materials – continue to be preferred for assembling manufactured products. According to the Industrial Fasteners Institute in Cleveland, Ohio, mechanical fastening accounts for 75% of the work performed in assembly plants; 45 000 employees manufacture fasteners in North America; 200 billion fastener pieces are produced annually. The key consumers of fasteners are the aircraft/aerospace, automotive, farm equipment, construction, electronics and the appliance industries. Fastener manufacturing consumes 2.2 billion lb (1 million tons) of material annually, 80% of which is carbon and alloy steel. While traditionally made in mass quantities, the trends characterizing the fasteners industry are decreasing lot sizes, decreasing delivery times and increasing demand for higher quality, customized design and materials. A similar trend is observed in electrical and electronics components which are becoming increasingly customized. Figure 10.2 illustrates a variety of terminals for electrical wires. As customer requirements change, it becomes more difficult to limit fasteners and terminals to a standard selection.

The Fastener Quality Act was passed by the US Congress in November 1990. It addresses issues of mismarked, substandard or counterfeit fasteners; however, final regulations to implement the Act have not yet been approved. The Act does mandate testing and certification procedures to assure the quality of high-grade fasteners.

10.2.4 Adhesives

Increasingly, adhesives are used to join a wide variety of parts in assembling products. The applications for adhesives in assembly are growing in number

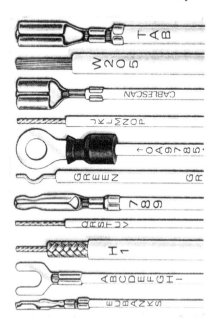

Fig. 10.2 Standard and customized wire terminals (courtesy of Eubanks Co.).

and in diversity (Benson and Iverson, 1994). But adhesives have only begun to replace the more traditional methods of fastening. Adhesive bonding is presently estimated to comprise about 1% of the world fastening market, partly because of concerns about long-term bond reliability and the need for non-destructive testing of joined parts.

An example of adhesive application for assembly is illustrated in Fig. 10.3. A manufacturer of bonding equipment advertises the cost-effectiveness and relative benefits of bonding by combining inert gas with hot-melt adhesive to generate the bonding foam. This method is offered as an alternative to other fasteners, particularly in automated high-production operations.

In the automotive, aircraft and electronics assembly industries, there are emerging applications for adhesives in welding bonding, structural tapes and films, structural cross-linking hot-melts and radiation-cured adhesive bonding. In a new trend, part of the assembly takes place in a developing country to take advantage of lower-cost labor or economic incentives. Often, assembly plants are located in several geographic areas to bring the assembly site closer to the

Fig. 10.3 Relative advantages of an adhesive material for assembly include less labor and time (courtesy of Nordson Corp.).

final customer. In those cases, adhesive technology is often viewed as relatively simpler to apply.

Global adhesive selection and availability for assembly will depend on three main factors: environmental and recycling regulations; development of new adhesive substrates; and raw material availability. However, the use of adhesives in assembly is expected to increase in the future because of improving performance, quality and convenience considerations.

10.2.5   *Assembly hardware and components*

General assembly components include bearings, couplings, formed parts, handles, knobs and hinges. Latches, seals, retaining rings, shock absorbers, slides, springs and strain-relief devices are also included in this group of parts which is essential to product assembly. Such components provide cost-saving and productivity improvement to typical product assembly operations. For instance, one retaining ring can replace multiple fasteners in holding a cover plate and reduce the total assembly cost. Hence, demand for such components is expected to continue a growth trend.

New and emerging assembly materials, equipment and components as described above are influencing the future of assembly. They improve the quality of the assembled product and at the same time increase the effectiveness of assembly work.

10.3   Emerging trends in the design of products and systems

The potential for flexibly automated assembly systems application will increase as more technical experience and guidelines for assembly automation are developed (Bullinger, Warnecke and Lentes, 1986; Heginbotham, 1986; Black, 1991). Trends in this area are described next.

10.3.1   *Product design*

Product design for assembly will continue to gain importance, as has already been the case since the mid-1980s, similar to part design for manufacture. Effectiveness measures for assembly-oriented product design increasingly include consideration of parts handling and joining.

10.3.2   *Modularity*

The modularity of assembled products is growing in importance with benefits both in product quality and assembly logistics. Design with a view to easier, less costly and better quality assembly means that designers do not only think in terms of 'parts'. In the future they will focus increasingly on 'modules' of which a product is composed. For instance, an assembly module of car brakes, axle, brake controls and shock absorbers will be considered, instead of designing each component for individual assembly.

10.3.3   Assembly systems

The range of future systems will essentially be characterized by combinations of present-day systems, including labor-intensive systems, hybrid systems with a combination of manual and automatic workstations, and flexibly automated systems. However, the relative distribution of these technologies will change. An important trend is the growing willingness of managers to justify automation investment by evaluating the impact on quality, in-process inventory and response to customer orders and design changes.

Labor-intensive assembly   Such systems will be used mainly for bulk-size products. The two conflicting planning issues for future labor-intensive systems will be (a) planning for timely available materials and components of required quality at the assembly site, and (b) achieving flexibility and responsiveness regarding fluctuations in order quantities and product type variations. The key to solving this conflict is planning through modular design of the product and the system.

Human-friendly assembly lines   The way to design future systems involving operators is the human-friendly assembly line (Makino and Arai, 1994). In studies in Japan, where a 30% shortage of assemblers is expected by the year 2000, automotive manufacturers plan to introduce redesigned, modular automobile structures. Short subassembly lines can be better suited for a certain level of automation and are more appealing to Japanese workers. The proposed features of a production facility include modular structures, parallel sub-lines, autonomous operations and attractive work areas.

10.3.4   Assembly robots

The increasing role of robots in assembly has been reported in many studies (e.g. Porter and Rossini (1988) Fig. 10.4). The problems of integrating robots and manual workstations will increase in the future. Methods for task allocation between humans and robots (Helander and Domas, 1986), for robot–human interfacing (Kreifeldt, 1988) and human–robot ergonomics (Nof, 1992a) are becoming useful for this purpose. For the hybrid assembly systems, several guidelines can be suggested.

1. Use redundant manual workstations for backup and flexibility.
2. Develop computer displays and control-decision support systems for better supervision of the robots.
3. Develop optimal layouts with regard to flexible material flow and to cooperation alternatives among human operators and robots.
4. Design modular, decoupled technological assembly subsystems.

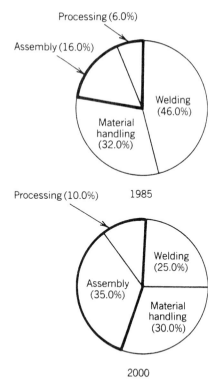

Fig. 10.4 The increasing role of robots in assembly, based on industrial robot sales, 1985 vs 2000 (Porter and Rossini, 1988).

10.3.5 Layout

The layout of future assembly systems with flexible automation is still subject to debate. The main advantages of work cell layout are the reduced material handling cost, reduced part-handling damage and relatively less direct labor cost. On the other hand, linear layout concepts imply that individual systems are lined up behind one another as with a transfer line, and the products to be assembled are conveyed continuously and efficiently by means of a transport system with workpiece carriers to the individual assembly areas or cells that comprise the line.

The possible linking of assembly systems is also being investigated by the automotive industry. For example, flexibly automated assembly systems are applied in which the assemblies are mounted and conveyed by an AGV system on pallets to individual programmable assembly cells or to special-purpose stations. Such linking offers even greater flexibility, particularly in view of the fact that some of the programmable cells can be redundant. When one cell is down, its operation is carried out by a neighboring cell, thereby eliminating the possibility of a total production standstill.

Added flexibility unfortunately entails greater costs. The degree of flexibility desired must therefore be weighed against the extra investment involved.

10.3.6 Knowledge-based systems planning

In planning future systems it will be possible to use the assistance of integrated CAD tools and expert systems (see Fig. 10.5 and Chapter 3, section 3.6). At present, for example, expert systems are being developed for planning assembly systems by the knowledge of experts, users, manufacturers and product developers. Knowledge-based support will relieve planners of routine tasks, simplify decision-making, learn from past experience and enable effects of change in surrounding conditions to be recognized early enough. Expert design systems of this nature with a sufficient data and knowledge base are already emerging and should become available in the near future as a common planning aid (Fig. 10.5(a)). A related direction is to automatically translate the plan into a control program, without human intervention, and activate assembly machines and robots (Fig. 10.5(b)).

No one can predict the future with exact precision, of course. Figure 10.6 includes some expert views as to likely developments in the 1990s. Indeed, by the mid-1990s these predictions have generally materialized.

10.3.7 Flexible facilities

Several examples illustrate emerging concepts of future flexible assembly systems. These and other similar examples have been proven in research laboratories and some have already been implemented in industry.

Dexterity The first example illustrates the increasing dexterity and accuracy of assembly robots. A wire-laying robot from Yamaha is capable of laying a wire harness, attaching connectors and providing wire terminals. An assembly system of this type was presented first in Japan in 1985, and precision assembly robotic systems have also been implemented (see examples and figures in Chapter 4). Similar systems have been developed since then in Europe and the US.

Flexible lines A flexible assembly line has been described in section 4.7.3. Oki-Elektrik produces several models of printer at its Tohuku plant. In a one-shift operation, 2000 printers per day are assembled automatically, with a typical batch size of about 1000 units. The cycle time of the system is 18 sec and the resetting time for all 127 stations is 15 min. The stations are linked by a double-belt conveyor and parts are delivered in magazines on pallets by an AGV system. Twelve operators carry out the resetting and repair any part-jamming or machine failure. In case of failure, several operators work to repair

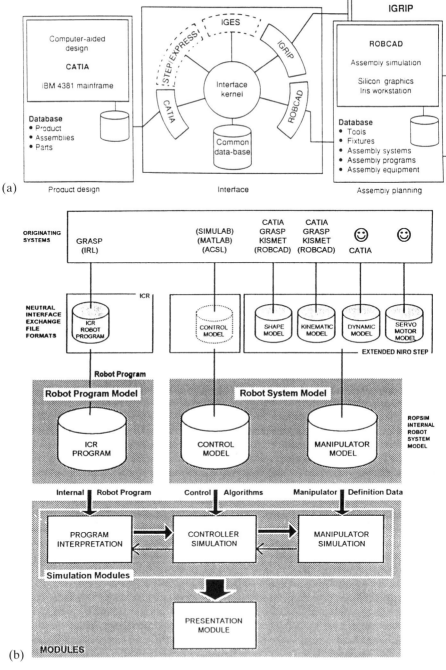

Fig. 10.5 Emerging integrated CAD, planning and control systems for assembly: (a) CIM interface between product design and assembly planning at the Institute for Production Systems and Design Technology (IPK), Berlin; (b) an integrated system using neutral interfaces between design information and motion-control devices at the Control Engineering Institute, Technical University of Denmark in Lyngby.

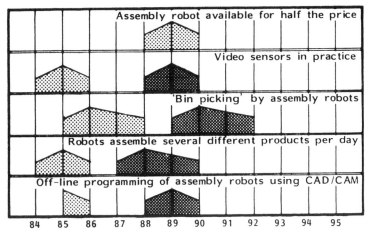

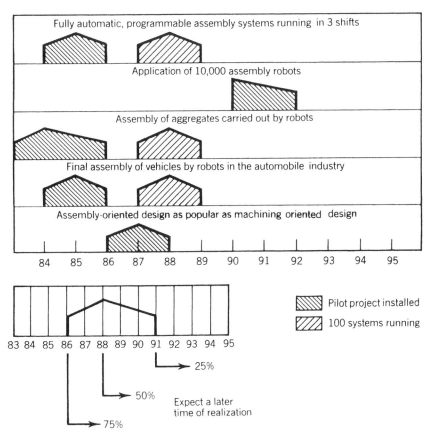

Fig. 10.6 Delphi studies on flexible assembly automation (top: Schraft (1988); bottom: Warnecke *et al.* (1992)).

the equipment, while others take over and perform manually the necessary assembly tasks so that the flow is maintained.

This example illustrates flexibility to accommodate changes in model, design and batch size. It also demonstrates the role of human operators in robotic assembly. The development of a system of such complexity would not be feasible, however, without consistent, assembly-optimized product design.

Examples of emerging flexible assembly facilities are shown in Fig. 10.7. In Fig. 10.7(a) a single robot station is depicted with the auxiliary equipment and interface to the cell or line. In Fig. 10.7(b), two robots with complementary tasks are shown in a cell. The laser-based calibration and metrology systems in this cell add further flexibility by integrating testing tasks.

Reconfigurability Another type of flexibility is the configuration adaptation of stations through which assemblies progress. A reconfigurable pilot cell is demonstrated in Fig. 10.7(c). Flow and task flexibility are enabled by an internal free-flow loop and six external by-pass loops to transfer pallets. The system control can divert different assemblies to different stations, depending on software instructions. Makino and Arai (1994) distinguish between this type of dynamically reconfigurable systems, i.e. by software, and static

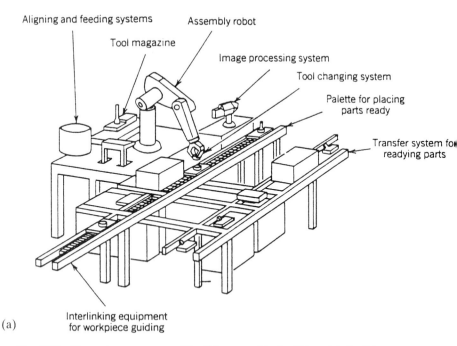

(a)

Fig. 10.7 Emerging concepts of flexible assembly facilities: (a) flexibly automated assembly station/cell (Warnecke *et al.*, 1992); (b) flexible assembly cell for small to medium size airframe components (courtesy of Rohr Co.); (c) dynamically reconfigurable assembly system (Valckenaers and Van Brussel, 1994).

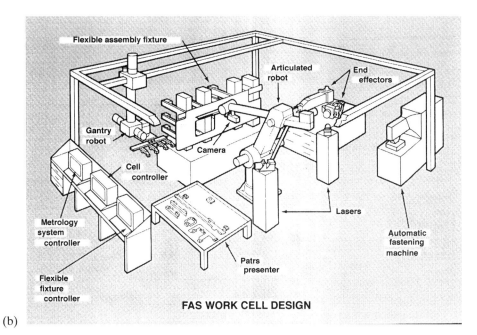

(b)

Flexible assembly fixture

Articulated robot

End effectors

Gantry robot

Camera

Cell controller

Metrology system controller

Lasers

Flexible fixture controller

Patrs presenter

Automatic fastening machine

FAS WORK CELL DESIGN

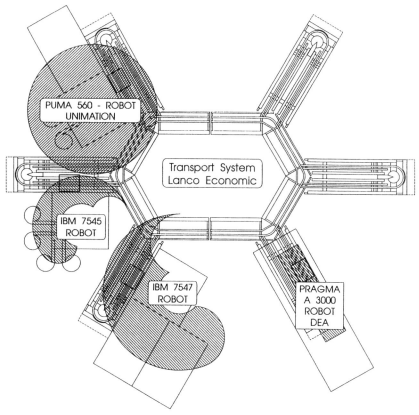

(c)

PUMA 560 - ROBOT
UNIMATION

Transport System
Lanco Economic

IBM 7545
ROBOT

IBM 7547
ROBOT

PRAGMA
A 3000
ROBOT
DEA

Fig. 10.7   *Continued*

reconfigurable systems, where configuration flexibility can be achieved by rearranging the equipment.

Even with these emerging concepts, a hybrid system with manual assemblers seems logical. Automation will displace human assemblers only where it can boost quality and productivity, with the highest potential displacement in the electronics industry. Indeed, a rapid relative growth in robots and vision systems implementation in the US has started in the mid-1990s with assembly being one of the fastest growing areas. Through analysis of future assembly requirements, including shorter product lifecycles, frequent design changes and smaller order quantities, the following predictions can be made.

1. Factories using mainly manual assembly systems will have to increase their product quality and productivity by integrating flexibly automated assembly equipment without compromising flexibility, reliability and cost-effectiveness.
2. Factories using rigid, special-purpose assembly automation will have to increase their flexibility by introducing flexible assembly automation.
3. Systematic methods of design for assembly, computer-aided assembly techniques and knowledge-based assembly task planning and control will increase in significance by supporting the design of cost-effective products and the selection and utilization of the most appropriate assembly technology.

Based on experience to date, design for assembly has convinced designers that about half of all products are not candidates for automation, at least in the near future. This conclusion is based on total order quantities that are too small or because automation cannot be cost-justified. But simplified assembly has already reduced significantly the total assembly costs even in manual or hybrid systems. An important future concept is modular product design coupled with modular design of assembly systems.

10.4 Assembly utilities

Assembly utilities can be defined as assembly service houses that specialize in certain technology, operations and equipment and serve as subcontractors. They are part of contract manufacturing, or outsourcing, which is a significant growing trend in American industry. This trend is fueled by the approaches of agile manufacturing and of lean production. Agile manufacturing and assembly (Kaplan, 1993) imply that companies compete in the global market by teaming up, only as needed, with the best available partners and subcontractors for the fulfillment of a given project. Usually, agility implies minimizing time to market and just-in-time deliveries (Strobel and Johnson, 1993). Lean production implies that instead of heavy investment in capital equipment and technology, manufacturers are more interested in efficiency, and allocate their scarce resources to core competencies, mainly R&D, design and marketing.

Hence, manufacturers are cutting costs, downsizing and reducing overheads, personnel and production capacity. The necessary assembly competencies, especially with specialized technologies, are then provided as services by utility companies.

In a recent survey by Coopers & Lybrand Co. of about 400 executives of companies identified as the fastest growing US firms in the 1990s, two-thirds use outsourcing and over three-quarters consider it a key factor in their company's success. Interestingly, in terms of outsourcing budget, assembly is found second only to payroll services. While respondents cite economic reasons for initially exploring outsourcing, 70% of those surveyed say outside providers are relatively more efficient and 45% say outsourcing allows them to focus on their core business products and growth.

An example of the services offered by an assembly utility is given in Table 10.1. The company specializes in electronic and electro-mechanical assembly

Table 10.1 Subcontract services offered (sample) by a utility for electronic and electro-mechanical assemblies

---

Surface mounted component technology
- Design review and guidelines
- Stenciling of solder paste
- Automatic solder paste application
- Manual or automatic component placement
- Infra-red, convection or vapor phase soldering
- In-line or ultrasonic cleaning

Chassis and product assembly
- Subassembly fabrication
- Light mechanical assembly
- Complete product assembly
- Assembly to standard requirements
- Turnkey assembly operations

Printed circuit board services, rigid or flexible
- Board cleaning and baking
- Manual or automatic assembly
- MIL spec (standard) assembly
- Manual or flow soldering
- In-line cleaning – aqueous or solvent
- Lead trimming
- Contamination inspection
- Gold tab replating
- 100% in-process and final inspection
- Printed circuit board rework and modification

Wire, cable, harness services
- Automatic wire cutting and stripping
- Automatic tubing cutting
- Automatic wire terminating
- Wire tinning
- Cable assembly
- Harness fabrication
- Automatic testing

Special services
- Manual and CAD design
- Purchasing components, materials
- Bare board testing
- In-circuit testing
- Functional testing
- Computer-aided inventory control
- Hardware/software design
- Certification

Courtesy of Altron Inc.

as **their** 'core competency'. They maintain an environmentally controlled facility with a variety of up-to-date equipment and technologies and with trained personnel (150–250 workers). The company offers a complete service including design review, purchase of components, building the assembly, testing and quality certification. Similar examples are wire harness assembly houses, and kitting and distribution supply companies. Common to all these examples is the preference of the client companies of such utilities to avoid having to handle costly in-house facilities, equipment, technology and inventory.

It is expected that more assembly utilities will emerge in the future, that more extensive design capabilities will be offered by first- and second-tier contract manufacturers and, following the modularity trend, assembly services will handle larger integrated modules. Because of competition, such utilities will be challenged to maintain state-of-the-art technology and provide consistent quality, just-in-time delivery and an expanded menu of services.

10.5   The influence of emerging information technologies

'What's tomorrow's most critical assembly technology?' asks the editor of *Assembly Engineering*, and answers: 'If you don't have time [to read his article] the answer is simply information.' The article then reviews information systems and technologies in support of assembly (Coleman, 1988). Many refer to the second half of this century as the 'Information Revolution' and the way information technologies have already impacted on assembly has been described throughout the previous chapters. Information technologies continue to evolve rapidly. In this section, several particular trends are summarized because their potential effect on industrial production is already evident with some unique influence on assembly.

10.5.1   *Artificial intelligence for assembly planning and control*

Artificial intelligence developments have intensified since the 1970s and improved assembly in two main areas: better machine performance, and better computer-aided design and control. Some of the important developments (see also Chapters 3, 5 and 9) include (Kusiak, 1990; Badiru, 1992):

1. design for assembly and assembly process planning by problem-solving and various reasoning techniques, such as search procedures, case-based reasoning, genetic algorithms and fuzzy logic;
2. assembly facility planning and design by various problem-solving and reasoning techniques;
3. assembly knowledge representation by object-oriented methods, state transition graphs, Petri nets and other methods as the basis for problem-solving;
4. expert systems and knowledge-based systems for planning, design and control;

5. quality, process control and diagnostic methods based on AI techniques;
6. automatic programming and languages to specify assembly design, operations and control;
7. learning techniques, particularly artificial neural network approaches for adaptive control and scheduling;
8. robot motion planning and control programs based on AI techniques;
9. AI-based vision algorithms and systems for vision-based assembly and inspection;
10. intelligent sensor systems (in addition to vision) based on AI techniques;
11. multimedia systems and improved graphic displays for virtual design and simulated assembly.

Further improvements of computing and communication hardware and software at lower cost will continue to play a major role in future assembly methods and facilities, as well as in their mathematical modeling. Three particular trends are discussed in the next three sections.

10.5.2 Software and tool integration

Most of the analytic models for planning, design and control of assembly which are described throughout the book become more useful when they are implemented in computerized decision-support systems. Better modeling software and computer interfaces will continue to improve the integration of models and information systems. Emerging assembly modeling techniques based on artificial intelligence methods such as neural networks, fuzzy logic, genetic algorithms, virtual reality and multimedia applications are all based on advanced software and hardware implementations.

Increasingly, software tools are integrated in line with the concurrent or lifecycle engineering trend. Two examples have been shown in Fig. 10.5 for the integration of product design and assembly planning, and for product design and assembly equipment control. In the area of facility design, for example, Fig. 10.8 depicts a prototype integration of previously separate modeling, control and design functions on a CAD workstation. Commercial CAD systems are also emerging with integrated tools, advocating 'seamless' transition among traditionally separate functions.

Tool integration offers designers two important advantages (Witzerman and Nof, 1995a, 1995b).

- **A common computational platform** is provided for geometric modeling, simulation playback, motion planning, cell control emulation and off-line device programming. A common platform facilitates the creation, operation and maintenance of virtual production system models.
- **Concurrent and collaborative engineering** is enhanced by reducing the effort associated with having to integrate the previously separate design,

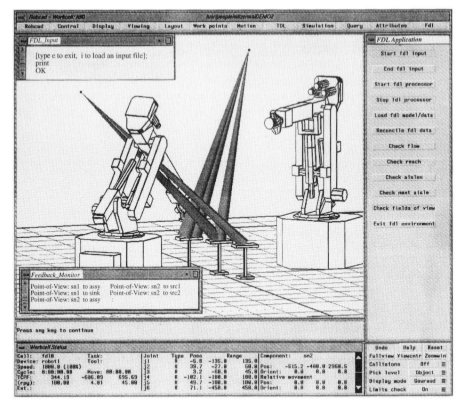

Fig. 10.8 Prototype Facility Description Language (FDL) application with integrated CAD tools on a ROBCAD workstation (Witzerman and Nof, 1995c, 1995d).

simulation, evaluation and integration of cell control and device task programs.

The development of an integrated assembly facility design tool environment, for instance, requires the following steps.

1. Determine communication protocols for the controller and peripheral devices.
2. Develop a logical representation for cell control.
3. Create cell control programs.
4. Model and develop programs for peripheral devices.
5. Integrate, test and refine programs and models using graphical emulation.

A related, emerging software development is in design languages. For example, a Facility Description Language (FDL) has been developed recently by Witzerman and Nof (1995c, 1995d) to serve as a common facility model for use by distributed collaborating designers. It is envisioned that a unified data and logic representation for concurrent design that specifies the geometry, control and material flow relationships in a cell will provide integrated information to drive various evaluation and design tools.

10.5.3 Collaboration among distributed processors

In response to the increasing complexity of industrial systems, companies resort to computer-supported integration, cooperation and collaboration. **Integration** is a process by which subsystems (such as computerized machines, workstations, information systems) share or combine tasks and resources, so that the whole system can produce better synergistic results. **Cooperation** is defined as the willingness and readiness of subsystems to share or combine their tasks and resources in what is called 'open systems'. But even with cooperation, there is still the issue of how to perform computer-supported integration effectively. **Collaboration** is defined as the active participation and work of the subsystems towards accomplishing integration. An important computer support function of collaborative integration is to prevent or overcome conflicts among the participating subsystems (Nof, 1995).

Collaboration in assembly can be found both at the machine level, discussed in this section, and at the human worker level, which is discussed in the next section. Multi-robot assembly work is becoming more efficient by advanced task and operating systems controls (Rodriguez, 1989; Rajan and Nof, 1994). For instance, a control approach called **collaborative coordination control** (CCC) synthesizes elements of distributed control with a collaborative strategy and has been developed and studied in recent years at Purdue University PRISM Program (Nof, 1992b).

The essence of CCC is its focus on collaboration: among machines, in sharing tasks and resources, and with human supervisors to obtain global information and external knowledge, and to plan revisions. Such control can be applied when the logic of physical assembly processes enables collaboration based on equivalence or parallelism of sub-tasks. Benefits include better reliability and quality by mutual backup, error recovery and rework; greater efficiency by dynamic reshuffling of tasks based on current workloads; and savings in critical resources, including floor space, that can be shared instead of duplicated. A good illustration of this emerging trend is assembly communication networks (described in Chapter 9).

10.5.4 Computer supported collaborative work

Computer supported collaborative (cooperative) work (CSCW) is a young field of research concerning systems that enable distributed groups of users to work more effectively. Software packages that have been developed for this purpose are called **groupware**. They are designed to facilitate human communication and cooperation. Groupware development requires an understanding of the fundamental rules and behaviors of cooperative human and organizational work.

A CSCW environment (Fig. 10.9) has to support interleaved routine work (i.e. asynchronous handling of multiple tasks and in different modes), dealing with frequent breakdowns, searching for a piece of information, etc. A natural switch between programs and systems has to be available, so that users can shift their attention quickly to current problems while coordinating individual activities, collaborating on a common activity or co-deciding (De Michelis, 1994).

At the foundation of a CSCW environment are message handling, an object-oriented database, a multimedia environment and various productivity tools. Over this foundation, there are a workflow manager to coordinate distributed procedures, an agenda manager to maintain timely commitments, and an information base that includes documents and information needed by users. Other components support procedure definition, communication, information search and a work center for each activity.

Collaborative design, or co-design, is a relevant application of CSCW for assembly. Co-design refers to integrated design activities. It is enabled by CSCW implemented on both hardware and software components. A CSCW system allows the integration and collaboration of distributed specialists in an environment where work with co-designs in manufacturing is essential. It is accomplished by integrating CAD and database applications, providing text and graphical representations for users.

A prototype of a co-design project has been developed at Purdue University with several design teams developing system components concurrently (Serna and Nof, 1994). The experiment is used to establish co-design protocols for concurrence control, error recovery, transaction management and information exchange. Protocols of information exchange handling are required to streamline and improve the efficiency of decision-making steps. The computer support system provides a graphical user interface to display windows for design

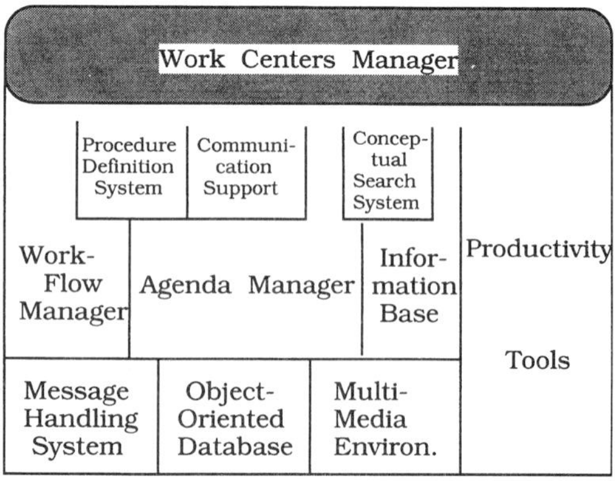

Fig. 10.9 A proposed organization of a CSCW environment (De Michelis, 1994).

drawings created using AUTOCAD and for discussion (on- or off-line) among distributed designers. The following design process steps have been studied in this project:

- **Conceptual discussions of a design project.** Data sharing allows up to four distributed designers at a time to view the same product or facility model on the CSCW graphics window. A group leader is assigned for purposes of file backup, time-out control and concurrence control when needed.
- **High-level conceptual design based on dialogues between engineers.** After initial model and problems review, specifications and tolerances must be discussed by manufacturing specialists and design engineers. A designer may leave the discussion, make necessary changes off-line and load a new model. Other parties must also evaluate the effects of any change on their area of functional responsibility.
- **Test and evaluation of models.** Error recovery and information exchange functions facilitate individual and joint evaluation. Older model versions may be retrieved for viewing. Current session changes are automatically stored. Files are stored until additional changes may be implemented. Tracking is based on files' save time.
- **Documentation.** The CSCW model allows engineers to exchange information and determine which of the models will be recommended. In the CSCW environment, all models may be displayed, but the most recent one is the first to be displayed and any pending changes are documented in the system's decision-window.

This research project has resulted in protocols that are useful for co-design and has demonstrated how they may be applied and integrated into a more effective CSCW design environment. Emerging and future developments of CSCW and co-design, for instance integration with virtual reality design studios, will enable better and potentially faster design and design verification. Further research is also required to understand human factors of co-design, e.g. trust, means for conflict resolution and issues of privacy and security.

10.6   Research issues in assembly

The economic significance of assembly and its challenging characteristics attract industry and academic researchers. In this section, recent research activities and directions in assembly research are summarized in three main areas: (1) material and information flow integration, (2) electronic assembly and (3) industrial perspectives on assembly research needs.

10.6.1   *Material and information flow integration*

Material and information flow issues have been summarized by an international panel with participants from government, industry and academia. The results are presented in Table 10.2. The left column lists four focal areas of issues and their associated problems. For each area, the right column includes research directions that have been identified by the panel as potentially promising to address those problems and issues. (Note that there is no particular correspondence for items listed on the same line; the correspondence is between the group of issues on the left and the group of research directions on the right.)

Table 10.2 Material and information flow integration

Issue/problem areas	Research directions for the area
1. **Human–machine interface**	
1.1 Task-oriented programming	• Graphics, 2-D and 3-D
1.2 Dynamic planning and rescheduling	• Graphic simulation; virtual reality
1.3 Dynamic facility reconfiguration	• Natural language specification, dialogues
1.4 On-line process tuning	• Windows for interactive collaboration
1.5 In-process knowledge acquisition	• Conflict prevention and resolution
2. **Flexible integration framework**	
2.1 Overall integration strategy	• Reference models for integration
2.2 Integrated design, planning, control	• Open software tools integration
2.3 Information management	• Open toolbox environments
2.4 Information access control	• Modular, configurable machines
2.5 In-process control, monitoring	• Reference specification languages
3. **Facility design and layout**	
3.1 Initial vs. adaptive design	• Knowledge-based planning
3.2 Capacity requirement planning	• New computational algorithms
3.3 Resource, material handling planning	• Neural networks and other learning methods
3.4 Technology selection	• Integration of supply chains
3.5 Exception handling	• Distributed and real-time control
4. **Implementation and economics**	
4.1 Portfolio of modifications	• Design simulation-to-specification
4.2 Small/medium company considerations	• Management simulation-to-specification
4.3 Equipment processing flexibility	• System decomposition to autonomous modules
4.4 Material flow flexibility	• Benchmarks for incremental implementation
4.5 Production management flexibility	• Representation of malfunctions
4.6 Economic justification	• System upgrading vs. a new system

Source: Nof and Woo (1991).

Some research directions listed have already started to develop over the last decade, as described in earlier chapters of this book. For instance, simulators, integrated design tools and knowledge-based planning have been developed and implemented to some extent. Emphasis on specific assembly issues and further refinement of techniques will require additional efforts.

10.6.2 Electronic assembly

A US research panel with participants from government, industry and academia reviewed issues related to electronic manufacturing. The conclusions that are relevant to assembly are summarized in Tables 10.3 and 10.4. A general conclusion is that a science base is needed for the assembly of circuit cards to better understand the fundamental processes. A unique feature of the electronics industry is its relatively complex processes. Integration and coordination are necessary because of this complexity, and because of the multiplant, multi-supplier and multi-echelon structure. Real-time control and

Table 10.3 Research needs in electronic assembly: design and operation models

Issues/problem areas	Research directions
System design methodologies	• Plant-, line- and cell-level methodologies, tools • Product- vs. process-focused design rules • Setup management design
Operational issues	• Circuit card assembly planning • Hierarchical scheduling; setup-dependent scheduling • Improved logistics • Optimization of purchasing
Modeling	• Multi-objective models of line balancing and scheduling • Improved means of verification • Strategic building block models for circuit card assembly
Enterprise strategy	• Inventory models throughout the supply chain • Models to facilitate negotiations, conflict resolution • Product design adequacy and cycle time prediction • Supplier–assembler relationship models
Organization and personnel	• Organizational structure, motivation and reward system • Management of culture change • Training

Source: Wilhelm and Fowler (1992).

Table 10.4 Research needs in electronic assembly: quality assurance

Issues/problem areas	Research directions
Total quality theory	• Total quality management models for electronic assembly • Measures of individual contributors to quality • Interdisciplinary teambuilding for quality
Concurrent engineering	• Interaction of product and process quality • Process capabilities and propagation of in-process errors • Fundamental effects of tolerance on product quality • Quality and learning curve of new product introduction • Impact of reduced cycle time on quality
System-wide quality modeling	• System-wide cost impact on quality • Models of cost by deviations from quality standards • Models and algorithms to predict quality measures • Circuit card design and quality dependencies • Strategies for locating test and rework operations
Quality and information systems	• Decision-support information for operators • Information on quality decisions for empowered operators • Displays of quality information

Source: Wilhelm and Fowler (1992).

scheduling, and emphasis on human resource management and training are two other common priority observations by this panel.

A Japanese review of emerging trends and needed research in assembly (Soni, 1990; Makino and Arai, 1994) is consistent with many of the panel's recommendations. Observed trends include:

1. methods evolving for quantitative evaluation of flexibility;
2. dedicated assembly machines being preferred for mass assembly;
3. dexterous assembly evolving rapidly;
4. manual assembly being still important;
5. main areas of consideration before introducing a flexible assembly process include:

   (a) techno-economic analysis;
   (b) task planning and configuration planning;
   (c) process of the physical assembly;

(d) modeling and simulation;
(e) programming and control;
(f) sensors for flexible assembly.

10.6.3 Industrial perspective on assembly research needs

A survey sponsored by the US Department of Defense (Martin-Vega *et al.*, 1995) includes 24 product lines (aircraft; missiles; microelectronics, communication and computer devices; electronic and mechanical assemblies; welded metal products) across companies ranging in size from $10 million to $2 billion in annual sales. The two main survey questions are:

- Can investments in research and development have a significant impact in either reducing the cost of and/or enhancing the effectiveness of assembly?
- If so, what research would lead to the highest return on investment in assembly R&D?

The survey results are given in Table 10.5. The research priorities of the 25 suggested directions according to industry experts are indicated first for the overall survey results, then for mechanically intense assembly, electrically intense assembly, electrical and mechanical assembly and for semiconductor companies.

Major findings of research needs, significance and the potential impact for all responding companies are as follows.

1. **Design for assembly (DFA).** This is the dominant, most important candidate for R&D investment. Design of products, tools and processes for ease of assembly is considered an opportunity for significant cost reduction and improved operational effectiveness.
2. **Manufacturing assembly integration.** The next consistent priority for assembly support activities is assigned to assembly planning and to quality assurance technology, followed by material flow management including JIT. These activities are considered integration-oriented functions. For instance, linking assembly schedules with fabrication and supply policies, planning pull-oriented assembly processes and integrating assembly with test and inspection will improve significantly the overall capacity and capability of a production company. The integration view of assembly research also emphasizes enterprise optimization, where assembly must be considered as part of the whole business.
3. **Flexible automation.** There is a desire in industry to automate many manual assembly operations. However, there is a need for standard, highly flexible and reusable components (with capabilities beyond present-day robots) that can be configured with relative ease into the automated support of assembly operations.
4. **Critical technologies.** Priority process activities in electronic assembly are soldering, interconnecting and SMT. Interconnect technology, for

Table 10.5 Industrial perspective on assembly research priorities

Research need	Overall priority	Mostly mechanical	Mostly electrical	Electrical/ mechanical	Semi conductor*
• Design for assembly	**1**	1	6	1	1
• Statistical process control	**2**	2	8	7	6
• Quality management	**3**	3	11	4	13
• Assembly system planning	**4**	8	14	10	7
• Assembly planning	**5**	5	5	6	5
• Wiring	**6**	4	18	14	–
• Fixturing	**7**	6	9	13	16
• Material flow management	**8**	7	2	8	15
• Cabling	**9**	11	20	12	–
• Line balancing	**10**	10	17	15	12
• Final assembly	**11**	9	13	11	–
• Just in time	**12**	15	12	5	8
• Surface mount technology	**13**	19	3	3	–
• Manual assembly	**14**	12	16	17	20
• Screw fastening	**15**	17	22	21	–
• Kitting	**16**	13	15	22	18
• Riveting	**17**	14	24	24	–
• Component placement	**18**	18	4	9	–
• Parts retrieval	**19**	16	10	16	19
• Soldering	**20**	25	1	2	–
• Through-hole insertion	**21**	20	7	19	–
• Press operations	**22**	21	23	23	–
• Sensors	**23**	23	21	18	14
• Robots	**24**	22	19	20	9
• Plug connection	**25**	24	25	25	–

*In semiconductor companies, other priority research needs are: 2 – Lead bond interconnect; 3 – Die attachment; 4 – Sealing; 10 – Die preparation; 11 – Environmental stress test; 17 – Visual inspection.

Source: Martin-Vega *et al.* (1995).

instance wiring in aircraft assembly and printed circuit boards, are considered by some industries as critical to their survival and require research for next-generation solutions.

5. **Environmental issues.** These are considered inhibitors of assembly and manufacturing in general. Typical concerns are the use of chemicals and

solvents, cleaning and degreasing, glues, plastics and epoxy. There are also concerns about unknown impacts on the environment of newly planned processes. Environmental issues must be considered during design for assembly. In particular, there is a need for a material substitution knowledge base.

10.7 Summary

Assembly, a significant element of production, is constantly undergoing change. Advances in technology and materials influence how assembly is performed and impact on the quality and reliability of assembled products and, as a result, the competitiveness of production companies. Another result, of course, is the effect on the quality of life of everyone who is using assembled products.

In this chapter, emerging trends in assembly are described in an effort to understand what changes have already appeared on the horizon. Trends are analysed and described in the components of assembly operations and systems, and in the design of products and facilities. An interesting trend is the emergence of assembly utilities which provide specialized services in response to the global market trends towards agile manufacturing and lean production.

Computers and information technologies have already transformed manufacturing and assembly, and their influence is also discussed. Assembly networks benefit from computer communication that enables the integration of assembly with other functions, such as test and inspection. The integration of computer-aided design tools and computer-support of collaborative design and planning work are other prominent trends. Emerging computing technologies and computer-supported design environments will also influence the modeling of assembly. For instance, design and planning models described in Chapters 3, 5 and 6 are being enhanced by better search techniques, while more powerful models of complex and transient assembly systems, as described in Chapters 7 and 8, can be developed and implemented for scheduling and flow control.

In response to changing requirements, research and development projects are meant to provide better, new methods and enabling technologies. Several recent panels and surveys have attempted to identify the high-priority assembly research needs and directions. Although there are different objectives and perspectives among academia and industry, the recommended priorities are generally consistent with the trends of the end of the twentieth century: emphasis on design for assembly consistent with the trend of lifecycle and concurrent engineering; emphasis on quality consistent with the quest for total quality management; interest in computer-based modeling and automation for flexibility and just-in-time flow management. There is even a renewed interest in flexible assembly line balancing, a technique that connects future assembly systems with the manual assembly lines that were the most prominent emerging trend in assembly during the beginning of this century.

10.8   Review questions

1. Out of the trends discussed in this chapter, list five that would also be most relevant for the assembly of:
    (a)   furniture;
    (b)   toys;
    (c)   home appliances;
    (d)   communication satellites;
    (e)   kitchen utensils.
2. Discuss emerging trends that are associated with the objective of flexibility in production.
3. Which emerging trends are associated with the following assembly functions?
    (a)   kitting in push systems
    (b)   just-in-time scheduling in pull systems
    (c)   cellular production/assembly systems
    (d)   design for ease-of-assembly
    (e)   design for ease-of-disassembly
4. Which specific assembly industry needs are reflected in the following emerging trends?
    (a)   assembly communication networks
    (b)   integrated CAD tools
    (c)   error diagnostics and recovery
    (d)   computer-supported collaborative work
    (e)   assembly with adhesives
5. Differentiate between 'assembly optimization' and 'enterprise optimization'.
6. What are the advantages and disadvantages of (a) customized fasteners, (b) adhesives?
7. Consider the design models described in Chapters 5 and 6 and explain which of them would be most appropriate for the design of an assembly utility facility.
8. What are the advantages and limitations of an assembly utility and how can it best fit into the supply chain?
9. Describe how production of the following products can benefit from the assembly utility outlined in Table 10.1:
    (a)   telephone answering machine;
    (b)   medical diagnostic kit;
    (c)   smoke detector.

10. What are some of the advantages provided by integrated CAD tools?
11. What are some of the advantages provided by computer-supported collaborative work?
12. (a) How do you envision assembly sequencing and process planning (as discussed in Chapters 3, 4 and 5) being performed with computer-supported collaborative work?
    (b) Consider the same for scheduling and time-managed flow control (as described in Chapters 7 and 8).
13. What are some of the advantages of an assembly facility design language? What features are required from such a language to achieve these advantages?
14. Explain the reasons for specific differences between the research priorities as perceived by different types of industries.
15. Explain several techniques and models discussed throughout the book that already attempt to address any of the first 10 research priorities suggested in:
    (a) material and information flow;
    (b) electronic assembly;
    (c) mechanical assembly.
16. Most of the research directions suggested by panels and surveys, as discussed in this chapter, address the need for solutions of recognized, existing or emerging problems. Can you think of other developments or discoveries that may significantly impact on future assembly, yet are not being developed in reaction to recognized needs?

References

Badiru, A.B. (1992) *Expert Systems Applications in Engineering and Manufacturing*, Prentice-Hall, Englewood Cliffs, NJ.

Benson, A.F. and Iverson, W.R. (1994) 'Special issue on buyers guide', *Assembly*, June.

Black, J.T. (1991) *The Design of a Factory with a Future*, McGraw-Hill, New York.

Bullinger, H.J., Warnecke, H.-J. and Lentes, H.P. (1986) 'Towards the factory of the future', *International Journal of Production Research*, Vol. 24, No. 4, pp. 697–741.

Coleman, J. (1988) 'What's tomorrow's most critical assembly technology?', *Assembly Engineering*, Vol. 31, No. 1, pp. 22–5.

De Michelis, G. (1994) 'A CSCW environment: some requirements', *Computer Supported Cooperative Work* (ed. A.R.S. Scribner), Avebury Technical, Hants, England, pp. 237–52.

Heginbotham, W.B. (1986) 'Programmable and robotic assembly', *Robotics and Material Flow* (ed. S.Y. Nof), Elsevier Science, Amsterdam, pp. 153–61.

Helander, M.G. and Domas, K. (1986) 'Task allocation between humans and robots in manufacturing', *Robotics and Material Flow* (ed. S.Y. Nof), Elsevier Science, Amsterdam, pp. 175–85.

Kaplan, G. (1993) 'Manufacturing à la carte: agile assembly lines, faster development cycles', *IEEE Spectrum*, Vol. 30, No. 9, pp. 24–6.

Kreifeldt, J.G. (1988) 'Ergonomics, human–robot interface', *International Encyclopedia of Robotics, Applications and Automation* (eds R.C. Dorf and S.Y. Nof), Wiley, New York, pp. 451–62.

Kusiak, A. (ed.) (1990) *Intelligent Design and Manufacturing*, Prentice-Hall, Englewood Cliffs, NJ.

Makino, H. and Arai, T. (1994) 'New developments in assembly systems', *Annals of CIRP*, Vol. 43, No. 2, pp. 501–12.

Martin-Vega, L.A., Brown, H.K., Shaw, W.H. and Sanders, T.J. (1995) 'Industrial perspective on research needs and opportunities in manufacturing assembly', *Journal of Manufacturing Systems*, Vol. 14, No. 1, pp. 45–58.

Nof, S.Y. (1992a) 'Industrial robotics', *Handbook of Industrial Engineering* (ed. G. Salvendy), 2nd edn, Wiley, New York, pp. 399–463.

Nof, S.Y. (1992b) 'Collaborative coordination control (CCC) of distributed multi-machine manufacturing', *Annals of CIRP*, Vol. 41, No. 1, pp. 441–5.

Nof, S.Y. (1995) 'Parallel and distributed models of integration in production systems', *Proceedings of the International Conference on Production Research*, Jerusalem.

Nof, S.Y. and Woo, T.C. (1991) 'Computer integrated engineering of material flow systems: key research issues', *International Journal of Computer Applications in Technology*, Vol. 4, No. 3, pp. 159–65.

Porter, A.L. and Rossini, F.A. (1988) 'Futurism and Robotics', *International Encyclopedia of Robotics, Applications and Automation* (eds R.C. Dorf and S.Y. Nof), Wiley, New York, pp. 565–78.

Rajan, V.N. and Nof, S.Y. (1994) 'Cooperation requirement planning for multiprocessors', *Information and Collaboration Models of Integration* (ed. S.Y. Nof), Kluwer Academic, Dordrecht, Netherlands, pp. 179–200.

Riley, F.J. (1988) 'The evolution of automatic assembly', *Assembly Engineering*, January, pp. 36–8.

Rodriguez, G. (1989) 'Recursive forward dynamics for multiple robot arms moving a common task object', *IEEE Transactions on Robotics and Automation*, Vol. 5, No. 4, pp. 510–21.

Schraft, R.D. (1988) 'Robots in Western Europe', *International Encyclopedia of Robotics, Applications and Automation* (eds R.C. Dorf and S.Y. Nof), Wiley, New York, pp. 1401–11.

Schwartz, W.H. (1988) 'An assembly hall of fame', *Assembly Engineering*, January, January, pp. 30–2.

Serna, V.M. and Nof, S.Y. (1994) *Codesign and Collaboration Protocols for Engineering Tasks Integration*, Research Memo 94-19, School of Industrial Engineering, Purdue University, West Lafayette, IN.

Soni, A.H. (1990) 'Research needs and opportunities in flexible assembly systems', *Proceedings of ASME 2nd Conference in Flexible Assembly Systems*, Chicago, pp. 221–8.

Strobel, R. and Johnson, A. (1993) 'Pocket pagers in lots of one', *IEEE Spectrum*, Vol. 30, No. 9, pp. 29–32.

Valckenaers, P. and Van Brussel, H. (1994) 'A theoretical model to preserve flexibility in FMS', *Information and Collaboration Models of Integration* (ed. S.Y. Nof), Kluwer Academic, Dordrecht, Netherlands, pp. 89–104.

Walker, C.R. and Guest, R.H. (1952) *The Man on the Assembly Line*, Harvard University Press, Cambridge, MA.

Warnecke, H.-J., Schweizer, M., Tamaki, K. and Nof, S.Y. (1992) 'Assembly', *Handbook of Industrial Engineering* (ed. G. Salvendy), 2nd edn, Wiley, New York, Ch. 19.

Wilhelm, W.E. and Fowler, J. (1992) 'Research directions in electronic manufacturing', *IIE Transactions*, Vol. 24, No. 4, pp. 6–17.

Witzerman, J.P. and Nof, S.Y. (1995a) 'Integration of simulation and emulation with graphical design for the development of cell control programs', *International Journal of Production Research*, Vol. 33, No. 11, pp. 3193–206.

Witzerman, J.P. and Nof, S.Y. (1995b) 'Tool integration for collaborative design of manufacturing cells', *International Journal of Production Economics*, Vol. 38, pp. 23–30.

Witzerman, J.P. and Nof, S.Y. (1995c) 'A Facility Description Language for concurrent design', *Proceedings of the 4th Industrial Engineering Research Conference*, May 1995, Nashville, TN, pp. 449–55.

Witzerman, J.P. and Nof, S.Y. (1995d) *Facility Description Language (Ver. 1.1) User Manual*, Research Memo 95-5, School of Industrial Engineering, Purdue University, West Lafayette, IN.

Index